Imaging in Clinical P

CLINICAL PRACTICE SERIES

Series Editor: Professor AC Kennedy PRCPS (Glas), MD, FRCP (Lond, Edin, Glas), FRCPI, FRSE, FACP (Hon), FRACP (Hon)

Muirhead Professor of Medicine, University Department of Medicine, Royal Infirmary, Glasgow

Imaging in Clinical Practice

Alan G Chalmers,
MB ChB, MRCP, FRCR

Consultant Radiologist, The General Infirmary, Leeds

James H McKillop,
MB ChB, PhD, FRCP

Senior Lecturer in Medicine and Honorary Consultant
Physician, Royal Infirmary, Glasgow

Philip JA Robinson,
MB BS, MRCP, FRCR

Consultant Radiologist, St James's University Hospital, Leeds

Edward Arnold
A division of Hodder & Stoughton
LONDON   BALTIMORE   MELBOURNE   AUCKLAND

© 1988 Alan G Chalmers, James H McKillop and Philip JA Robinson

First published in Great Britain 1988

British Library Cataloguing in Publication Data

Chalmers, Alan G.
  Imaging in clinical practice.
  1. Man. Diagnosis. Radiography—For surgery
  I. Title   II. McKillop, James H.
  III. Robinson, Philip J.A. (Philip Joseph Andrew)   IV. Series
  617'.07572

  ISBN 0–340–41214–3

Typeset in 10/12pt Times Roman by Wearside Tradespools, Fulwell, Sunderland. Printed and bound in Great Britain for Edward Arnold, the educational, academic and medical publishing division of Hodder and Stoughton Limited, 41 Bedford Square, London WC1B 3DQ by Butler and Tanner Ltd, Frome and London.

SERIES EDITOR'S PREFACE

This new series is planned to give an account of current clinical practice with authoritative guidance on investigation, diagnosis and management. The series is aimed principally at candidates preparing for the MRCP(UK) or those who are gaining further experience in the specialty of their choice after acquisition of the MRCP(UK) diploma. The texts will also be useful to senior registrars and consultants who are seeking to keep themselves up to date across the various subspecialties in medicine and who also prepare lectures and seminars for younger postgraduates. Some of the more gifted senior undergraduates should also be interested in consulting the books. These comments apply also to overseas postgraduates both while working in this country and after their return home. Most of the countries in the Middle East, Africa, India, Pakistan and the Far East are still very much in tune with the British practical clinical approach to patient management, finding highly technological approaches less appropriate to their local needs. The books are not extensively referenced since an important objective is to keep them to a size that is easily handled but lists of key references for further reading are provided.

Imaging in Clinical Practice is the second title in the series. It has been written by Dr Chalmers and Dr Robinson, two consultant radiologists with complementary interests in imaging procedures, and by Dr James McKillop, the consultant physician with special expertise in nuclear medicine. All three are active clinicians and teachers with well-established reputations in their respective fields. The book is designed and planned for clinicians (surgeons as well as physicians) and the authors have pursued throughout a practical clinical approach based on a sound but not overelaborate groundwork of underlying scientific principles. All of the major systems of the body are included and the book is remarkably up to date in a rapidly changing field.

Glasgow, 1988   AC Kennedy

AUTHORS' PREFACE

The introduction of a large number of imaging techniques over the last 25 years has led to considerable confusion as to the role of individual procedures in clinical practice. Our aim in this book has been to provide a guide to the best utilization of imaging procedures in a wide variety of clinical situations. As such the book is intended primarily for clinicians. We believe it will be useful to doctors preparing for clinical examinations such as the MRCP or FRCS, and to older clinicians who wish to update their knowledge — particularly in areas outwith their immediate specialty interest. Senior undergraduates should find in it a useful overview of the bewildering array of alternative methods.

The first chapter deals with some general concepts which apply to imaging techniques and gives a brief account of the scientific basis of the various imaging modalities utilized in medicine. It also discusses some of the general principles which lie behind their use in clinical practice. Chapters 2–9 deal in turn with individual organ systems. Special imaging techniques are described at the beginning of each chapter followed by an account of the role of imaging in the investigation of clinical problems relating to that organ system. Chapter 10 is a summary of the use of imaging in oncology, but also includes a section on imaging in acquired immune deficiency syndrome.

As the book is not intended to be a manual to the performance of imaging procedures, technical details have been included only in so far as they are necessary to understand the nature of the investigations and to appreciate their strengths and limitations. Similarly, the book does not aim to teach image interpretation, and illustrations have been limited in number.

Each author has contributed to some extent to every chapter. As we have particular interests in different aspects of medical imaging, we hope the combined approach has allowed us to produce a balanced account of the most effective use of this group of techniques in clinical practice.

Leeds and Glasgow, 1988
Alan G Chalmers
James H McKillop
Philip JA Robinson

CONTENTS

LIST OF ABBREVIATIONS

ACTH	Adrenocorticotrophic hormone
ADH	Anti diuretic hormone
AIDS	Acquired immunodeficiency syndrome
AP	Antero-posterior
ARC	AIDS related complex
ASD	Atrial septal defect
ATN	Acute tubular necrosis
^{195}Au^m	Gold-195m
AVM	Arteriovenous malformation
CDH	Congenital dislocation of the hip
CMV	Cytomegalovirus
^{51}Cr	Chromium-51
CRST	Calcinosis, Raynaud's, sclerodactyly, telangiectasia
CSF	Cerebrospinal fluid
CT	Computed tomography
DISIDA	Di-isopropyl iminodiacetic acid
DMSA	Di-mercapto succinic acid
DSA	Digital subtraction angiography
DTPA	Diethylenetriamine pentacetic acid
DVI	Digital vascular imaging
DVT	Deep venous thrombosis
EDTA	Ethylenediaminetetraacetate
EF	Ejection fraction
ERCP	Endoscopic retrograde cholangiopancreatography
^{67}Ga	Gallium-67
GI	Gastrointestinal
Gy	Gray
HIDA	Dimethyl imminodiacetic acid
HIV	Human immunodeficiency virus
HMPAO	Hexamethylpropyleneamineoxime

HOCM	Hypertrophic obstructive cardiomyopathy
HPOA	Hypertrophic pulmonary osteoarthropathy
^{123}I	Iodine-123
^{125}I	Iodine-125
^{131}I	Iodine-131
ICP	Intracranial pressure
ICRP	International Commission on Radiological Protection
IDA	Imminodiacetic acid
^{111}In	Indium-111
IUCD	Intra-uterine contraceptive device
IV	Intravenous
IVC	Intravenous cholangiography
IVU	Intravenous urogram
KUB	Kidneys, ureters and bladder
LIP	Lymphocytic interstitial pneumonia
MAI	Mycobacterium avium intracellulare
MBq	MegaBecquerel
MHz	MegaHertz
mIBG	Meta-iodobenzylguanidine
mGy	MilliGray
MRI	Magnetic resonance imaging
MS	Multiple sclerosis
NRPB	National Radiological Protection Board
OCG	Oral cholecystography
PA	Postero-anterior
PE	Pulmonary embolism
PET	Positron emission tomography
PGL	Persistent generalized lymphadenopathy
PTC	Percutaneous transhepatic cholangiography
PUO	Pyrexia of unknown origin
PYP	Pyrophosphate
SeHCAT	23-selena-25-homotaurocholate
SIADH	Syndrome of inappropriate ADH secretion
SVCO	Superior vena caval obstruction
^{99}Tc^m	Technetium-99m
^{201}Tl	Thallium-201
TSH	Thyroid stimulating hormone
VIP	Vasoactive interstitial polypeptide
V/Q	Ventilation/perfusion
VSD	Ventricular septal defect

1

INTRODUCTION TO CLINICAL IMAGING

The Development of Imaging Methods

In the 1860s, Sir James Young Simpson, the famous Edinburgh obstetrician, is said to have predicted that 'a light to which opaque screens are not impervious' would soon allow physicians to inspect their patients' internal anatomy without the intervention of the surgical knife. Thirty years later, Professor Roentgen published his first radiograph, an image of his wife's hand, and his discovery was soon taken up all over Europe and America.

Radiographs

Conventional radiographs show detailed anatomy of the bony structures and also give a clear demonstration of the lungs by virtue of the air contained within them. However, the ability of X-rays to discriminate between different types of soft tissue is very limited. Over the last sixty years, many different techniques have been developed for increasing the contrast between different soft tissues. This is generally achieved by the injection or ingestion of materials which are relatively opaque to X-rays, such as barium sulphate and the iodinated parenteral contrast media, or by the injection of air or carbon dioxide. Some of these techniques, such as arteriography and lymphography, carry small risks of toxicity and are not particularly pleasant for the patient. Their replacement by non-invasive methods seems desirable.

Scintigraphy

In the late 1940s and early 1950s, scintillation probes were used to detect the distribution of previously administered radioactive tracers within the body. At first, rectilinear scanners and gamma cameras were used with iodine-labelled agents but, with the development during the early 1960s of a wide range of new radiopharmaceuticals labelled with technetium, it became possible to visualize many different organs and physiological processes. The more recent addition of dedicated mini- or microcomputers has allowed the development of methods for visualizing local blood flow and other functional forms of imaging, e.g. individual kidney function and cardiac ejection fraction measurement.

Ultrasound

Initially developed for use in obstetrics, this technique now has wide applications. Because ionizing radiations are not used, multiple repeat examinations are possible. This is useful in monitoring the effects of treatment. Ultrasound is also more effective than other methods in detecting disease at certain sites. Recently, 'real-time' scanners have been introduced. These give a continuously renewed image of the area being examined and are less dependent on the skill of the operator than earlier machines.

Computed Tomography (CT)

The theory of how to reconstruct images from a series of observations made from the periphery of an object has long been established but only computers have made it feasible to apply this technique to medicine. The first operational computed (or computerized) tomography (CT) scanner was designed and built in the late 1960s by an English Nobel laureate, Godfrey Hounsfield. By 1975, CT brain scanners were installed in many of the neurosurgical centres throughout the UK and the first two prototype whole-body scanners were in operation. Over the last ten years, technical improvements have resulted in faster and better quality scans.

Magnetic Resonance Imaging (MRI)

The most recently developed major imaging technique is magnetic resonance. This technique offers not only the same level of anatomical detail as CT but also has exciting possibilities for tissue characterization using combinations of different radiofrequency pulse sequences to detect biochemical variations between different tissues.

Objectives of Imaging

Excluding pure research applications, the main purposes of imaging may be summarized as follows:

1. *Screening*
   Large volume population studies, particularly when focused on high risk groups, may allow the earlier detection of disease and lead to improved survival (e.g. mammographic screening for breast cancer).

2. *Detecting anatomical abnormalities*
   Imaging procedures may be used to visualize individual organs or areas of the body which are suspected clinically of harbouring the lesion responsible for causing illness.

3. *Characterizing lesions*
   Where an abnormality is apparent either clinically or on initial imaging tests, further procedures may help to discriminate between various possible pathologies.

4. *Defining the anatomical extent of lesions*
   Once local pathology has been recognized, it is often important, particularly where surgical treatment is planned, to define as clearly as possible the extent of the lesion itself, its effect on surrounding organs and the presence of metastatic disease in cases of malignancy.

5. *Assessing the response to treatment*
   After treatment, whether by surgery, drugs or radiotherapy, imaging procedures may need to be repeated to determine the effect of treatment on local pathology. In cases of malignant disease, it is often important to repeat imaging procedures at intervals to look for evidence of recurrent or residual disease, particularly in patients whose symptoms reappear.

6. *Excluding disease*
   With our present state of knowledge some clinical diagnoses can only be reached by the exclusion of other possibilities. Imaging often plays a major role in this type of investigation.

This list is not meant to be exhaustive but to indicate the type of question which imaging procedures are aimed at answering. No doubt readers will be able to think of examples of usage that do not fall into any of the above categories.

Using Imaging Departments

Choosing the Most Appropriate Test

The range of procedures available and the speed of development of new techniques is such that many physicians and surgeons requesting imaging procedures find it impossible to keep up-to-date with the imaging sciences. The traditional role of the radiologist was to carry out X-ray procedures and report on the findings; nowadays imaging consultants must take on the additional responsibility of helping to guide their clinical colleagues in choosing the most appropriate procedures for their patients. Making the correct choice requires a degree of dialogue between the specialities: the clinician knows what questions he wants answered, whereas the imaging specialist knows the possibilities and limitations of his techniques.

Requisitions

The radiologist carrying out an investigation without clinical data is in the same position as a physician or surgeon who is invited to examine a patient without being allowed to take a history. Imaging procedures are much more likely to be fruitful if they are carried out with the knowledge of the patient's clinical presentation. The clinical data presented on the requisition form is used by the imaging specialist firstly to ensure that the correct procedure is being asked for, secondly to tailor the examination where necessary to the particular clinical problem, and thirdly to evaluate the results of the procedure as accurately as possible in the light of clinical findings. Great detail is not required except occasionally for specialized procedures, but it is, very often, valuable to know about previous surgery and the results of previous imaging investigations.

Preparing the Patient

For procedures for which the patient needs to starve overnight or undergo bowel cleansing, detailed instructions will be sent out by the imaging department. However, in addition to the physical preparation of the patient, it is both helpful and humane to warn the patient what to expect when he or she attends for investigation. An explanation in person is more effective than a letter and the referring clinician is in the best position to do it. A patient who arrives in the X-ray department having been told that he is just going to have 'a simple X-ray of his small bowel' and then finds himself undergoing duodenal intubation is not likely to relish a return visit. Patients also like to know when and how they will receive news of the results of their tests. There is a feeling among imaging specialists that since their contact with patients is usually brief, they are not placed the best to discuss the implications of test results with the patients. It is understandable that when patients arrive for imaging procedures not knowing what to expect and leave with no knowledge of the results, they may regard the imaging department as an unfriendly or impersonal environment. A little prior explanation can save a lot of unnecessary anguish.

Imaging Techniques

Plain Films and Contrast Radiology

X-rays are produced when a beam of high-energy electrons strikes a suitable target, usually metallic tungsten. The electron beam is generated from a heated wire filament by applying a very high voltage (usually in the range of 60 000–120 000 volts) produced by a step-up transformer from the mains supply. X-rays cannot be focused so the direction of the X-ray beam is controlled by using lead shields to absorb the unwanted rays.

The absorption of X-rays by tissue varies accordingly with the energy of the radiation, the density of the tissue and its average atomic number. Normally expanded lung consists mostly of air and so produces only a little attenuation of the X-ray beam. Bone and other heavily calcified tissues absorb much more of the beam whereas absorption by soft tissues is intermediate. Adipose tissue has a distinctly lower mean atomic number than other soft tissues so layers of fat are often discernible on plain films. The density and mean atomic number of blood, muscle and the solid organs are too similar to produce visible differences on plain films. The demonstration of detailed anatomy of these soft tissues can often be improved by using radiographic contrast media.

Contrast media are non-toxic compounds which include a component of high atomic number. An aqueous suspension of barium sulphate is used for the majority of gastrointestinal examinations, whereas iodinated derivatives of aniline dyes are used for intravascular administration. Iodinated oils are used for lymphography and bronchography. The injection of air or carbon dioxide provides a further variety of contrast medium with limited application.

After the X-ray beam has been attenuated to various degrees by the different tissues through which it passes, the resulting heterogeneous beam is allowed to strike the detector surface which may be either a photographic film, a fluoroscopic screen, or one of the newer, reusable detectors. The image produced is thus a two-dimensional summation of the three-dimensional structure through which the X-ray beam has passed. The majority of X-ray images are recorded directly on conventional film which consists of a celluloid base coated with a layer of photographic emulsion. Fluoroscopy is used with some procedures involving contrast administration — partly in order to obtain the best position for recording the images on film, and partly to obtain a visual impression of motility and other functions of the organs being investigated. Fluoroscopic images are viewed by means of a television monitor coupled to an image intensifier, a device for magnifying the brightness of the image from the fluorescent screen itself. Permanent records are made either by photographing the image from the TV chain, by video recording, or, more recently, by digitizing the image and storing it on computer disc. Direct digital acquisition of the radiographic image is a recent development which may eventually replace the use of photographic film in radiology.

Conventional and Computed Tomography

Tomograms are images of the anatomy within a predetermined section through a part of the body. The sections are of finite thickness so are often referred to as 'slices'. Conventional tomograms are obtained by simultaneous movement of the X-ray tube and the film in opposite directions during the exposure. The extent and direction of movement are linked mechanically so that the fulcrum lies within the plane of interest in the patient. Structures which lie outside the plane of interest are blurred, whilst those within that plane remain sharply defined. The movement may be linear, circular, elliptical, spiral or hypocycloidal; the more extensive the movement, the thinner will be the focal plane of the image. However, the X-ray beam always passes through the full thickness of the body so the resolution of the image is always degraded by the contribution of X-ray absorption by overlapping tissues.

Computed tomography (CT) operates on quite a different principle. In this technique, the X-ray tube rotates around the outside of the patient. With most devices the exposure is made over a 360° rotation. The detectors may be fixed in a ring around the patient or may rotate isocentrically with the X-ray tube. Each of many detectors registers a large number of individual measurements of X-ray attenuation during a single scan which may take 4–5 seconds. The computer then calculates an attenuation value for each point within the section of anatomy being scanned. This is then assigned a value on an arbitrary scale of X-ray absorption (Hounsfield units) and displayed as a two-dimensional image.

The unique contribution of CT scanning to radiological diagnosis arises from three technical features. First, the slices are 'pure' tomograms with no contamination from overlapping tissues; the detail obtained is thus much clearer than with conventional tomograms. Second, the detectors used are much more efficient and precise in measuring X-ray attenuation than are conventional X-ray films or fluoroscopic screens; this means that minor differences in X-ray absorption by different tissues allow the recognition of detailed soft-tissue anatomy. Third, the axial view provides an anatomical dimension which is often missing from conventional radiographic studies.

For CT examinations of the abdomen and pelvis, it is helpful to opacify the lumen of the bowel using dilute oral (and occasionally rectal) contrast medium. Intravenous contrast medium is often helpful in improving the delineation of vascular structures and this manoeuvre also shows most lesions of the parenchymal solid organs more clearly. Most CT scanners now offer a range of slice thickness — the thinner slices being used to demonstrate more detailed anatomy, although the radiation dose is higher with thinner slices. If a series of contiguous slices is obtained, a computer reconstruction of sagittal, coronal and, sometimes, oblique planes of anatomy can be produced. The resultant images however are limited in resolution by the thickness of the original axial slices.

As with any other radiological procedure, optimum results are obtained if the techique is tailored to the particular clinical problem under investigation.

Scintigraphy

Scintigraphy (nuclear medicine) is based on the principle that various elements and radioactively labelled compounds (radiopharmaceuticals), usually administered by intravenous injection, will localize predominantly in a particular organ or group of organs. In diagnostic tests the emitted radiation is utilized to measure the uptake, distribution or turnover of the agent within the organs of interest.

The substances administered for nuclear medicine tests consist of either radioactive isotopes of elements or non-radioactive compounds labelled with a radioisotope. The radioisotopes used in scintigraphy emit gamma-rays or, less commonly, low energy X-rays as they decay. Most of the radioisotopes used in diagnostic nuclear medicine have half-lives of hours or a few days. In addition to the physical half-life, the radiopharmaceuticals have a biological half-life related to the metabolism and excretion of the compound. The physical and biological half-lives combine to give an effective half-life which is important in determining the radiation dose delivered to the patient.

Radiopharmaceuticals localize in the organ of interest as a result of one or more of the organ's functions. The emitted radiation is then used to build up a map of concentration of the radiopharmaceutical within the target organ. This map is the nuclear medicine image and is produced using a radiation detector such as a gamma camera.

The detection of radiation by the gamma camera is dependent upon a sodium iodide crystal. To ensure that only gamma rays travelling in a limited range of directions reach the crystal, a lead collimator is placed in front of it. The collimator consists of a lead sheet with many, usually parallel, holes in it. When gamma rays strike the sodium iodide crystal they produce a flash of light which is converted to an electrical pulse and amplified by photomultiplier tubes. Positioning electronics are then used to determine the X and Y co-ordinates of the light flash on the crystal. In this way the distribution of the radiopharmaceutical in the organ under study is determined and displayed, usually as an image showing the relative uptake in different areas. Similar information can be obtained from a rectilinear scanner where a radiation detecting probe is moved backwards and forwards across the region of interest and the radioactivity over a number of points is registered. The gamma camera has the advantage over the scanner of allowing multiple views to be obtained easily. In addition, a gamma camera is almost always required for dynamic studies in which the change in radiopharmaceutical distribution over a period of time is recorded.

Computers are increasingly used in nuclear medicine. They are essential for the most dynamic imaging studies where quantitative measurements of changes in radiopharmaceutical distribution are produced. In static nuclear medicine studies, i.e. those in which the distribution of radiopharmaceutical at a single time-point is recorded, computers have a role in image processing which may make detection of abnormalities easier.

Scintigraphic tests are more dependent on physiology or biochemistry (function) than on structure (anatomy). Functional and anatomical changes in disease are related but often occur at different rates. Scintigraphy will give information different from that obtained with ultrasound or radiology which predominantly reflects structural changes; in particular, functional changes may occur at an earlier stage in a disease process than gross structural changes. Thus, scintigraphy may be positive before structural changes can be detected.

A second advantage of scintigraphic procedures is that they are non-invasive. With very few exceptions they involve no more than an intra-venous injection of a very small amount of the radiopharmaceutical. Reactions to radiopharmaceuticals are uncommon and usually minor. The radiation dose resulting from administration of the isotope is usually lower than from corresponding radiological studies.

The major disadvantage of scintigraphy is its low level of specificity in most applications. Different pathologies frequently affect organ function in similar ways and lead to the same scintigraphic result. In some cases, the scintigraphic findings plus the clinical context will make a particular diagnosis so likely that treatment can proceed without further investigations. On other occasions, scintigraphy serves as a screening procedure to be followed by other tests such as biopsy or invasive radiology directed at the area of abnormality demonstrated by scintigraphy.

One approach to increasing lesion detection by scintigraphy has been the development of tomographic techniques. Single photon emission (computed) tomography (SPECT or SPET) uses standard radiopharmaceuticals and a conventional gamma camera detector which is capable of rotating round the patient and images are obtained at regular (usually 6°) intervals. The gamma camera is interfaced to a computer which stores the raw data. Tomographic slices can then be reconstructed by various mathematical techniques. SPECT techniques have been especially useful when the area of interest lies deep in the body or when the ratio of activity between the lesion and surrounding tissue (target to background ratio) is not high. SPECT studies have established clinical roles in various organ systems, notably in cerebral perfusion studies, myocardial perfusion studies and in tumour imaging. Positron emission tomography (PET) is limited in its clinical application by the need to have a dedicated cyclotron close by, and by the highly sophisticated imaging and computing system which is employed. PET has exciting possibilities for research into biochemical processes *in vivo* because of its ability to utilize short-lived tracers of biologically important atoms such as carbon, oxygen and nitrogen.

The resolution of nuclear medicine images is inherently limited and although technological advances will give some improvement, the degree of anatomical detail possible will always be less than that achieved by conventional radiology, ultrasound and computed tomography.

Ultrasound

The human ear can detect sound with frequencies up to 20 000 c/s or 20 kHz. Frequencies above this level cannot be heard and are collectively described as 'ultrasound'. The ultrasound beam is produced by a material which is structurally deformed when a voltage is applied to its surface. Such piezoelectric materials not only resonate and emit ultrasound waves when a voltage is applied, but also produce a voltage at their surface when a mechanical stress is applied. Transducer substances therefore have the ability to both emit and detect ultrasound waves, the effects produced at the transducer being directly proportional to the voltage or mechanical stress applied. By varying the size of the piezoelectric disc, the frequency of the ultrasound beam produced can be varied.

Transducers emit pulses of ultrasound lasting only a few microseconds. This is then followed by a pause to allow the detection of returning echoes before the next pulse is emitted. The ultrasound beam is attenuated or reflected depending on the physical properties of the tissues under examination. The intensity of the returning echoes is translated into a grey scale display on a television monitor. This gives a tomographic image of the anatomy under study. Early ultrasound equipment produced a variety of 'static' images. Modern machines produce a 'real-time' display. 'Real-time' is, in effect, a rapid sequence of ultrasound images continuously displayed on a television monitor. This not only allows the detection of movement, e.g. vessel pulsation, but provides the operator with a continuous sequence of tomographic sections. By changing the orientation of the ultrasound probe, sections in the sagittal, transverse or oblique planes can be acquired. By varying the frequency and the focusing of the ultrasound beam, structures at varying depths can be imaged. The higher the frequency of the ultrasound beam, the better the anatomical resolution — but this is obtained at the cost of reduced tissue penetration.

The construction of the ultrasound probe determines the shape of the image produced. Sector images appear triangular and are generally produced by the transducer sweeping the beam to and fro through an arc of approximately 90°. Linear array scans are produced by a probe which contains a sequence of elements which are fired sequentially and rectangular images result. With modern display software, two adjacent linear array images can be joined together on the television monitor to give an extended field of view. This is useful when documenting large masses and is particularly helpful in obstetric scanning. It also allows the measurement of structures which are too large for a single field of view. The triangular sector scans give a more restricted field of view but are superior to linear array transducers when access is limited, e.g. when scanning through the ribs or between surgical dressings or drain sites.

'Hard copy' documentation is achieved by freezing the real-time display and recording the static image on photographic film. Alternatively, a video recording of the continuous real-time images can be made. This can be particularly useful in imaging the heart.

Doppler ultrasound is used for the assessment of blood flow. This technique uses the principle that when an ultrasound wave strikes an object which is moving, the reflected beam is of a different frequency to the incident beam. Either continuous or pulsed Doppler ultrasound can be used to assess intravascular blood flow, whether it be in the peripheral vascular tree, the intra-abdominal vessels, extracranial neck vessels or the heart.

The advantages of ultrasound imaging are that it is quick and simple to perform, and harmless to the patient. Very little preparation is required other than fasting prior to an abdominal study and a full bladder for a pelvic examination. As ultrasound is reflected by gas, scans can be severely impaired by excessive intestinal gas. Prior bowel preparation has only limited effect on the volume of gas within the bowel loops. Ultrasound is also absorbed readily by bone, so bony structures act as a barrier to the technique.

The interpretation of ultrasound images is less straightforward than conventional radiography or CT scanning. This means that there is a degree of operator dependency. This problem is less marked with modern real-time equipment than with the older static scanners.

Magnetic Resonance Imaging (MRI)

Magnetic resonance is a new form of imaging which provides a display of sectional anatomy without the use of ionizing radiation. The technique is based on the principle that when tissue is placed within a strong magnetic field, protons align themselves with respect to the forces applied by that field. Pulsed radiofrequency waves are then applied to disturb this alignment.

During the intervals between the pulses, the protons will emit radiation of characteristic frequencies as they return to their previous alignment. These emissions demonstrate the distribution and density of protons within the tissue under examination. By varying the radiofrequency pulses, and by observing differing characteristics of proton behaviour in response to these variations, MRI shows considerable versatility in the images which can be produced. Currently, this versatility comes at a price: namely the length of time taken to produce a set of images. This effectively limits the throughput of patients. At the time of writing, the technique is expensive, costs being roughly three times that of comparable CT studies. Current developments in MRI technology anticipate a reduction in the time taken to acquire images.

Although still in the experimental stages, image enhancement by intravenous contrast media also appears to have a role in MR imaging. Radiographic contrast agents are not appropriate, but agents which contain paramagnetic ions and therefore create their own magnetic fields are currently being assessed. The agent most widely used at the time of writing is gadolinium-labelled diethylenetriamine pentacetic acid (DTPA). Following intravenous injection, gadolinium-DTPA is distributed in the extracellular fluid space and excreted by glomerular filtration.

Unlike CT, MR imaging is not restricted to the axial plane. Anatomical images can be generated in coronal, sagittal and oblique planes as well as in axial section. A further advantage of MR over CT is the absence of signal from bone which eliminates the streak artefacts often seen on CT images around the base of the skull.

Magnetic resonance is many times more sensitive than CT in detecting pathological changes in grey and white matter, and its ability to image in multiple planes improves the demonstration of brain anatomy. The use of surface coils enables detailed images of superficial structures to be obtained so that orbital disease is most elegantly shown by this technique. Longitudinal demonstration of the spine and the spinal cord by MRI is a major advantage. Both normal and abnormal intervertebral discs are well shown, in addition to disorders of the spinal cord itself.

Electrocardiograph-gated MR studies are able to demonstrate the thickness of the myocardium and to differentiate between infarcted and normal muscle, a distinction that has not previously been possible by anatomical methods. Applications of MR in abdominal disease arise largely from the improved contrast resolution this technique offers so that early detection of parenchymal disorders of the liver and kidney may be possible. The use of sagittal and coronal image planes is also likely to be valuable in the detection, staging and monitoring of pelvic malignancy.

Radiation Dosimetry and Risk Assessment

The Biological Effects of Ionizing Radiation

Biological tissues absorb X-rays by means of interactions between the individual photons of the X-ray beam and atomic nuclei or their electron shells within the tissues. Tissue damage at atomic level may lead to cell death or to altered biological characteristics, including genetic mutation and the appearance of malignant growth patterns.

The absorption of ultrasound and radiofrequency waves in the body results in the transfer of energy with a rise in local temperature. With very high energy ultrasound radiation, mechanical disruption can occur at cellular level. However, the energies used for diagnostic ultrasound and MR imaging are several orders of magnitude lower than the levels needed to produce any biological effects, so these techniques are generally regarded as being free of risk.

The harmful effects of ionizing radiation manifest themselves in one of three ways: local tissue damage, teratogenesis and genetic effects.

Local tissue damage

This category includes the erythema, desquamation of epithelial surfaces, vascular damage and subsequent fibrosis which is seen very frequently after radiation therapy. These changes only occur with relatively high doses of radiation and tissue damage of this type should never be associated with diagnostic procedures. The one possible exception is the lens of the eye which is relatively radiation sensitive; special care needs to be taken in cases where multiple X-ray exposures of the eye are needed, e.g. in thin section tomography or CT of the orbit.

Teratogenesis

The fact that ionizing radiation can induce malignancy after a long latent period is well known; what is not known with any degree of confidence is the probability that a given dose of irradiation will result in malignant trans-formation. It is not yet certain whether the teratogenic effects of irradiation are dose-related, even at very low dose rates, or whether there is a threshold below which low levels of radiation are risk-free. For practical purposes we assume the worst, i.e. that even very low levels of radiation carry a risk. However, there is no clear information on whether the relationship between the risk of malignancy and the dose absorbed is linear or non-linear. A number of different theoretical models have been proposed but there is so little good data available on low level radiation exposure and the incidence of subsequent malignancy that the data fits all of the models equally badly. The risk estimates used in clinical practice are derived by extrapolation from data on the incidence of malignancy in patients exposed to high radiation levels in the aftermath of the atomic bombing of Japan.

Radiations of different energies carry a varying degree of risk in relation to the energy they impart to the tissue. This variation is small in the range of X-ray energies used in diagnostic procedures, so for practical purposes this factor can be ignored in risk assessment. The risk of subsequent malignancy is much higher when the tissues being irradiated are dividing rapidly, e.g. in the developing fetus. Irradiation of the embryo at an early stage of development (during the phase of organogenesis) also carries a potential risk of inducing somatic abnormalities in the fetus itself. For these reasons, special precautions must be taken to minimize irradiation of the abdomen and pelvis of pregnant women (see below).

Genetic effects

It is known from experimental studies that irradiation of the gonads can produce chromosomal damage resulting in genetic mutations which may appear in the offspring or in subsequent generations. Since the cause of 'spontaneous' mutations, which occur fairly frequently, remains unknown, it is virtually impossible to measure from a given radiation dose the risk of inducing mutations in subsequent generations. However, the general principle is accepted that we must, wherever possible, avoid irradiating the ovaries and testes of children and of adults in the reproductive age group.

Absorbed Doses from Diagnostic Procedures

The dose of radiation absorbed by the body is measured in terms of the energy that is imparted to the tissues by the radiation. The unit of measurement of absorbed dose is the Gray (Gy) which is defined as the amount of ionizing radiation which results in the deposition of 1 J of energy in 1 kg of the material being irradiated. Diagnostic X-ray procedures generally result in only part of the body being irradiated and so the biological effects are confined to this particular area. With diagnostic use of radionuclides however, the radiation dose is delivered differently depending upon the mechanism for handling the particular radiopharmaceutical (see earlier).

Radiation doses from procedures involving radiographs of different parts of the body and radioisotope tests using different radiopharmaceuticals may be compared by estimating the mean dose to the entire active bone marrow — the most sensitive target for whole body radiation. The estimated levels of the whole body (bone marrow) dose resulting from various diagnostic procedures are listed in Table 1.1. It must be emphasized that there may be considerable deviation from these figures in individual patients depending upon the patient's size, the number of X-ray exposures, the choice of X-ray energy, the type of film/screen combination used and the care taken in minimizing scattered radiation. In the case of radioisotope procedures, variations may occur according to the size of the patient, the administered dose of radiopharmaceutical, the rate at which the pharmaceutical is

Table 1.1 Typical whole body bone marrow doses from diagnostic tests

Dose	Tests
≤0·2 mGy	Lung perfusion/ventilation scintigraphy Thyroid scintigraphy using $^{99}Tc^m$ pertechnetate Chest X-ray Dental X-ray
0·3–1 mGy	Colloid liver/spleen scintigraphy HIDA scintigraphy Thyroid scintigraphy using ^{123}I Renal scintigraphy using DMSA Skull X-ray Pelvic X-ray (AP view only)
1–5 mGy	Bone, brain and multigated cardiac scintigraphy Renal scintigraphy with DTPA Thallium scintigraphy Cholecystogram Thoracic spine X-ray Intravenous urogram Lumbar spine X-ray (AP views)
5–10 mGy	Gallium scintigraphy CT scan Barium meal Barium enema X-ray pelvimetry

metabolized by different individuals and the presence of hepatic or renal failure.

Imaging in Women of Childbearing Age

Until recently, the use of imaging tests involving ionizing radiation (i.e. radiology and nuclear medicine but not ultrasound) in women of childbearing age was guided by the 'ten-day rule'. This rule was aimed at minimizing the risk of inadvertent irradiation of a fetus and is based on the fact that pregnancy is very unlikely to have occurred in the ten days after the start of the last menstrual period. It was recommended that any non-urgent procedure involving ionizing radiation should be carried out within this ten day period, waiting until after the next period if necessary. In the case of urgent tests, the ten-day rule could be waived. The ten-day rule could be inconvenient for patients, delaying investigations and perhaps necessitating additional trips to the hospital outpatient department. The rule also affected the functioning of departments, especially when booking investigations for which special preparation was required. Recent statements from the International Commission on Radiological Protection (ICRP) indicate that the risks to a fetus of irradiation *in utero* during the first two weeks of life are very small. The ICRP now believe that there need be no special limitations on radiological procedures during this period. Accordingly, the ten-day rule is no longer employed in many areas and the National Radiological Protection Board has suggested three new guidelines:

1. Any woman having a radiological procedure should be considered to be pregnant when a menstrual period is known to be overdue or missed, unless there is information indicating the absence of pregnancy.
2. Any woman being considered for radiological procedures where the uterus may be in or near to the primary beam should be asked 'Are you, or might you be pregnant?' Unless the answer is an unequivocal 'No', the patient should be regarded as if she were pregnant.
3. When a woman who is, or who is regarded as being, pregnant requires a radiographic examination in which the primary beam irradiates the fetus, special care should be taken to ascertain that the examination is indeed needed. The risk of irradiating the fetus may be much less than that of not making a proper diagnosis, so the examination should still be made if properly indicated, albeit with particular care to minimize the dose to the fetus.

At the time of writing, the NRPB recommendations apply only to radiological procedures but it is expected that similar guidelines will apply to radionuclide investigations.

As a result of the NRPB statement many areas now use a '28-day rule'. This states that non-urgent radiological procedures may be performed in women of reproductive age if it is not more than 28 days since the beginning of the last menstrual period. This strategy is clearly much more flexible than the ten-day rule and should result in many fewer investigations being postponed.

Unlike ionizing radiation, there is no evidence that diagnostic ultrasound causes fetal damage. No restriction is placed on the use of ultrasound in women of childbearing age, and in fact the technique is widely employed at all stages of pregnancy for indications such as determination of gestational age, diagnosis of twin pregnancy, study of fetal malformations, diagnosis of placental abnormalities and measurement of fetal size.

Risk Assessment

Virtually all medical and surgical interventions carry a degree of risk; the role of the individual practitioner is to ensure the best possible outcome for his or her patients. The risks associated with irradiation for diagnostic purposes might well be considered in the same light as the risks of morbidity or mortality associated with the use of drugs and with surgical procedures. Although we have a substantial volume of data regarding the qualitative aspects of biological damage from irradiation, the only good quantitative data is derived from animal experiments and from long-term studies of populations exposed to relatively high doses of radiation, i.e. Japanese survivors of the atomic bombs.

By extrapolation from the A-bomb survival data, it has been estimated that the lifetime risk of death from radiation-induced cancer in a patient exposed to a whole body dose of one mGy is 1:100 000. This dose is about the middle of the range most frequently encountered with diagnostic procedures. Similar levels of lifetime risk of death have been estimated to result from the smoking of 15 cigarettes/day or travelling 500 miles by motor car.

In terms of the population exposure to ionizing radiation, the total dose absorbed from diagnostic procedures by far exceeds the dose from all other man-made sources of radiation. However, even in countries with very highly developed and active health services, the absorbed dose to the population from medical use forms only a small fraction of the total annual dose. By far the major component of dose is derived from inevitable exposure to naturally occurring radioactive elements which are present in the minerals which make up the earth's crust.

Comparisons between the risks arising from the diagnostic use of ionizing radiation and the risks of other human activities may help to put the potential danger of medical radiation into perspective. However, even when quantitative risk estimates can be obtained and compared, it is clear that some risks are more acceptable than others. For example, the risk of maternal morbidity associated with normal pregnancy is socially and emotionally acceptable whereas the very much smaller risks associated with the use of oral contraceptives are much less readily accepted. The risks associated with driving motor vehicles and with moderate alcohol consumption tend to be accepted whilst the relatively infinitesimal risks arising from atmospheric nuclear fallout or food additives receive much more attention.

Counter-balancing risk and potential benefit is a complex problem with no easy answers. It is the role of the imaging professionals to ensure that the required diagnostic information is obtained whilst keeping any radiation exposure as low as reasonably achievable. The physician or surgeon who sends the patient for investigations involving irradiation should be confident that the information obtained by the investigation will contribute towards the management of the patient in some positive way.

Assessing the Utility of Diagnostic Tests

The value of diagnostic aids in clinical practice depends on a number of factors. Cost is important though it is often difficult to determine the precise cost of an individual test. Assuming an acceptable cost, the choice between various investigations depends on the utility and reliability of their results. If a test yields a numerical result it is usually relatively simple to define a 'normal' or reference range and to consider test results falling outside this range as abnormal or positive. For tests which do not yield a numerical result, and this is the case in most imaging tests, the borderline between positive (abnormal) and negative (normal) results is more difficult to define. It should be recognized that results obtained from this type of investigation are more subjective as they involve interpretation by an observer and are prone to variation between observers.

Furthermore, it is often an over-simplification to treat the outcome of imaging procedures as dichotomous, i.e. producing clear-cut positive or negative results. Image interpretation may be regarded more properly as a series of continuous variables, more or less skewed according to the clinical context. On the one hand, the distinction between 'fracture' and 'no fracture' on a radiograph is quite clear-cut in most cases; on the other hand, the presence of pulmonary venous congestion on a chest film is a finding of continuously variable severity. Biological variation between individual patients may be such that a particular feature on a diagnostic image may fall within the normal range for the population, although it may be abnormal for that particular patient. For example, the transverse diameter of the heart shadow on a chest film may normally be as great as 15·5 cm in a large muscular patient, whereas a heart of 13 cm size may be abnormal in a patient with six previous chest films showing the heart size to be constantly 11 cm in diameter.

Difficulties arise not only from variations in the degree of abnormality of particular image features, but also in the level of certainty with which the abnormality is detected. Uncertainty in diagnostic imaging arises from three main sources: perceptual difficulty (e.g. does this chest film show faint miliary mottling or is it normal?), interpretative difficulty (e.g. is the small filling defect on the barium enema caused by faeces or by a polyp?), and technical difficulty (e.g. is the quality of the examination good enough to show the lesion we are looking for, if it is present?). Studies which compare the relative contribution of different diagnostic procedures often fail to take these factors into account. The results of such assessments are also influenced by the selection of patients to be investigated (see below) and by the method of verification of the diagnosis, particularly in respect of patients in whom the imaging tests are negative. How can we be sure that these patients really do *not* have the lesion we are looking for? In many cases it is only by prolonged follow-up or detailed pathological correlation.

Bayes' Theorem

Bayes' theorem, as applied to diagnostic tests, states that the likelihood of a disease being present after the test result is known (post-test probability) is dependent on the pretest probability of the diagnosis. The pretest probability of the disease is also referred to as the prevalence in the population under consideration.

The use of Bayes' theorem can best be illustrated by working through an example. Assume that a test has both a sensitivity of 90 per cent and a specificity of 90 per cent for a particular disease. Then 90 per cent of patients with the disease will have a positive test result and 90 per cent of patients without the disease will have a negative test result. In clinical terms this would represent quite a respectable performance. If the test is then applied to a population of 100 patients of whom ten have the disease and 90 do not (10 per cent pretest probability or prevalence), nine positive test results will be obtained from patients who have the disease (true-positive and a further nine from those who do not (false-positive). Thus, at this disease prevalence the post-test probability of the disease being present with a positive result (also known as the positive predictive value) is 50 per cent (9/18). Negative test results will be obtained in one patient with the disease (false-negative) and 81 patients who do not have the disease. The post-test probability of the disease with a negative result is thus 1·2 per cent (1/82). The post-test probabilities for positive and negative results at various disease prevalences, and assuming both sensitivity and specificity of 90 per cent, are noted in Table 1.2. It can be seen that the significance of either a negative or a positive test result changes greatly according to disease prevalence in the group under study.

The application of Bayes' theorem has certain difficulties in practice. It is usually assumed when applying it, that the sensitivity and specificity of the test remains constant at varying prevalences. This is not necessarily true. For example, it might be that a symptom which defined a group with a low prevalence of disease was also associated with mild disease. In this case the sensitivity of the test for the disease might be lower than in a high prevalence group who also, on average, had more severe disease. There are difficulties

Table 1.2 Bayes' theorem
Post-test probabilities of disease being present with positive and negative test results at various pretest probabilities (prevalence).
A test sensitivity of 90% and a specificity of 90% are assumed.

Prevalence (%) (pretest probability)	Post-test probability (%) Test positive	Test negative
10	50	1·2
30	79·4	4·6
50	90	10
70	95·4	20·6
90	98·8	50

in calculating precise pretest probabilities in many cases and it is often difficult to obtain accurate values for sensitivity and specificity of a particular test. Nevertheless, the use of Bayes' theorem and of similar analytical approaches is an essential step in assessing the diagnostic utility of investigative procedures.

Further Reading

Dixon, A. (1983). *Body CT — a Handbook.* Churchill Livingstone, Edinburgh.

Maisey, M.N., Britton, K.E. and Gilday, D.L. (1983). *Clinical Nuclear Medicine.* Chapman and Hall, London.

·Meire, H.B. and Farrant, P. (1982). *Basic Clinical Ultrasound.* Teaching Series **4**. British Institute of Radiology, London.

Merrick, M.V. (1984). *Essentials of Nuclear Medicine.* Churchill Livingstone, Edinburgh.

Squire, L.F. (1982). *Fundamentals of Radiology.* Harvard University Press, Boston.

2

RESPIRATORY DISORDERS

Imaging Techniques

The Plain Chest Radiograph

With the exception of adipose tissue, the various soft-tissue structures in the body attenuate X-rays to a fairly uniform degree so that conventional radiographs are unable to distinguish between them. Uniquely, the fine structure of the lungs is readily demonstrated by radiographs because the soft-tissue components are normally outlined by air. This allows radiographic demonstration of detailed anatomy of the lung which cannot be achieved in any other organ without contrast injection. The amount of anatomical information available, the simplicity of the procedure, and the requirement for only the least costly and unsophisticated equipment have brought the plain chest film to prime position in the investigation of not just cardiothoracic disorders but also of virtually all patients who need investigating. No diagnostic test can replace careful clinical assessment of a patient, but the plain chest film probably has more influence on the further investigation and on the immediate management of patients than any other diagnostic procedure.

It is impossible to discuss the use of other imaging procedures in chest disease without considering an approach to the interpretation of plain chest films, and this is done later in this chapter.

The information on a single chest film is purely anatomical: interpretation in physiological terms depends upon correlation with past experience and experimental data. Sequential changes on two or more films however, can allow physiological data to be deduced more directly. For example, lateral movement of the mediastinum between inspiratory and expiratory films will indicate unequal pressure in the two hemithoraces as with a large pneumothorax; or, the gradual disappearance of septal lines on consecutive chest films obtained during the treatment of left heart failure may be interpreted in terms of falling left atrial pressure.

The limitations of the chest film will be discussed more fully in the second part of this chapter but a few basic reminders are worth pointing out. Firstly, soft tissues are generally indistinguishable on the chest film — i.e. blood, tumour, effusion, pus, myocardium and lung vessels all show the same density; secondly, the amount of air present within a lung indicates no more about the ventilation of that area than the heart size does about cardiac output; thirdly, although the range of causative factors in producing lung disease is very wide indeed, the ways in which the lung can respond to injury are limited. Although it is often possible to categorize the type of disturbance which is going on in the lung (i.e. consolidation, congestion, effusion, etc.), pin-pointing the aetiology of the lesion is possible much less frequently.

Conventional and Computed Tomography

One of the major disadvantages of the conventional radiograph is that three-dimensional anatomy is projected onto a two-dimensional image with superimposition of overlapping structures. Conventional tomography was developed over 50 years ago as a method for improving the anatomical detail available on radiographs by reducing as much as possible the effect of overlapping structures. The method involves the simultaneous movement, in diametrically opposite directions, of the X-ray tube and the film during the exposure. This movement, which may be linear, circular, elliptical or in the form of a more complex closed curve, has the effect of blurring out the images of structures which are outside the 'focal plane' which is the level in which the fulcrum of tube/film movement lies. The extent of movement during the exposure determines the effective thickness of the focal plane. Tomograms can be obtained in the anteroposterior, lateral or oblique planes.

Tomography may be used to provide more detailed localization of a lesion shown on plain films, e.g. whether or not a tumour has crossed a fissure. It may be used to try to establish whether or not a lesion is present, when the chest film is equivocal. A further application is to characterize a lesion which has already been shown, e.g. to show the margin of the lesion more clearly, to demonstrate the presence of calcification within it, etc. Tomography may also be used to bolster the validity of the 'exclusion' chest film, e.g. in searching for small lung metastases in patients with primary malignancy, or in providing detail of the main bronchi in patients presenting with haemoptysis.

In all its applications, conventional tomography has been superseded by computerized tomography where the latter facility is available. The ability of CT to resolve subtle differences in the X-ray attenuation of different tissues, e.g. its clear discrimination between pleural fluid and consolidated lung, gives CT a distinct advantage over linear tomography. A further advantage of CT is its complete freedom from overlapping structures (except within the volume of the slice). The axial plane in which scans are routinely obtained is ideal for examining the mediastinum since most of the major structures run longitudinally, and it also clarifies the anatomical relationships of peripheral lung lesions with the chest wall and pleura.

The radiation doses involved are of the same scale for both linear and computed tomography so, in all aspects except cost and availability, CT is preferable.

Ventilation/Perfusion Scintigraphy

Scintigraphic studies of lung function are usually referred to by the symbols 'V' for ventilation and 'Q' as the generally accepted physiological symbol for flow. Although the major application of lung scintigraphy is in the investigation of patients with suspected pulmonary embolism, knowledge of the distribution of lung function is often of clinical interest in patients with other cardiothoracic problems.

Lung perfusion scintigraphy

The scintigraphic assessment of regional lung perfusion is achieved by gamma camera imaging after the intravenous injection of labelled particles. The radiopharmaceutical usually used is technetium-labelled albumin in the form of microspheres or macroaggregates. These particles range in size from 5 to 100 micrometres (μm). When injected intravenously they pass through the right side of the heart and are distributed into the pulmonary circulation. They are too large to pass through the capillaries and most of the particles impact in precapillary arterioles. Assuming that the particles are well mixed with blood during their passage through the right heart, their distribution within the lung is proportional to regional blood flow. An area of impaired perfusion will have fewer impacted particles and will appear as a 'cold' area on the gamma camera images. Over the few hours after injection the particles break up and are cleared from the pulmonary circulation by macrophages.

In adults, the number of administered particles is about 100 000–200 000 which will occlude roughly 0·1 per cent of the precapillary arterioles in the normal lung. No changes in pulmonary haemodyamics or in lung function have been observed in patients with normal pulmonary artery pressures after injection of perfusion scanning agents. In severe pulmonary hypertension, especially if it has been long standing, the increased pulmonary resistance produced by particle embolization of the already compromised lung circulation may be enough to cause right heart failure. A few cases of sudden death shortly after injection of labelled particles have been reported in patients with critically severe pulmonary hypertension. In such patients it is suggested that the number of injected particles should be reduced to one-third or one-half of the usual dose. In patients with right-to-left shunts, some systemic distribution of the particles will occur. Although this is theoretically hazardous, no adverse consequences have been found in practice.

Normal perfusion scintigraphy shows almost uniform distribution of radioactivity. There may be diminished perfusion at the apices if injection is performed in the upright position reflecting the influence of gravity on regional perfusion. Normally, anterior, posterior, lateral and/or oblique views are obtained. This allows segmental perfusion to be assessed accurately.

Lung ventilation scintigraphy

Gamma camera assessment of lung ventilation is achieved by having the patient inhale radioactive inert gas or aerosol. The gases used most widely are xenon and krypton. Xenon-127 (half-life 28 days) and xenon-133 (half-life 5·3 days) require closed-circuit administration. The gas is inhaled through a face mask from a spirometer, the exhaled gas being collected in a large reservoir. Images of inhalation (single breath) and equilibration (steady state) are first obtained, then the source of xenon is shut off and consecutive images are obtained during the wash out phase to demonstrate regional ventilation. Because of the prolonged half-life of xenon isotopes, ventilation can be easily imaged in only one projection. Imaging in multiple projections is possible when the inhaled gas is very short-lived, e.g. krypton-81m (half-life 13 seconds). The use of multiple projections allows more precise matching of ventilation and perfusion. The use of krypton-81m is also technically simpler since it does not require a closed circuit for administration and the radiation dose to the patient is also much less than with the xenon nuclides. Unfortunately, the availability of krypton-81m is very limited.

Multiple-view ventilation studies are also feasible using inhaled aerosols of technetium-labelled compounds such as $^{99}Tc^m$ — DTPA. The apparatus for producing aerosols is relatively simple and the pharmaceutical can be prepared easily in kit form. The major disadvantages of this technique are that if the same radionuclide (i.e. technetium-99m) is used for both ventilation and perfusion, then the two examinations will have to be separated in time by a few hours; normally the aerosol procedure will be carried out first, DTPA being gradually absorbed from the bronchial mucosa and excreted by the kidney so that perfusion imaging can be carried out 1–2 hours later. A further disadvantage is that patients with airways disease show abnormal deposition of aerosol particles in the major bronchi which casts uncertainty over the interpretation of regional ventilation in these patients.

Whereas perfusion imaging can be carried out on any patient who is able to keep reasonably still for a few minutes, ventilation imaging, particularly the xenon and aerosol techniques which require closed circuit breathing, do need a degree of co-operation from the patient. Patients with severe dyspnoea or chest pain may find the inhalation procedures difficult or uncomfortable and the quality of the results may be compromised in this group.

The interpretation of lung scintigraphy will be discussed later in more detail in relation to pulmonary embolism, but a few general points may be made here. First, classifying the perfusion and ventilation of each region of the lung as 'normal' or 'abnormal' is a subjective exercise. Several studies have shown variations between different observers viewing the same images. Disagreements are least likely to occur where images are thought to be normal and where clear-cut abnormalities of gross degree are present. Disagreements (and therefore, by inference, errors) are much more likely where subtle changes are noticed and particularly when attempts are made to grade the severity and/or extent of the lesions on some sort of quantitative scale.

Other Scintigraphic Studies

Bone scintigraphy, described in more detail in Chapter 8, will be useful in investigating chest wall lesions. The age of rib fractures shown on a chest film, the presence of fractures not visible on the chest film, and evidence for the extension of primary tumours into the chest wall can all be shown on bone scintigraphy.

Scintigraphic studies of the oesophagus should be considered when investigating patients with unexplained chest pain and those in whom the chest film shows evidence of recurrent aspiration pneumonia. Radionuclide studies of oesophageal transit and gastroesophageal reflux are described elsewhere in this book.

Thyroid scintigraphy, also described in more detail elsewhere, has an important role in the assessment of patients with masses in the neck and/or superior mediastinum.

Scintigraphy using gallium citrate is occasionally helpful in evaluating patients with chronic inflammatory diseases of the lung. Gallium uptake can be used as an indicator of the level of activity in, for example, sarcoidosis and cystic fibrosis when the plain chest films show the anatomical disturbances but do not necessarily reflect the activity of the disease.

Ultrasound

Because ultrasound waves are scattered in air and absorbed rapidly in bone, applications for this technique in the thorax are limited. However, scanning between the ribs allows ultrasonic visualization of the pleural space and effusions can be readily detected. Ultrasound also has a major role in cardiac diagnosis, described in Chapter 3. Masses in the lower mediastinum can sometimes be demonstrated ultrasonically by scanning from below the costal margin and lesions in the superior mediastinum can sometimes be reached by scanning through the thoracic inlet.

Contrast Radiology

The oesophagus

Contrast examination using either barium or low-osmolality water-soluble media will be helpful in patients in whom oesophageal pathology may be responsible for pulmonary symptoms. Where the plain chest film suggests a mediastinal abnormality, contrast radiology of the oesophagus is often useful in showing the relationship of the oesophagus to other mediastinal structures, particularly mediastinal mass lesions.

Bronchography

Detailed endobronchial anatomy can be demonstrated radiographically after the bronchial walls have been coated with an opaque medium. Usually an inert oil-based iodine-containing compound is used. The material is introduced via endotracheal intubation with direct injection into the trachea immediately below the vocal cords or by trickling it through the larynx over the base of the tongue. All of these methods require care and skill but are safe in practised hands. The objective is to coat the bronchi down to fourth or fifth division but not to fill the alveolar air spaces. Radiographs are obtained in anterior, lateral and/or oblique views to demonstrate the bronchi of the two lungs separately. Transient impairment of gas exchange and of mucociliary clearance is inevitable with this technique, so it should be used with great caution in those patients whose respiratory reserve is already compromised.

   With the development of fibreoptic bronchoscopy, bronchography is now infrequently used for diagnosis. Its main application is as a preoperative technique in patients being considered for local resection, e.g. with patchy bronchiectasis, and occasionally to demonstrate segmental anatomy in patients with congenital anomalies of bronchial division.

Angiography

The vascular anatomy of the lung may be studied by pulmonary or systemic arteriography. Pulmonary angiography is a relatively straightforward technique with low morbidity requiring only the puncture of a femoral or antecubital vein for the introduction of a catheter via the right heart into left or right pulmonary artery. Where necessary, segmental catheterization can be carried out using catheter guidance systems. Pump injection of a substantial volume of contrast medium is accompanied by rapid sequence filming of the lung fields. Usually anterior and lateral views of each lung are obtained consecutively. Complications of the procedure are uncommon, the main hazard being that of inducing arrhythmias during passage of the catheter through the right heart, particularly in patients with an unstable myocardium or those with abnormalities of cardiac anatomy. Pulmonary angiography has a role in confirming or refuting the diagnosis of pulmonary embolism in cases where V/Q scintigraphy is not conclusive (Fig. 2.1). Other applications include investigation of suspected pulmonary arteriovenous malformations, fistulae, and as an adjunct to cardiac catheterization in patients with congenital anomalies; the latter subject is discussed in Chapter 3. The introduction of digital subtraction angiography could further reduce the morbidity of vascular studies of the lung but, at the time of writing, the place of DSA in the lung has not been established.

Selective catheterization of the bronchial arteries which arise from the

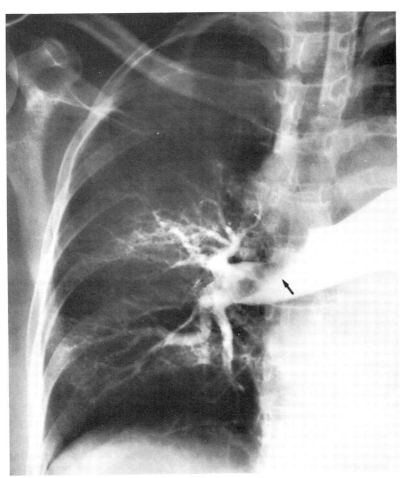

Fig. 2.1   Selective right pulmonary arteriogram showing an embolus protruding from the orifice of the main upper lobe artery.

thoracic aorta is occasionally used to try to localize the site of bleeding in patients with persistent haemoptysis. Therapeutic embolization has had some success in such patients. The technique is not easy and should not be undertaken lightly since there is a theoretical risk at least of damaging arterial branches to the spinal cord.

Angiographic demonstration of the great vessels of the mediastinum is relevant in patients with thoracic aneurysms or with clinical evidence of vena caval obstruction. In most cases a diagnostic assessment of the circulation can be made using non-invasive techniques such as CT and scintigraphy. Invasive angiographic procedures which require direct arterial or venous catheterization can be reserved for the detailed assessment required preoperatively in patients being considered for surgical treatment.

Endoscopy

Laryngoscopy and bronchoscopy are usually carried out by physicians and surgeons and are largely outside the scope of this book. However, it is important to consider the relationship of the endoscopic methods to other imaging techniques — though, clearly, direct visualization of the mucosal surface of the larynx, trachea and bronchial tree cannot be improved upon by indirect imaging procedures. The facility to obtain biopsies or bronchial washings directly from the affected areas is of great diagnostic value. However, it should be remembered that the whole of the lung parenchyma is outside the field of view of the bronchoscopist, that many peripheral lesions will produce no endobronchial abnormality, and that those hilar and mediastinal lesions which arise extrabronchially will sometimes be undetectable endoscopically.

Interpreting the Chest Radiograph

Assessment of Technique

The accuracy of interpretation of a chest film is greatly dependent upon correct positioning of the patient and suitable exposure factors being used. Conventionally, chest films are obtained with the patient standing and the anterior surface of the chest adjacent to the film cassette so that the X-ray beam travels in the posteroanterior (PA) direction. This minimizes the area of lung masked by the mediastinal structures since the bulk of the heart is in the anterior part of the chest cavity. Routine films are obtained in deep inspiration. Films obtained using mobile apparatus on patients too ill to attend the X-ray department are generally taken in anteroposterior projection, either supine or sitting. The heart and other mediastinal structures appear relatively larger on AP films than on PA films and a greater proportion of the lung is obscured. Ill patients are also often unable or unwilling to hold a deep inspiration so that the lung bases are also less well shown. Movement blur may further degrade the visible lung detail. Straight positioning of the patient during the exposure must be checked by observing symmetrical placement of the ends of the two clavicles relative to the upper dorsal vertebral bodies. Even a little rotation of the patient may produce different densities in the two lungs and apparent mediastinal shift. Over-exposure may conceal subtle abnormalities in the lung fields, whereas underexposure may prevent visualization of more central abnormalities, particularly the area of lung overlapped by the heart. On a correctly exposed film, it should be just possible to discern the disc spaces between the thoracic vertebral bodies and also to see the posterior ribs faintly through the heart. In a normal subject good inspiratory effort will bring the diaphragm down to the level of the anterior ends of the fifth or sixth rib and to about the tenth rib posteriorly.

Doubt and Certainty in Interpretation

The plain chest film probably contains more information about detailed anatomy than any other single image using any technique available to us. Traditionally, this information is analysed by a subjective visual inspection of the film. With few exceptions, the inferences which are drawn are qualitative. This is not to say that particular features of a chest film (such as the transverse diameter of the heart) could not be measured easily, but that the detection of the majority of chest film lesions depends on seeing features which should not normally be present, and not seeing features which are normally visible. Some features lend themselves readily to this type of qualitative interpretation, e.g. for practical purposes a pneumothorax is either present or absent. Other types of abnormality appear with different degrees of certainty, e.g. even with sequential films it may be difficult to discern precisely when a patch of consolidation or a fine nodular pattern first became visible. Some types of abnormality are much more specific than others. For example, a homogeneous opacity within which the bronchi are outlined by air ('air bronchogram') clearly indicates that the lung is consolidated — i.e. the alveolar air spaces are filled with fluid, but the bronchi remain patent. An opacity which is entirely similar except for the absence of the air bronchogram could be due to either consolidation of the lung with the bronchi also full of fluid, or to collapse of the lung with or without an overlying pleural effusion.

Degrees of certainty in interpretation are different in different parts of the image. For example, small mass lesions are seen more easily in the periphery of the mid-zones than close to the hila or in the apices (where they may be overlapped by the ends of the clavicles and first ribs), or in the posterior basal segment of the left lower lobe behind the heart. Finally, the confidence with which a chest film can be interpreted is underpinned by its technical adequacy. The factors described in the preceding section should be taken into account when assigning confidence levels to X-ray reports. For example, the exclusion value of a good quality film correctly positioned and exposed is far better than that of a film obtained using mobile apparatus on an uncooperative patient under difficult circumstances.

Radiologists often use descriptive terms (adjectives and adverbs) to give some indication of the levels of certainty in their interpretation of the film. The rationale behind this needs explaining. 'The heart is enlarged' is a commonplace interpretation, but one which logically assumes that the size of the heart has a normal range which is well known. Since heart size is a biological variable with a (probably) Gaussian distribution, only in cases of gross enlargement will the heart size clearly be well outside the normal distribution curve. Since heart size is also known to vary with the size and build of the patient, it would be sensible to take these factors into account before drawing any conclusions about an individual film. Of course, it is possible to circumvent these difficulties by reporting that 'the transverse diameter of the heart measures 16 cm', a statement which immediately allows comparison of that measurement with known data from groups of patients with and without heart disease. However, most radiologists will try to relate the film upon which they are commenting to their memory of other films they have seen so that the report 'the heart is enlarged' actually means 'in my opinion, the heart is larger than it should be in a patient of this age, sex and build'. The reporting radiologist may reinforce the confidence of his or her interpretation by adding further qualifying words. Reporting that 'the heart is grossly enlarged' carries the same inference as 'the heart is definitely enlarged' — i.e. the heart is so large that it is virtually impossible for it to be normal. Conversely, reporting that 'the heart is slightly enlarged' or 'minimally enlarged' actually means 'I am not sure whether this heart is enlarged or not'.

Finally, the confidence with which a particular diagnosis can be made or refuted on the chest film can sometimes be increased by varying the technique to suit the particular clinical problem. For example, in order to find out whether or not a pneumothorax is present, a film taken on expiration is desirable since the contrast between the lung and air in the pleural space is enhanced at lower lung volumes; to detect free pleural fluid, a lateral view or a PA view with the patient lying on the affected side is more likely to detect a small volume of fluid than the conventional erect PA film.

Radiographic Appearances

It is not the purpose of this book to give a detailed account of the interpretation of radiographic appearances. However, some knowledge of the major types of chest film abnormality and their pathological correlations is useful to any clinician dealing with acutely ill patients.

Homogeneous opacities

Total collapse of a lobe or a lung produces a homogeneous density in the affected area together with movements of landmarks within the thorax (hemidiaphragms, hila, lung vessels, fissures and mediastinum) which differ and so are characteristic for each of the lobes. Consolidation produces similar but larger densities without the volume-loss associated with collapse. Air-filled bronchi visible within an area of uniform opacity ('air broncho-gram') are an indication of pneumonic consolidation. The absence of a normally visible interface between lung and adjacent solid structures — e.g. heart borders, hemidiaphragms — is a sign of collapse or consolidation in the adjacent part of the lung. Free pleural fluid is seen surrounding the lung bases in the conventional erect film: it produces a meniscus around the pleural surface of the lung analogous to the effect of pushing a football into a half-filled bucket of water. The mobility of pleural fluid can be shown by obtaining an additional film in the supine or lateral decubitus position. A variety of different pathologies can produce the appearance of a solitary mass lesion on the chest film. Occasionally, the appearances will be specific — e.g. a sharp-edged mass of 3 cm size containing 'popcorn' calcification will be a hamartoma of the lung. Sometimes the combination of clinical and biochemical data with the chest film appearance and data from further imaging procedures will allow a confident diagnosis to be reached. In other cases, the appearances may remain non-specific and histology of the lesion then becomes essential for definitive diagnosis (Fig. 2.2). If the chest film shows multiple mass lesions a different list of likely causes must be considered, but the same principles of differential diagnosis still apply.

Patchy opacities

Heterogeneous areas of opacity with ill-defined edges constitute patchy consolidation, usually a result of infection or oedema. Patchy infiltration with lymphomatous or leukaemic tissue or alveolar cell carcinoma may also produce a similar appearance. Patchy oedema is most often caused by left heart failure but can also result from allergic, toxic or ischaemic insults to the lung. Clues for differential diagnosis of patchy lesions may be found in the state of the pulmonary vessels, the presence of pleural fluid or mediastinal abnormalities. However, it is often the evolution of the chest film over the course of the illness which provides the most useful information.

Mottling

Densities which are numerous, approximately uniform in size, and which have sharper outlines than the patchy opacities described above, combine to produce the appearance of 'mottling'. The causes of mottling are rather

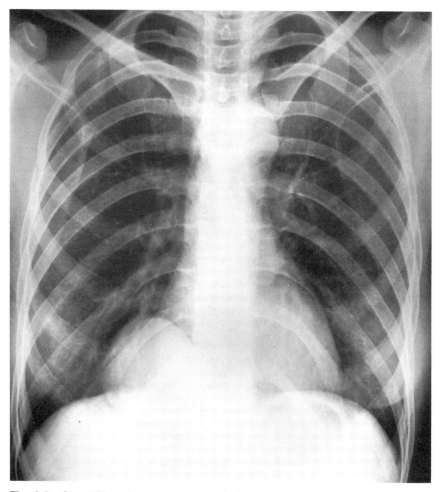

Fig. 2.2 Chest film of an asymptomatic immigrant. Differential diagnosis of the sharply defined mass overlying the right cardiophrenic angle includes springwater cyst, pericardial defect, diaphragmatic hernia, foregut duplication cyst etc. At surgery, a hydatid cyst was removed.

more varied than the range of radiographic appearances so it is often impossible to do more than narrow down the differential diagnosis by critical review of the chest film. Useful clues may be obtained by recording the size of the lesions, their distribution, their density, the presence of associated linear or ring-shaped densities, the state of the lung vessels, the presence of pleural fluid and of hilar or mediastinal lymph node enlargement. For example, miliary tuberculosis produces a diffuse pattern of very fine nodules typically without pleural or hilar lymph node abnormality. The nodules of sarcoidosis on the other hand are typically larger and more well defined, and usually there is concurrent enlargement of hilar lymph nodes. The combination of nodules and basal septal lines should prompt an enquiry for an occupational history predisposing to pneumoconiosis, although the same combination with the addition of pleural fluid is more suggestive of malignancy. However, if these features are combined with cardiac enlargement and pulmonary venous congestion, left heart failure could explain the entire picture.

A correct interpretation of the chest film showing miliary mottling is more likely if clinical and laboratory findings are taken into account along with the radiographic changes. The most important additional imaging procedure is to obtain sequential chest films and, wherever possible, to obtain previous chest films.

Lines and rings

Short linear densities usually result from interstitial oedema or thickening of the interlobular septa (Kerley B-lines) and are seen in patients with raised pulmonary venous pressure, lymphatic obstruction, pneumoconiosis or interstitial fibrosis. Long, straight or curved lines are usually pleural in origin except for the rarely seen Kerley A-lines radiating from the hilum in patients with carcinomatous lymphangitis or heart failure. Parallel line shadows suggest the bronchial wall thickening seen in bronchiectasis. Broad tubular densities in the lower zones are seen in bronchiectasis and usually caused by dilated bronchi full of secretions; linear band-like opacities seen in the periphery of the lungs result from either local atelectasis or fibrosis following infection or infarction. Large ring shadows surround cysts, bullae or cavitating lung lesions. Smaller ring shadows, particularly in clusters, usually indicate areas of cystic bronchiectasis. A widespread coarse reticular ('honeycomb') pattern is seen in children with histiocytosis X, some cases of tuberose sclerosis and neurofibromatosis, but also occurs in some patients with end-stage sarcoid or pneumoconioses. A fine reticular pattern is seen in patients with interstitial fibrosis resulting from chronic fibrosing alveolitis, sarcoid, systemic sclerosis and other collagen disorders, asbestosis, and some forms of drug reaction.

Hypertransradiancies

If there appears to be hypertransradiancy affecting the whole of one lung, it is important to check for rotation or scoliosis first. Films exposed with the patient rotated will show increased blackening (hypertransradiancy) on the side of the chest rotated away from the film cassette. Dorsal scoliosis is associated with asymmetry of the rib cage which can produce similar discrepancies in transradiancy between the two sides. Once these factors are excluded, it is usually not difficult to decide whether one lung is abnormally opaque, or the other lung is abnormally transradiant. If there is difficulty in making this distinction, the simplest next step is to obtain films in inspiration and expiration. The lung which changes most between inspiration and expiration is the normal one. If there is still doubt, V/Q scintigraphy will show which lung is functioning most normally.

Increased transradiancy of an entire lung is most commonly the result of air-trapping due to check valve obstruction of the main bronchus by either a foreign body or, less commonly, by an endobronchial tumour. Increased transradiancy without air-trapping is seen in McLeod's syndrome. Localized hypertransradiancies result from local collections of air either within the lung (cysts, bullae, lobar or segmental emphysema) or within the pleural space (pneumothorax).

Calcifications and other high densities

Calcified deposits within the lung are most often the result of previous tuberculous infection. In North America, histoplasmosis and coccidioidomycosis also commonly leave calcified residues in the lung. Dense flecks of calcification are seen in some hamartomas and in bone-forming metastases from osteosarcoma. Primary carcinomas of the lung do not calcify but may encompass an area of pre-existing calcification. Widespread nodules of calcific density occur — rarely — in patients with rheumatic heart disease (particularly after recurrent attacks of rheumatic fever) and are also an occasional sequel to chickenpox pneumonia. Widespread lung nodules which appear dense because of their heavy metal content may be seen following industrial exposure to metallic iron or tin. Pulmonary haemosiderosis, either primary or resulting from mitral valve disease, is also characterized by high density nodules. Calcification in the cartilages of the bronchial tree is an occasional incidental finding in the aged; arterial calcification in the pulmonary vessels is usually associated with severe pulmonary hypertension. Lymph node calcification is most often a sequel of infection but is occasionally seen in patients with silicosis and occurs after chemotherapy or radiotherapy in some cases of lymphoma. Calcification in the pleura and in mediastinal lesions is discussed below.

Cavitating lesions

It is important to recognize excavation within a lung lesion since this almost invariably indicates active progressive disease. Cavitation can occur in almost any lesion within the lung field, including inflammatory, granulomatous and neoplastic lesions. Pointers towards the underlying pathology are derived from observation of the site of the lesion, the thickness and regularity of the cavity wall, the presence of other lesions in the lung, and the time sequence of progressive changes. Cavities arise through central necrosis in areas of consolidation (tuberculous or pyogenic pneumonia), in areas of infarction, within primary or secondary tumours (particularly squamous cell carcinoma), or within granulomatous or fibrotic masses (rheumatoid nodules, Wegener's, progressive massive fibrosis). A fluid level within the cavity indicates the formation of an abscess or the aseptic liquefaction of a tumour or infarct. Solid material within a cavity could be necrotic debris but should also raise the suspicion of mycetoma due to secondary fungal infection of a pre-existing cavity. Multiple cavitating lesions should suggest tuberculous pneumonia but also occur in rheumatoid disease, Wegener's granulomatosis, and occasional patients with lung metastases.

Lung vessels

Inspection of the plain film may show alterations in the size and number of vessels — either locally or throughout the lung, crowding or displacement of adjacent vessels, alterations in the relative sizes of upper and lower zone vessels, and abnormalities of the branching pattern. In the periphery of the lung, it is virtually impossible to distinguish arteries from veins. Centrally, the arteries are seen to arise from the position of the main pulmonary artery while the veins converge on the left atrium; thus the veins tend to cross the arteries in the hilar regions and so the vessels can often be distinguished here. The arteries also tend to follow the branching pattern of the bronchi more closely so lobar and segmental arteries may be recognized particularly when the bronchi are seen end on. It is important to remember that the size of lung vessels is an indication of the volume of blood within them. In the case of the arteries, there is a rough correlation between volume and flow, but this does not apply to the pulmonary veins. Generalized enlargement of pulmonary arteries indicates a left-to-right shunt with increased pulmonary blood flow; generalized reduction in the size of pulmonary arteries (oligaemia) indicates reduced flow, most often associated with right-to-left cardiac shunts. Reduction in the size and/or number of peripheral arteries with enlargement of the pulmonary trunk indicates pulmonary arterial hypertension. A sudden change in the calibre of vessels locally within a lung is seen in some patients with pulmonary embolism although in most cases the chest film is normal. Enlargement of the pulmonary veins occurs in patients with left-to-right shunts but is much more commonly a result of raised left atrial pressure resulting from left heart failure or valvular heart disease. In the standing or sitting position, the lower lobes carry about three times as much blood flow as the upper lobes so that on the chest film the upper lobe vessels appear distinctly smaller than lower lobe vessels at the same distance from the hilum. This perfusion gradient is lost in patients with basal lung disease and in those with left heart failure where vasoconstriction in the lower lobe vessels diverts blood to the upper lobes producing a recognizable alteration in the relative size of vessels. The perfusion gradient is lost in patients with left-to-right shunts and other high-output states.

The pleura and chest wall

Free fluid in the pleural space gravitates to the base of the lung and so is first seen on the PA film as a homogeneous density obliterating the costophrenic angle. This feature usually indicates the presence of at least 300–400 ml fluid; smaller amounts may only be visible on the lateral view where the fluid is first seen in the posterior costophrenic sulcus. If films taken in different positions show no movement of the pleural opacity, it is impossible to distinguish between loculated fluid and solid material. Air within the pleural space allows the lung to become separated from the chest wall so that the lung edge becomes visible. With both air and fluid in the pleural space a characteristic horizontal interface (fluid level) is visible. Calcification in extensive areas of pleural thickening is a consequence of previous tuberculous pleurisy, empyema, or traumatic haemothorax. Small irregular plaques of calcification, particularly over the diaphragmatic surfaces of the pleura, are seen in patients previously exposed to asbestos. Peripheral opacities with a sharply defined medial margin usually arise in the pleura or chest wall. Bone destruction in the adjacent ribs indicates a malignant pathology (usually myeloma or metastatic disease) or rarely an abscess.

The diaphragm

The position of the diaphragm is indicated by the edges of the adjacent aerated lung. Where there is basal consolidation or pleural fluid on the right side, the position of the hemidiaphragm cannot be assessed. On the left side, air in the gastric fundus and/or splenic flexure of the colon provide useful indicators for the level of the hemidiaphragm. Free intraperitoneal air, when present, outlines the undersurface of the diaphragm on erect films. The high position of a hemidiaphragm indicates paralysis, eventration or shrinkage of the overlying lung. Depressed diaphragms indicate overinflation, usually due to airways disease.

The mediastinum

Abnormal mediastinal contours are most frequently the result of heart disease (discussed in Chapter 3). Diffuse widening of the mediastinum can be the result of oedema, fibrosis, fatty infiltration, superior vena caval obstruction, or extensive tumour. Gas visible in the mediastinum usually indicates traumatic damage to the oesophagus, trachea or extrapleural bronchi, but may also result from the migration of gas from the pleural space in patients with pneumothorax and in infants or small children with airways obstruction. Localized opacities arising in or adjacent to the mediastinum will almost always require further imaging investigations, but careful inspection of the PA and lateral chest films should considerably narrow down the differential diagnosis by localizing the site of origin of the lesion.

Further Investigation

Lesions in the Lung

Lung lesions shown on the plain chest film can be characterized to a greater or lesser degree by additional anatomical information obtained from tomography or CT, and by the physiological data obtained by scintigraphy. However, in many cases the most important dimension to the differential diagnosis of lung lesions is that of time, and the most useful further procedure one can perform is to obtain the patient's previous chest films and to take further follow-up films. No investigation is more cost-effective than the plain chest radiograph and sequential films taken over a time scale appropriate to the duration of the pathology under suspicion are very often enough to reach a firm diagnosis.

Imaging techniques

When there is difficulty in establishing the cause of a large homogeneous or patchy opacity, CT or linear tomography will help sometimes by adding further anatomical information. In particular, the axial view of the chest provided by CT shows the relationship of lung lesions to the chest wall, bronchi and major vessels very clearly. CT may show an air bronchogram within a lesion which looks homogeneous on the chest film. In other cases, CT may reveal a central fluid component indicating abscess formation or the breakdown of a necrotic tumour (Fig. 2.3). Segmental distribution of lung lesions is elegantly shown by CT and either CT or linear tomography can show whether a lesion has crossed one of the major fissures. The question of bronchial obstruction often arises in patients with peripheral lung consolidation and either CT or linear tomography can be used to outline the lobar and segmental bronchi. Bronchoscopy gives a more direct view of endobronchial disease and has the advantage that biopsies may be obtained under direct vision. CT has the advantage of showing not only the bronchi but all the surrounding structures as well. In patients with peripheral opacities or band shadows, the question of pulmonary infarction will arise. V/Q scintigraphy will distinguish between infection and infarction in some but not all cases. The area of opacity will be underperfused, whatever the cause. Areas of ischaemia in regions of the lung which appear normal on the chest film and are normally ventilated indicate the strong probability of multiple emboli (Fig. 2.4). A perfusion deficit which is less extensive than the abnormality on the chest film (and less extensive than the disturbance of ventilation which accompanies it) is typically found in infective consolidation. The limit of resolution for lung lesions by scintigraphy is 2–3 cm so scans will be normal at the site of band opacities unless the surrounding area of lung is also ischaemic.

The management of solitary mass lesions in the lung often comes to focus

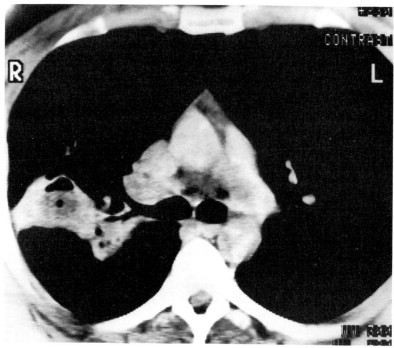

Fig. 2.3 CT scan through an area of dense consolidation in the right lung shown on plain chest films. Within the solid area of lung is a central fluid component with an air/fluid level indicating abscess or necrotic tumour.

on a decision of whether or not to carry out a thoracotomy and resection. Additional imaging data is often helpful, particularly if supplemented by histology or cytology from guided needle biopsy. CT or linear tomography will illustrate the margin of the lesion: a spiculated border being typical for primary carcinoma, whilst a clear smooth margin suggests a benign lesion. The presence of calcification within the mass will again suggest a benign cause except in the rare instance of a carcinoma arising close to an old tuberculous scar. In the latter case, the surrounding lung may be expected to show signs of pre-existing disease. Improved localization of the mass lesion and particularly its relation to the pleural surfaces both at chest wall and at the fissures, and to the major bronchi and vessels will all be helpful to the surgeon. Where the nature of the mass is still dubious, percutaneous needle biopsy may be guided fluoroscopically. The question of whether a single nodule detected on a PA chest film is truly solitary is one which is best answered by CT. The detection of small nodules in other regions of the chest will have a major impact on the future management of the patient. The discovery of a clear-cut mass lesion with a 'stalk' joining it to the hilum should arouse suspicion of an arteriovenous malformation. Rapid sequence CT with contrast enhancement may be conclusive in these cases, but

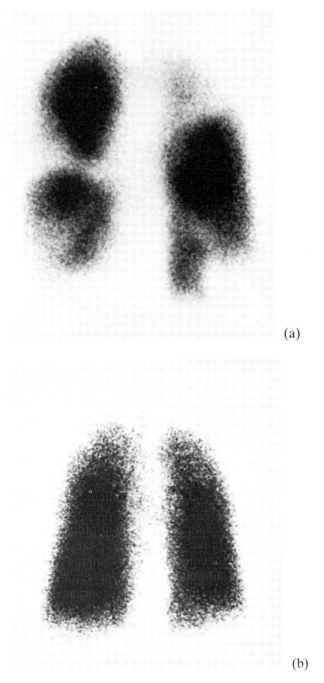

(a)

(b)

Fig. 2.4 (a) Perfusion and (b) ventilation lung images from a patient with postoperative dyspnoea and a normal chest X-ray. There is a high probability of multiple emboli.

pulmonary arteriography should be carried out before resection since these lesions are often multiple.

As with other diffuse lung diseases, the use of CT will demonstrate the extent and distribution of interstitial fibrosis and of emphysema more precisely than the plain chest film. The early changes of these conditions are also shown more easily by CT. However, at the time of writing, CT has not made much impact on the differential diagnosis of diffuse lung disorders. Radiographic methods may give little idea of the level of disease activity in patients with established chronic inflammatory lung disorders and in such cases gallium scintigraphy may be a useful indicator. Studies in children with cystic fibrosis and adults with sarcoidosis have shown good correlations between gallium uptake in the lung and other evidence of progressive disease activity.

Most patients presenting with bronchiectasis appear to have localized lesions on the chest film but CT will often show the disease to be more extensive. If surgery is seriously being considered, bronchography is required for a complete demonstration of segmental bronchial anatomy. V/Q scintigraphy may also be useful to establish the relative contribution of the two lungs to overall function and also to highlight further unsuspected areas of abnormality.

Similar considerations apply in patients with bullous emphysema. CT will often show the extent of the disease to be greater than was suspected from the chest film, and CT should certainly be undertaken before deciding on surgical resection of an apparently single bullous area. V/Q scintigraphy is again likely to make a useful contribution by indicating the relative function of different regions of the lung.

Investigation of the hyperlucent lung is also simplified by V/Q scintigraphy. If doubt exists as to which of the two lungs is diseased, it can be assumed that the lung which shows the better function on V/Q scintigraphy is the normal one. Hyperinflation affecting a lobe or a whole lung will often be the result of endobronchial obstruction by foreign body or tumour, and requires bronchoscopy for both diagnosis and treatment.

Lesions of the Chest Wall, Pleura and Diaphragm

It is often difficult to see on the plain film whether a peripheral lesion is arising within the lung or from the chest wall. This distinction can usually be made by CT or by linear tomograms taken in a projection which is tangential to the chest wall nearest to the lesion. Of course, many peripheral lung lesions spread to the pleura, e.g. carcinoma, whilst some lesions of pleural origin infiltrate the lung, e.g. mesothelioma. When pulmonary and pleural components coexist, it is sometimes impossible to decide which arose first.

Invasion of the chest wall by lung tumours reduces the chances of successful resection and so is worth knowing about preoperatively (Fig. 2.5). Sometimes invasion of the chest wall can be recognized on CT but in other

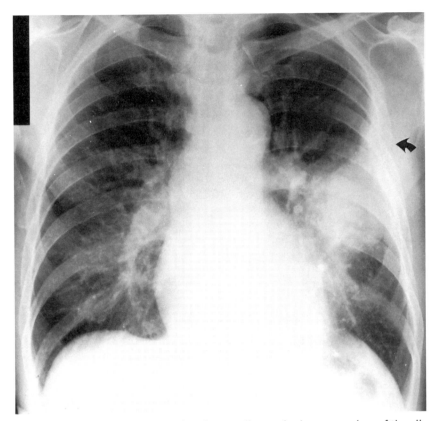

Fig. 2.5 Bronchial carcinoma: the chest radiograph shows erosion of the ribs adjacent to the mass indicating chest wall invasion (arrow).

cases it is impossible to distinguish between contact and invasion. Bone destruction shown on the chest film or on oblique rib radiographs is a good indicator of non-resectability. If symptoms point to chest wall invasion but the plain films show no bone destruction, bone scintigraphy is worthwhile, as this technique will usually show the first signs of bone reaction to invading tumour.

In the absence of free fluid, pleural opacities on the plain film may be either liquid or solid. The distinction is generally readily made by CT (Fig. 2.6). This method can also be used to identify appropriate sites for aspiration or biopsy of pleural lesions. Similarly, ultrasound scanning will detect pleural fluid as long as there is a clear pathway of access for the scanning probe. This is achieved round the periphery of the lung by scanning in the intercostal spaces or at the lung bases by scanning from below the costal margin (Fig. 2.7). Because of the feasibility of approaching from almost any angle, the adaptability and mobility of real-time machines, and the absence of radiation, ultrasound is the method of choice for guiding

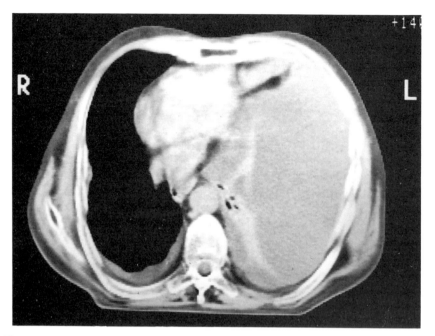

Fig. 2.6 CT scan through the lower chest of a patient whose plain radiographs showed a total homogeneous opacity of the left hemithorax. The totally collapsed lung is clearly seen within the pleural fluid.

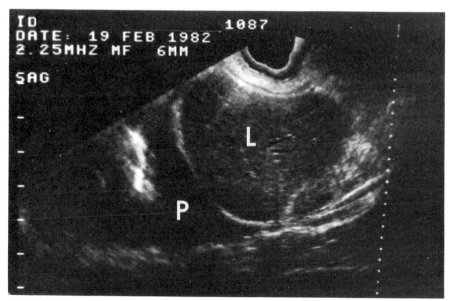

Fig. 2.7 Sagittal ultrasound scan through the right upper quadrant of abdomen and lower chest showing the liver substance (L) bounded by the diaphragm above which is a collection of pleural fluid (P). (Courtesy of Dr H Irving).

percutaneous biopsy and aspiration procedures into the chest wall and pleura.

Deciding whether an elevated diaphragm is paralysed or simply displaced can be achieved by observing it in real-time. If the position of the diaphragm is visible on the plain film (i.e. if the lung base is clear or if there is gas under the diaphragm) fluoroscopy during deep breathing will show whether diaphragmatic movement is normal, reduced, absent or paradoxical. In the presence of pleural effusion or basal lung consolidation, fluoroscopic screening of the diaphragm is much less fruitful and, in these circumstances, ultrasound scanning is indicated. On ultrasound, a normal diaphragm is strongly reflective and can still be readily identified when there is fluid either above or below it. This method is also the most likely imaging technique to succeed in detecting subphrenic fluid in the presence of basal lung consolidation and effusion. Virtually all patients with subphrenic abscesses will have a basal pleural effusion and some loss of volume in the lower lobe of the lung on the affected side, and if there is no gas in the abscess, plain films and fluoroscopy will be unrewarding. Ultrasound should, in most cases, be able to indicate the relative proportions of fluid above and below the diaphragm and thus the likely site of primary pathology. Access for ultrasound is much better on the right side (scanning through the liver) than on the left where the gas-filled colon and the gastric fundus will reduce the available access. In difficult cases, CT will usually distinguish subphrenic from subpulmonary fluid by its distribution relative to the lung base but this is not always easy. The diaphragm is only clearly visualized on CT when there is fluid both above and below it.

The Mediastinum

Careful scrutiny of the plain PA and lateral radiographs, with penetrated PA and/or oblique views, if necessary, will allow the site of origin of most mediastinal masses to be classified into anterior, middle and posterior compartments. Local views of the spine and ribs are useful when the mass arises in the posterior mediastinum. Neurogenic tumours commonly arise on intercostal nerves or nerve roots; they may erode or displace the adjacent ribs and widen the exit foramen of the affected nerve roots. Fore-gut duplication cysts are almost invariably associated with bony anomalies in the spine whereas symmetrical paraspinal soft-tissue masses may arise from infection or tumour within a vertebral body.

Masses anterior to the spine in the superior mediastinum may displace or compress the trachea, the usual causes being thyroid enlargement or lymph node disease. Thyroid scintigraphy is worthwhile in patients with superior mediastinal masses. Iodine-123 should be used rather than pertechnetate since the latter gives a high background count in the mediastinum. Not only may scintigraphy show functioning thyroid tissue within the mediastinum but it will also indicate the condition of the cervical thyroid; this is relevant since the vast majority of mediastinal thyroid masses are inferior extensions of multinodular goitres in the neck. If surgery is contemplated, CT is necessary to show the extent of the intrathoracic thyroid and its relation to the great vessels.

Thymomas

A soft-tissue mass arising anterior to the heart and close to the midline should raise the suspicion of thymoma. CT will confirm the location and surrounding anatomical relations of the mass. Benign thymomas, once they are large enough to obliterate the normal thymus, are indistinguishable from lymph node masses which may arise in the same site. Malignant thymomas have a characteristically heterogeneous texture with areas of necrosis or liquefaction and tend to invade the pericardium, pleura and chest wall producing effusions. The normal adult thymus, although not visible on plain films, is recognizable on CT up to the fourth decade. Small tumours and subtle degrees of thymic enlargement can be detected on CT when the chest film looks entirely normal. It is, therefore, worth carrying out this procedure in patients with myasthenia before considering thoracotomy. Germinal-cell tumours are also most typically located in the anterior mediastinum. Demonstration of fat densities within the mass on CT is characteristic (Fig. 2.8), but more likely to be seen in benign than in malignant teratomas.

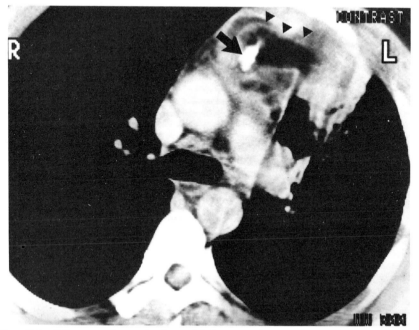

Fig. 2.8 CT scan through the mid-thorax of a patient with an anterior mediastinal mass showing an area of dense calcification (arrow) representing a tooth and fatty tissue (arrow heads). Diagnosis: dermoid cyst.

Superior vena caval obstruction

The clinical features of superior vena caval obstruction (SVCO) are seen in some patients with diffuse widening of the superior mediastinum (Fig. 2.9). The site and extent of venous obstruction can be confirmed by contrast venography performed by arm vein injections, but this procedure is not without discomfort to the patient and there may be difficulties in controlling the bleeding at the puncture sites if catheters are introduced. Radionuclide venography can be carried out with fine needle injections into the hand veins, carries no risk of toxicity or discomfort to the patient, and gives an

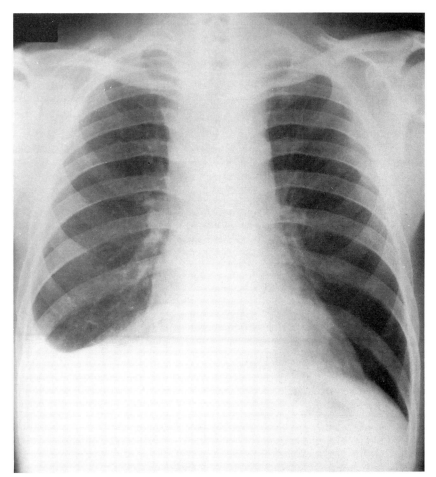

Fig. 2.9 Chest film of a patient presenting with a superior mediastinal obstruction showing diffuse widening of the superior mediastinum and a right pleural effusion. The mediastinal appearance is non-specific and could be caused by oedema, fatty infiltration, or tumour. CT with contrast enhancement showed occlusion of the SVC and innominate veins with venous engorgement and oedema of the mediastinum.

elegant demonstration of the presence and site of venous obstruction and the pattern of collateral drainage. Once the site of obstruction has been shown, the abnormal anatomy is best demonstrated by CT. Usually it is not difficult to distinguish SVCO due to a mediastinal tumour from the appearance of chronic inflammatory mediastinitis (usually tuberculous) and from idiopathic mediastinal fibrosis. A similar chest film appearance may result from fatty infiltration of the mediastinum — often associated with Cushing's syndrome — or from diffuse mediastinal oedema resulting from cardiac tamponade or tricuspid stenosis. Again, these conditions should be readily identified by CT.

Air-fluid levels

The presence of an air-fluid level in the mediastinum on the erect chest film indicates the need to examine the oesophagus. If oesophageal perforation is clinically suspected, water-soluble contrast medium of low osmolality should be used. In all other circumstances, barium is preferable. A rounded density in the superior mediastinum with a single fluid level in it is likely to be a pharyngeal pouch and if this is confirmed on barium swallow, no further tests need be done. A dilated oesophagus containing an air-fluid level is typically seen in the late stages of achalasia, but occasionally slow-growing carcinomas may cause enough stasis and proximal dilatation to produce an air-fluid level on the chest film. The diagnosis can be confirmed by barium swallow, with endoscopic biopsy if there is any doubt. Barium swallow is also useful in patients with middle mediastinal masses, firstly to indicate the anatomical relation of the oesophagus to the mass and secondly to look for evidence of invasion of the oesophagus which may influence surgical management.

Mid-mediastinal masses

Bronchoscopy is an essential step in the investigation of middle mediastinal masses since direct spread from a primary bronchial carcinoma is common. However, CT is more informative about the extrabronchial extent of the mass, its relation to surrounding structures, and the presence of lymph node enlargement elsewhere in the mediastinum. Developmental bronchial cysts are usually closely associated with the main bronchi or trachea but occasionally lie more peripherally or in the posterior mediastinum. CT demonstration of a clear-cut margin and fluid content is helpful in defining these lesions but in some cases the contents may be so viscid as to appear solid on CT. Oesophageal and neuro-enteric cysts usually lie more posteriorly and are associated with bony anomalies of the spine. Sequestration cysts also occur most commonly at the lung bases. These lesions typically obtain arterial blood supply from the upper abdominal aorta, so once a suspicion of this diagnosis has been raised, aortography should be considered as an aid to planning surgery.

The lymph nodes

Lymph nodes are present in all areas of the mediastinum. Any or all of them may be enlarged, giving rise to local or generalized mediastinal mass disease. CT is the preferred method for demonstrating lymph node enlargement (Fig. 2.10). Lymphomas and most metastatic tumours in lymph nodes appear as uniform soft-tissue masses. The lobular shape of clusters of lymph nodes is usually preserved but may be lost in rapidly growing lymphomas or metastases, particularly when only a single site is involved.

Lymph nodes can be identified in the anterior mediastinal fat, around the aortic arch, surrounding the trachea and main bronchi, in the posterior mediastinum immediately below the carina and to the right of the oesophagus, in the cardiophrenic angles anteriorly, and in the retrocrural area alongside the aorta.

Aortic aneurysm

Saccular or fusiform aneurysms of the thoracic aorta can often be recognized from the plain film. CT typically shows dilatation of the affected part of the

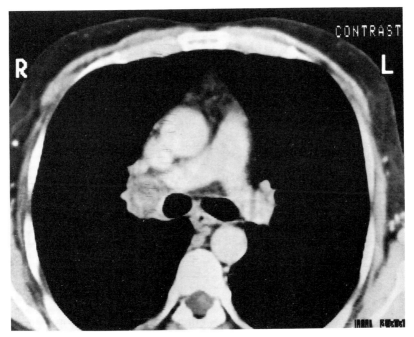

Fig. 2.10 CT scan through the mid-thorax of a patient with a peripheral bronchial carcinoma. The aorta, SVC and main left and right pulmonary arteries are visualized with contrast enhancement; immediately lateral to the right main bronchus is a soft-tissue mass of tumour density. Diagnosis: hilar node metastasis.

aorta with calcification in the wall and thrombus partly filling the lumen. Soft-tissue thickening around the aorta, mediastinal oedema and pleural fluid are all pointers to recent leakage from an aneurysm. With rapid sequence CT after contrast injection, anatomical detail of the aorta may be adequate for surgical appraisal and the presence of aortic regurgitation can be shown non-invasively by ultrasound. Aortography may still be needed however, particularly combined with coronary artery catheterization since severe coronary disease may preclude surgical treatment of a thoracic aneurysm. Acute aortic dissection is a more difficult problem. Most patients in whom this diagnosis is clinically suspected are acutely ill and in poor physical condition. Often only supine chest films can be obtained; this makes it difficult to draw conclusions about the width of the mediastinum. Rapid sequence CT with contrast enhancement will show the majority of dissection flaps and also has the advantage of detecting evidence of leakage into the mediastinum or pleural spaces. Dissections in the ascending aorta are usually detectable by ultrasound. Nevertheless, at the time of writing, aortography is still regarded as necessary in patients with suspected acute dissection. The use of digital subtraction angiography allows direct aortography to be carried out with fine gauge arterial catheters using relatively small amounts of contrast media, so the risks of this technique are minimized.

Other mediastinal masses

Mediastinal masses which are adjacent to the diaphragm can often be explored by ultrasound scanning from below the costal margin. The most common site for a pleuropericardial ('spring-water') cyst is in the right cardiophrenic angle; ultrasound will distinguish between a simple cyst and the more common lipoma or large fat pad seen at this site which can produce a similar chest film appearance. Fatty tissue is shown by CT *par excellence* and any doubt about the nature of presumed fat pads in the cardiophrenic angles can be resolved easily by this method. Hiatal hernia presenting as a mediastinal mass is usually diagnosable on plain films — with a mouthful of barium if there is any difficulty. Other forms of diaphragmatic hernia, particularly congenital anterior (Morgagni) defects and the posterior (Bochdalek) hernia may require CT and/or barium enema examination to identify the contents. If there is doubt about the nature of soft tissue herniated through the anterior part of the right hemidiaphragm, colloid scintigraphy of the liver will show whether there is functioning hepatic tissue within the hernia.

With posterior mediastinal masses, bone abnormalities on the plain films point to neural tumour or neurenteric cyst. The extraspinal component of the mass is then best shown by CT but myelography, or CT with dilute contrast in the subarachnoid space, may be needed to show the intrathecal extent of a dumbell-type lesion.

In summary, the further investigation of mediastinal abnormalities shown on the plain film depends largely upon the site of the lesion. In most circumstances CT is the method of choice with considerable contributions made by contrast studies of the oesophagus, aorta and spinal canal in appropriate cases. Thyroid scintigraphy is useful for masses in the superior mediastinum while ultrasound scanning will help to sort out lesions abutting against the diaphragm.

Specific Clinical Problems

Pleuritic Chest Pain

The lesion responsible for the pain will often be visible on the plain chest film. If a pneumothorax is suspected clinically, an expiratory film should be obtained in addition to the film taken on deep inspiration since small pneumothoraces are more easily visible on expiration. A peripheral area of consolidation with or without a pleural effusion can indicate either infection or infarction. V/Q scintigraphy will be abnormal in either case but may show evidence of emboli elsewhere in the chest. The majority of emboli do not cause infarction and it is doubtful whether pulmonary embolism itself is responsible for pleuritic pain. If the chest radiograph is normal, the diagnosis of infection or infarction is not excluded in patients with recent onset of symptoms since the radiographic changes tend to lag behind the clinical events. In such cases, a further film after 24 hours will usually show the lesion. If the initial film shows a large pleural collection, a further film in the supine or lateral decubitus position will show whether the fluid is freely mobile and often also allows more of the underlying lung to be visualized. Ultrasound or CT can be used to confirm that peripheral densities are fluid rather than solid and to localize suitable sites for diagnostic aspiration. Rib fractures or bone erosion from chest wall tumours may be visible on the plain film or on oblique views of the chest wall. However, if these films do not help and there is clinical suspicion of a chest wall lesion, bone scintigraphy is indicated. Scintigraphy will also identify bone lesions in the dorsal spine which may be responsible for girdle pain. Acute inflammatory processes in the upper abdomen may also present with pleuritic-type pain in the lower chest. Clinical clues will point towards abdominal disease in most, but not all, cases. Even when the chest film shows an abnormality at the lung base, the primary pathology may be below the diaphragm, so it is worth considering investigating this area. Initial imaging procedures should be a plain radiograph of the upper abdomen followed, if necessary, by ultrasound scanning of the liver, gall bladder and subphrenic regions.

Haemoptysis

In many cases the initial chest film will show the responsible lesion. A central or peripheral lung tumour, an area of consolidation, basal bronchiectasis, and acute or chronic pulmonary oedema are common causes. Difficulties arise when the chest film is normal, when there is doubt whether the lesion shown on the chest film is the one causing the haemoptysis, and when bleeding is so profuse that its source needs to be found and dealt with urgently. There is evidence that pulmonary embolism may result in localized bleeding into the lung without going on to infarction, so V/Q scintigraphy is worthwhile if the chest film is normal. Bronchoscopy may also be carried out to exclude a bleeding endobronchial lesion. In patients with pre-existing lung disease the same further investigations are suggested, although the results are often more difficult to interpret in the absence of data from previous examinations. In some cases, the diagnosis can be reached only retrospectively after treatment and resolution of the acute problem. In patients with profuse haemoptysis, bronchoscopy will usually identify which segment the bleeding is coming from. This is not always possible however, and in difficult cases scintigraphy using sulphur colloid may be used to find the approximate site of extravasation; this technique is mostly used for the localization of gastrointestinal bleeding and it is described in Chapter 4. As a last resort, selective bronchial arteriography can be tried for localization and therapeutic embolization of the bleeding vessels, but this is a difficult and potentially dangerous technique which should not be undertaken lightly.

Dyspnoea

As with chest pain and haemoptysis, most lesions causing dyspnoea will be identifiable on the initial chest film. Once a radiographic abnormality has been recognized, its further investigation can follow the lines discussed previously. If the chest film is normal, or if the degree of dyspnoea seems out of proportion to the abnormalities shown on the plain film, V/Q scintigraphy may be valuable since multiple pulmonary emboli frequently cause no visible alteration to the chest radiograph. Where lung function tests show a restrictive or obstructive defect yet the chest film appears normal, CT may show subtle changes of emphysema or interstitial fibrosis. CT may also be useful in patients with established lung disease whose dyspnoea is worsening. For example, in patients with asbestos-related pleural disease, the development of an intrapulmonary component spreading from the pleural lesion is suggestive of mesothelioma. Progressive activity in chronic lung disease may be indicated by abnormal uptake on gallium scintigraphy.

Suspected Hilar Mass

Assessment of the lung hila is one of the more difficult aspects of chest film interpretation and there are two situations where one's index of suspicion must be particularly high. First, in patients suspected of harbouring a primary bronchial tumour — e.g. those presenting with unexplained haemoptysis or with evidence of metastatic disease elsewhere — and second, patients in whom the presence of enlarged hilar lymph nodes will be an important influential factor in diagnosis and/or staging — e.g. those with peripheral lung tumours, suspected lymphoma or sarcoidosis. If an endobronchial tumour is suspected, bronchoscopy is the logical step. If examination of the hilar lymph nodes is required, linear tomography or CT should be performed. Linear tomography carried out in steep oblique or lateral projections has the advantage of giving a longitudinal demonstration of the lobar and segmental bronchi. Thin section CT with contrast enhancement can also be used to show detailed bronchial anatomy and also to illustrate the surrounding vessels and lymph nodes very clearly.

Postoperative Chest Problems

Patients with lobar or segmental atelectasis in the first few days after surgery are more in need of treatment than of investigation. Persisting lobar collapse should raise the question of bronchial obstruction for which bronchoscopy is the appropriate vehicle for diagnosis and treatment. The distinction between infection and pulmonary infarction in the postoperative chest can sometimes be made by V/Q scintigraphy but it is important to remember that these two conditions can coexist and that pulmonary embolism is commonly asymptomatic in the perioperative period. However, V/Q scintigraphy is certainly indicated in patients with unexplained dyspnoea or haemoptysis. Surgery to the upper abdomen is often followed by the appearance of basal consolidation and pleural effusion and it may be difficult to decide whether the primary cause lies above or below the diaphragm. Screening the diaphragm is of little help in these patients unless there is visible subphrenic gas since reduced diaphragmatic movement occurs with basal pneumonia as well as with subphrenic abscess. This question can usually be decided by ultrasound scanning supplemented by CT if there is technical difficulty in obtaining access due to gaseous distention, wounds, dressings, etc.

Chest Trauma

Many patients with chest trauma are successfully managed on plain films alone but, in selected cases, contrast radiology, CT and scintigraphy can all contribute. Where oesophageal perforation is suspected, a careful swallow using low osmolality water-soluble contrast medium is required. Mediastinal haematoma can be shown by CT but, in cases of blunt trauma, bleeding is just as likely to come from small mediastinal veins as from aortic leakage. If a traumatic dissection or partial rupture of the aorta is suspected, urgent aortography is preferable. CT may also show small areas of lung contusion, pleural effusions and even small pneumothoraces which can be readily missed on a supine chest film in an acutely distressed patient; these additional findings, however, are unlikely to make much impact on the clinical management of the patient. Rib fractures, if not visible on plain films, will be picked up by bone scintigraphy. This technique will also differentiate between recent and old vertebral fractures in patients with wedged dorsal vertebrae shown on radiographs. Probably the most valuable contribution of CT in acute thoracic trauma is in the demonstration of spinal fractures which may be difficult to pick up on radiographs, particularly when the vertebral appendages are involved. Bone fragments which may impinge on the spinal canal are most readily identified in the cross-sectional view provided by CT. In patients with suspected diaphragmatic injury where the lung base is obscured on the plain film, ultrasound will usually establish the position of the diaphragm. Ultrasound is also the first-choice technique for demonstrating the presence of pericardial fluid which may be suspected in patients showing an acute increase in heart size.

Lung Tumours: Staging and Resectability

The first consideration in patients with lung tumours is the site and extent of the primary growth — particularly its relation to pleural surfaces, the mediastinum and the chest wall. If these relationships are not clear from plain films, linear tomography or CT is needed. In cases of doubtful chest wall invasion, bone scintigraphy can be helpful. The second consideration is that of the presence of lymph node metastases. Hilar nodes may be shown on plain films or linear tomograms but subcarinal nodes are often difficult to detect and are much better shown by CT. The third consideration relates to the presence of metastatic disease elsewhere. Several studies have shown that prognosis in patients with primary lung cancer is strongly influenced by the presence of metastatic disease, particularly in the liver, adrenals and brain, and some authorities suggest CT scanning of the head and upper abdomen before embarking on thoracotomy.

Finally, lobectomy or pneumonectomy may be particularly hazardous in patients with pre-existing chronic lung disease and, in these cases, V/Q scintigraphy will indicate the relative function of the two lungs. The functional contribution of a lung which harbours a tumour, particularly one close to the hilum, is often surprisingly reduced, even when the tumour is quite small and the lung apparently fully inflated on the chest film. If the patient has already lost most of the function of the affected lung, its removal may be expected to produce little further deficit in functional capacity.

Looking for Lung Metastases

Lung metastases are sought in patients with primary malignancy for two reasons. First, the presence of lung metastases will usually influence how the primary tumour is treated, and second, if metastases are sufficiently few in number, it may be worth resecting them. It is well established that linear tomography is more likely to detect small metastases than the plain chest film and that CT is more sensitive still in detecting small nodules. However, it is impractical to subject every patient with primary malignancy to a CT examination of the lungs and unreasonable to delay treatment of the primary tumour. In patients with tumours which metastasize to the lung only rarely or very late in their course — e.g. most tumours of the central nervous system and transitional cell tumours of the urinary tract — the use of CT is unlikely to be cost-effective. On the other hand, the detection of pulmonary metastases is of vital importance in the management of patients with testicular tumours, melanoma, and osteosarcoma, all of which commonly metastasize to the lung at a fairly early stage. Where surgical treatment implies a major resection, e.g. removal of hepatocellular carcinoma, a careful search for lung metastases should be made. In patients with adenocarcinoma of the kidney, the presence of lung nodules may be used as an indication for therapeutic embolization of the kidney as an alternative to surgery. Where surgical excision of an apparently solitary lung metastasis is contemplated, as in some patients with renal cell carcinoma or osteosarcoma, the absence of other lesions should be confirmed by the most sensitive method available, i.e. CT scanning.

Chronic Lung Disease: Is It Active?

This question arises most frequently in the patient whose chest film shows evidence of previous tuberculous infection and who is now presenting with signs of acute disease. Inactive disease is usually represented on the chest film by thick linear densities enclosing air spaces representing areas of fibrosis and emphysema, together with calcification and pleural thickening. It is virtually impossible to be certain that this type of lesion is not active, but radiographic features which point positively towards active disease include alveolar consolidation, pleural fluid and cavity formation, especially if a fluid level is present. The most decisive radiographic sign, however, is the development of new lesions detected on comparison with previous films. If clinical and bacteriological examination are unhelpful and no old films are available, linear tomography or, preferably, CT may be used to look for cavities which are particularly significant if containing fluid. The use of gallium scintigraphy for indicating the level of activity in cystic fibrosis and sarcoidosis has been mentioned already.

'Routine' Chest Radiograph

Radiography of the chest is carried out far more frequently than any other imaging investigation. The test is relatively simple, inexpensive, informative, and without detectable hazard to the patient. Nevertheless, it is still possible to have too much of a good thing. The value of routine preoperative chest films in patients younger than 40 years undergoing minor surgery is in doubt. The value of consecutive chest films has been emphasized already, but it must be pointed out that the interval between successive chest films should be appropriate to the rate of evolution of the disease process. For example, no one would doubt the value of films repeated at intervals of a few hours immediately after thoracic surgery, but obtaining daily chest films after uncomplicated abdominal surgery or during the treatment of pneumonia seems prodigal. Finally, it is emphasized, again, that the exclusion value of the chest radiograph is very largely dependent upon its technical quality. A supine film obtained with mobile X-ray apparatus on a collapsed patient during the excitement of resuscitation is likely to be far less informative than a departmental film positioned and exposed carefully and with the patient's co-operation. Films which are non-diagnostic should be repeated; clinical decisions should not be based on inadequate radiographs.

Further Reading

Naidich, D.P., Zerhouni, E.A. and Siegelman, S.S. (1984). *Computed Tomography of the Thorax*. Raven Press, New York.

Putnam, C.E. (1981). *Pulmonary Diagnosis — Imaging and Other Techniques*. Appleton Century Crofts, New York.

Sostman, H.D., Rapaport, S., Gottschalk, A. and Greenspan, R.H. (1986). Imaging of pulmonary embolism. *Investigative Radiology* **21**, 443.

Waxman, A.D. (1986). The role of nuclear medicine in pulmonary neoplastic processes. *Seminars in Nuclear Medicine* **16**, 285.

3

THE CARDIOVASCULAR SYSTEM

Cardiac Imaging

A variety of imaging procedures are used in the assessment of cardiac disease. They can be conveniently divided into non-invasive studies, which do not require cardiac catheterization, and invasive studies which do.

Non-Invasive Studies

Plain X-rays

Straight X-rays of the chest can provide considerable information about the heart. The standard view is the posteroanterior (PA) chest X-ray taken with the X-ray tube 2 m from the film. This standard distance avoids distortion of the cardiac image due to magnification factors. Anteroposterior (AP) films, especially portable films, are unreliable in interpreting cardiac size and shape because of great variations in magnification factors. In addition, lateral or oblique views may be obtained on occasions.

The overall heart size is assessed from the PA chest film by measuring the maximum transverse diameter of the heart. As a rough guide the maximum cardiac transverse diameter should not be greater than 50 per cent of the maximum diameter of the thorax (Fig. 3.1). This measurement is often used in standard clinical practice. For more accurate assessment of cardiac size, tables are available which give normal values of the cardiac diameter for different heights and weights for both sexes. A further method is to assess cardiac volume from the length, width and depth of the heart as assessed on PA and lateral chest X-rays. The last method is rarely employed in clinical practice.

An increase in cardiac size may be due to enlargement of one or more of the cardiac chambers or to a pericardial effusion. In pericardial effusion the heart, characteristically, has a globular shape, but this may also be seen in

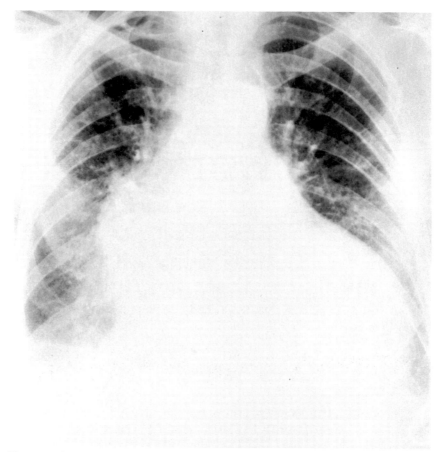

Fig. 3.1 Gross cardiomegaly and pulmonary vascular congestion in a patient with cardiac failure secondary to long-standing hypertension and ischaemic heart disease.

congestive cardiomyopathy and the presence of a pericardial effusion is best diagnosed by echocardiography (ultrasound).

The cardiac shape is also assessed from the plain X-ray. In the normal individual, approximately two-thirds of the cardiac shadow lies to the left of the midline. On the PA chest film the left border of the heart is normally produced (from above downwards) by the aortic knuckle, the pulmonary outflow tract, the left atrial appendage (which is not usually detectable as a separate bulge), and the left ventricle. The right border of the heart is formed by the superior vena cava or aorta superiorly — the aorta becoming more prominent with increasing age — and the right atrium below.

Enlargement of the heart to the right of the midline may be due to the increased size of either the right atrium or right ventricle. If the enlargement extends to the superior vena cava then it is likely that the right atrium is enlarged. Enlargement of the heart to the left of the midline, implies enlargement of one or both of the ventricles. A normal transverse diameter of the heart does not exclude ventricular enlargement as right ventricular enlargement initially occurs anteriorly, while the early stages of left ventricular enlargement may produce increased convexity of the left ventricular border without any change in transverse diameter. It is often not possible to say from the PA film whether an increased cardiac shadow is due to right or left ventricular enlargement. A left lateral view may help. In right ventricular enlargement, the anterior border of the heart maintains contact with the sternum beyond the normal upper point of midway between the diaphragm and the angle of Louis. In left ventricular enlargement, the main feature on the left lateral view is a prominence of the convexity of the lower posterior heart border behind the inferior vena cava. It is also important to distinguish between dilatation of the ventricular cavities, where the cavity size is enlarged, and hypertrophy in which the ventricular muscle mass is increased. The two states commonly coexist, but pure hypertrophy does not cause an increased cardiac size until it is advanced. By contrast, ventricular dilatation produces an enlarged cardiac shadow at an earlier stage. The presence of ventricular hypertrophy can be determined by ECG or echocardiographic studies.

Enlargement of the left atrium is detected from the PA chest film as a distinct bulge on the left heart border due to the left atrial appendage. The left heart border thus loses some of its concavity in the upper part and becomes straighter. Additionally, the X-ray density of the middle of the heart is increased in left atrial enlargement and a double edge may become visible on the right heart border. Lesser degrees of left atrial enlargement may be more easily detected on lateral X-rays as an increase in diameter of the upper part of the cardiac shadow. In the past, left atrial size was often estimated from lateral and oblique views by the indentation of the barium-filled oesophagus. This procedure has now been almost entirely replaced by echocardiography.

The chest film should be examined for evidence of calcification in the cardiac valves, coronary arteries and aorta, and in the ventricular myocardium.

The lung fields are also assessed on plain X-ray (Fig. 3.1). The presence of pulmonary venous hypertension produces a characteristic change in the lung fields. Normally, the vessels in the upper lung fields are smaller than those in the lower lungs in the erect position. With pulmonary venous hypertension, however, the upper zone vessels become more prominent and exceed the diameter of the lower zone vessels. This appearance results from a relative diversion of pulmonary blood flow from the lower to the upper zones in venous hypertension. If these signs are accompanied by an enlarged pulmonary trunk it is likely that the patient also has pulmonary arterial hypertension.

Pulmonary oedema may produce a variety of X-ray changes including enlargement of the hilar vessels with a hazy and indistinct surrounding edge (the so-called 'bat's wing appearance'), and widespread mottled pulmonary shadows. The latter appearance may be difficult to differentiate from bronchopneumonia. Kerley's B lines may develop. These indicate fluid within the interlobular septa of the lung and consist of fine, dense horizontal lines which extend to the periphery, usually most easily visible in the costophrenic angles. Kerley's B lines occur most often in pulmonary oedema but may be produced by other conditions such as lymphangitis carcinomatosa, sarcoidosis and occupational lung disease. Pleural effusions may develop in pulmonary oedema, either bilaterally or unilaterally, though more commonly on the right side.

Echocardiography

The use of ultrasound to examine the heart (echocardiography) has become increasingly important. In M-mode echocardiography a representation is obtained of the motion of the various cardiac structures in one dimension. The structures visualized will depend on the level at which the transducer is placed. Three levels are characteristically employed: that of the aorta and left atrium, the level of the anterior cusp of the mitral valve and the level of the left ventricular cavity. Increasingly, M-mode echocardiography is being replaced by two-dimensional echocardiography which allows better study of cardiac anatomy. The views obtained are limited by the bony structures of the chest wall and by the poor passage of the ultrasound beam through lung tissue. This latter fact may make it impossible to obtain echocardiographic studies in some patients with chronic obstructive airways disease and hyperinflation of the lungs. The two-dimensional studies are performed in four standard projections: the parasternal, apical, suprasternal and subcostal projections.

Echocardiography is an accurate method of determining the dimensions of the cardiac chambers, especially the left atrium and left ventricle. The thickness of the left ventricular free wall and the interventricular septum can also be assessed with considerable precision. Left ventricular function can be assessed from echocardiography. Various parameters have been derived for quantitating overall ventricular function from echocardiography. The ventricular ejection fraction can be estimated from echocardiography by measuring changes in dimensions of the ventricle during the cardiac cycle and converting these to volume changes.

Thus:

Ventricular ejection fraction =
$$\frac{\text{End diastolic volume} - \text{End systolic volume}}{\text{End diastolic volume}} \times 100$$

These methods yield clinically useful results though they are less reliable in very abnormal hearts where some of the geometric assumptions used in their calculation become invalid. Regional ventricular function is evaluated from echocardiography, with the demonstration of areas of reduced or absent or even paradoxical ventricular wall motion. Attempts have been made to obtain echocardiographic studies of ventricular function during exercise but in general only the resting state can be successfully studied.

The mitral and aortic valves can be studied well by echocardiography with respect to calcification of valve cusps, normality of valve cusp motion during the cardiac cycle and the presence of valvular regurgitation. It is also possible to detect the presence of vegetations on the valves in patients with infective endocarditis. The study of right heart valves is possible in some patients but is more difficult.

Echocardiography is the technique of choice for the demonstration of pericardial effusion (Fig. 3.2) and echo techniques are widely used for the non-invasive diagnosis of intracavitary masses such as mural thrombi, thrombi in the left atrial appendage (Fig. 3.3) or, less often, intracardiac tumours.

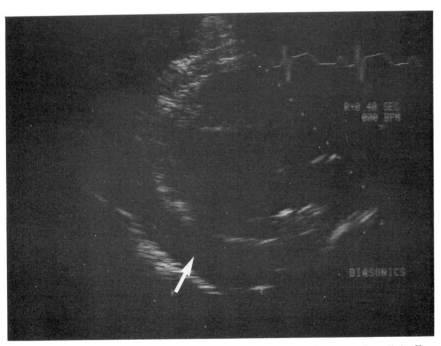

Fig. 3.2 Two-dimensional echocardiogram demonstrating a pericardial effusion as an echo-free area (arrowed) adjacent to the cardiac cavity.

Doppler ultrasound studies are also employed in the heart. They rely upon the fact that when an ultrasound wave is reflected from a moving structure, such as blood, the shift in the change in the frequency of the ultrasound wave is proportional to the velocity of the blood flow — among other factors. By combining Doppler and other echocardiographic methods

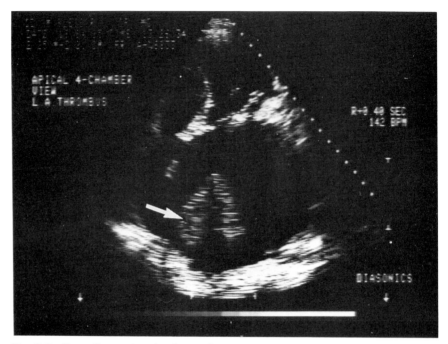

Fig. 3.3 Two-dimensional echocardiogram demonstrating a left atrial thrombus (arrowed) in a patient with long-standing atrial fibrillation.

it is now possible to obtain reliable, non-invasive, estimates of the severity of valvular stenosis.

In paediatric cardiology, where precise detail of cardiac anatomy is of paramount importance, the use of two-dimensional echocardiography and Doppler ultrasound have greatly enhanced non-invasive diagnosis. Ultrasound is now an essential investigation in all patients with congenital heart disease. As a result of echocardiography, many of these patients can undergo cardiac surgery without the need for extensive invasive investigation.

Computerized tomography

At present CT plays little or no part in the routine clinical assessment of cardiac disorders, although it is valuable in the investigation of abnormalities of the aorta and in pericardial tumours. Recently introduced, fast CT techniques may have a role in the future in assessing bypass graft patency and cardiac aneurysms.

Nuclear medicine

Nuclear medicine procedures are now widely accepted for the non-invasive assessment of myocardial perfusion and ventricular function. Scintigraphic techniques can also be used for the detection of myocardial infarction and for the diagnosis and quantification of intracardiac shunts.

Myocardial perfusion

The standard non-invasive scintigraphic technique for the assessment of myocardial perfusion uses the monovalent cation thallium-201 (^{201}Tl). When present in tracer amounts ^{201}Tl behaves physiologically like potassium. After intravenous injection, the tracer is widely distributed in the body, with around 4 per cent of the injected dose accumulating in the myocardium. In clinical practice, the major determinant of ^{201}Tl uptake in the myocardium is regional blood flow, but having been delivered to the myocardial cell the tracer must be trapped by it. This process is thought to involve the ATP-ase-dependent sodium–potassium pump. Thus, myocardial images obtained with ^{201}Tl are a composite of blood flow and myocardial cell function, with the former being more important in most cases.

Thallium-201 imaging is most often employed in patients with established or suspected coronary artery disease and involves the technique of stress and redistribution imaging. If the isotope is injected (in a dose of 60–80 MBq) at peak stress, the initial distribution will reflect regional myocardial blood flow. In the normal stress study, uniform distribution of uptake is seen in the left ventricular myocardium which appears as a horseshoe or a doughnut shape. Some right ventricular uptake is also often seen. Areas of stress-induced ischaemia have reduced ^{201}Tl uptake and appear as cold areas on images obtained immediately after stress. Areas of fibrotic or infarcted myocardium will also show reduced uptake. Over the few hours after stress injection, the uptake within areas of ischaemia will improve, so that images obtained several hours after stress without injection of further tracer will show improvement of ^{201}Tl uptake within ischaemic areas. Such abnormalities are classified as 'reversible' (Fig. 3.4). By contrast, areas of fibrosis or infarction will not change significantly between stress and redistribution — a 'fixed defect'. The differentiation between reversible and fixed defects can also be made by comparing stress images with those obtained after injecting a second dose of ^{201}Tl at rest. This doubles the radiation dose to the patient and requires hospital attendance on two occasions several days apart. Stress and redistribution imaging is therefore usually employed in preference to rest and stress imaging.

Best results from myocardial imaging are obtained by exercising patients to peak stress or until they have evidence of myocardial ischaemia, such as angina. This is potentially hazardous in patients with coronary heart disease. Continuous monitoring of ECG and blood pressure is essential and full resuscitation facilities must be available in the room where the test is being

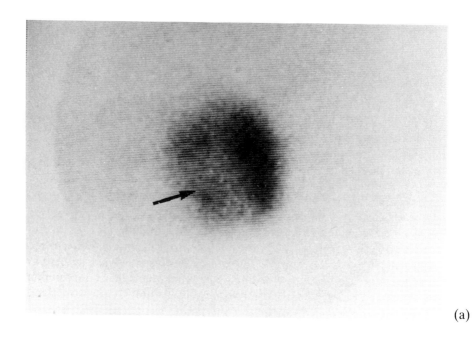

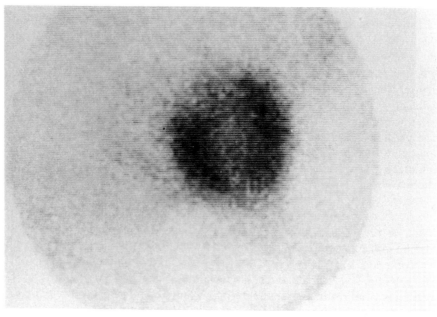

Fig. 3.4 (a) Stress ^{201}Tl myocardial image (45° LAO projection showing decreased tracer uptake in the interventricular septum (arrowed). (b) Normal 4 h post-injection ('redistribution') image showing complete reversal of the defect. The patient had a 90 per cent left anterior descending coronary artery stenosis at coronary angiography.

conducted. Stress myocardial imaging does not appear to be intrinsically more dangerous than standard stress electrocardiography.

In patients who are unable to exercise, the injection of ^{201}Tl after induction of coronary vasodilation by intravenous dipyridamole gives similar information.

Once images are obtained, they may be assessed either visually or by quantitative techniques involving computers. Most recent studies suggest that the quantitative methods improve accuracy when compared to visual analysis.

Thallium-201 has a number of shortcomings as an agent for myocardial imaging including a long half-life, resulting in a relatively high radiation dose to the patient, a low fractional uptake by the myocardium (4 per cent of the injected activity) and low energy photons which are not ideal for the gamma camera. No superior compound has yet become routinely available but considerable effort is being expended in the development of alternative radiopharmaceuticals — notably certain ^{99}Tc^{m}-labelled agents and ^{123}I-labelled fatty acids. It is possible that the newer agents will give information biased more towards myocardial cell function (extraction) rather than blood flow.

Ventricular function studies
Radioisotopic assessment of left ventricular and, to a lesser extent, right ventricular function is now well established in clinical practice. Two main approaches are possible: the first-pass technique and the ECG-gated (equilibrium) technique.

In first-pass nuclear angiocardiography a bolus of tracer is injected into a peripheral vein and its initial passage through the central circulation is monitored by a gamma camera interfaced to a computer. The tracer utilized for these studies is most often ^{99}Tc^{m} either as pertechnetate or bound to a variety of other compounds. The usual injected activity is 600–800 MBq. Various projections may be utilized such as left anterior oblique (LAO), anterior (A) or right anterior oblique (RAO). The passage of the bolus through the left ventricle is observed with high temporal resolution and an activity time curve can be generated from the region of the left ventricle using the computer system. With a good bolus, the passage of activity through the left ventricle takes 5–10 cardiac cycles, during which cyclical fluctuation of activity within the ventricle is seen. For the purposes of data analysis, the activity time curves derived from the cycles encompassing the first pass of the bolus are summed to yield a composite cycle.

In the ECG-gated (equilibrium) method a tracer is utilized which remains intravascular for some hours after injection. The compounds usually employed are ^{99}Tc^{m} human serum albumin (HSA) or ^{99}Tc^{m}-labelled autologous red blood cells. Labelling of the patient's own blood cells is most commonly done by an *in vivo* method involving the intravenous injection of cold (i.e. non-radioactive) stannous pyrophosphate followed 20–30 minutes later by 600–800 MBq $^{99}Tc^m$ pertechnetate. Red blood cells can also be labelled *in vitro* by incubating a sample of blood with stannous pyrophosphate initially, and then with $^{99}Tc^m$.

With the equilibrium technique, activity is present simultaneously in both ventricles. They are best separated in the left anterior oblique projection and this is utilized when quantitative evaluation of ventricular function is being performed. Other views may be employed for the analysis of regional left ventricular wall motion. In equilibrium angiocardiography, data is collected from multiple cycles (more than 100). The R wave of the ECG is used to signal the beginning of each cycle and each R–R interval is broken-up into a number of segments, usually 16–32. Image data from each of these segments is stored in a different part of the computer memory, with corresponding segments from each cycle being superimposed. At the end of acquisition, a composite cycle is obtained containing data from each of the individual cycles occurring during the period of study. Because data from a large number of cycles is being summated it is important to ensure that irregular beats are excluded as these will have a different length from the average cycle. It is usual to exclude in some way cycles which have an R–R interval which is more than 10 per cent different from the average value.

The quantitative analysis of both first-pass and equilibrium studies involves calculation of the left ventricular ejection fraction. It is assumed that the activity within the ventricle is proportional to the volume of blood within the ventricle so that the activity time curve derived from the ventricle is equivalent to the volume time curve and thus:

$$EF\,(\%) = \frac{\text{end diastolic activity} - \text{end systolic activity}}{\text{end diastolic activity}} \times 100$$

The main problems in deriving the ventricular activity time curve relate to identifying the borders of the ventricle, especially around the valve planes and in removing counts derived from overlying and underlying structures (background). Various computer programs can be employed to do this successfully. Values of EF calculated from nuclide techniques correlate well with those obtained from the standard method of contrast ventriculography. The isotope method is less hazardous to the patient and is less expensive. In addition, measurement of the ejection fraction from contrast studies makes certain assumptions about the shape of the ventricle which may be invalid in patients with diseased hearts. The nuclear techniques are largely independent of geometric assumptions of this kind.

Regional motion of the ventricular wall can also be assessed from radionuclide studies, either by visual analysis or by a variety of quantitative or mathematical approaches. Particularly when quantitative methods are used, the radionuclide angiocardiogram in multiple projections is accurate in differentiating normal regional wall motion from reduced wall motion (hypokinesia) and paradoxical wall motion (dyskinesia).

The nuclear angiocardiogram can be used to derive values for ventricular volumes and to evaluate parameters of ventricular filling and emptying, but specialized techniques are required which are not routinely available.

In the study of patients with possible coronary artery disease it is usual to carry out nuclide angiocardiography both at rest and during peak bicycle exercise — performed either in the erect or supine position and imaged throughout. In the normal patient, the left ventricular ejection fraction will be normal (>50 per cent) at rest and will rise during exercise. In patients with coronary disease, the resting ejection fraction and regional wall motion will usually be normal unless there has been prior infarction. On exercise however, the ejection fraction falls, characteristically, and new regional wall motion abnormalities may appear. The precautions described for exercise ^{201}Tl imaging should also be observed during exercise nuclide angiocardiography.

Alternative forms of stress such as hand-grip exercise or placing the patient's hand in ice cold water (the 'cold pressor test') have been used for nuclide ventriculography but none is a completely satisfactory substitute for dynamic exercise.

Rest and exercise studies are most commonly carried out using the equilibrium approach because of the ease with which repeated measurements of ventricular function can be made. The first-pass technique has the advantage of allowing data to be collected over a few seconds at true peak exercise rather than over a period of a few minutes of maximal sustainable exercise as in the equilibrium approach. Because the first-pass technique requires the injection of a separate bolus for each observation of ventricular function, accumulated radiation dose to the patient limits the number of levels of exercise at which data can be acquired. By comparison, the equilibrium method makes it possible to obtain information from all exercise levels performed. This limitation of first-pass studies may be solved by the development of suitable short-lived tracers of which the generator produced gold-195m (^{195}Au^m) is a notable recent example. The first-pass technique may also have problems in the patient with a paroxysmal arrhythmia. If the passage of the bolus coincides with a run of abnormal beats, the data obtained will not reflect true ventricular function. If a single first-pass study is being performed in this type of patient, stannous pyrophosphate should be given 30 minutes prior to the injection of the ^{99}Tc^m bolus so that an equilibrium study can be obtained if the first-pass study is invalidated by an arrhythmia. In patients with severe, sustained, arrhythmias neither a first-pass nor equilibrium study may be possible.

Detection of acute myocardial infarction

A clinical diagnosis of acute myocardial infarction can usually be confirmed or refuted by serial electrocardiograms and serial estimations of serum levels of appropriate enzymes. In some cases, however, the diagnosis remains in doubt. The findings of a defect on a rest ^{201}Tl scan or of a resting regional wall motion abnormality on a nuclide angiocardiogram may indicate the presence of acute infarction but may also be due to old infarction or to other pathologies. An alternative approach is to use $^{99}Tc^m$-labelled pyrophosphate ($^{99}Tc^m$-PYP) which will accumulate preferentially within acutely infarcted myocardium — so-called 'hot spot' imaging.

Technetium-99m pyrophosphate is a bone-scanning agent. The mechanism of uptake within acute infarction is uncertain but appears to involve binding to denatured protein within the infarct. The radiopharmaceutical is injected in a dose of 600–800 MBq and images are obtained $1\frac{1}{2}$–2 hours later in anterior, left anterior oblique and left lateral projections. In the normal subject, no blood pool or myocardial activity is seen and only the skeleton is visualized. In acute myocardial infarction, localized uptake is seen in the myocardium. It is possible to localize the site of infarction using this technique, but assessment of the precise size of the infarct is not possible.

The sensitivity of $^{99}Tc^m$-PYP for detecting acute myocardial infarction is around 95 per cent overall, but is lower for subendocardial compared to transmural lesions. The scan is rarely positive within 12 hours of the onset of symptoms and peak uptake of the tracer usually occurs 48–96 hours after the infarct. Thereafter the intensity of uptake fades. The finding of myocardial uptake of $^{99}Tc^m$-PYP is usually specific for infarction but abnormal images occasionally result from aneurysms, unstable angina, and cardiomyopathy.

Recent work on hot spot imaging of acute myocardial infarction has concentrated on attempts to develop attaching labelled monoclonal antibodies to fragments of the myocardial contractile protein myosin.

Shunt assessment

Radionuclide techniques have been applied to the evaluation of left to right intracardiac shunts. A bolus of activity is administered in the same fashion as for first-pass ventricular function studies. A region of interest is then placed over the lung and an activity time curve is generated. By employing various mathematical formulae (which are beyond the scope of this book), it is possible to look for the presence of early recirculation of activity to the lungs, demonstrating the presence of an intracardiac shunt. A good bolus of activity is essential for these studies and this should always be monitored in the superior vena cava or right heart.

When technically adequate, isotope studies are sensitive to the presence of a left-to-right shunt and the calculated magnitude of the shunt correlates well with the results obtained from more standard techniques such as cardiac catheterization and oximetry. Recent advances in echocardiography and Doppler ultrasound have resulted in the scintigraphic left-to-right shunt technique being employed less often.

The presence of a right-to-left shunt can be suspected from first-pass studies by the simultaneous appearance of activity within the right and left sides of the heart. This approach, however does not lend itself to quantification of the size of the shunt. Estimation of the size of the right-to-left shunt can be obtained by injecting $^{99}Tc^m$-microspheres or macroaggregates intravenously in the same fashion as for a perfusion lung scan. In the absence of a right-to-left shunt almost all of the labelled particles will impact in the pulmonary circulation. A very small proportion (usually less than 4 per cent) will reach the systemic circulation due to physiological shunting through the bronchial veins and will be deposited in capillary beds outside the lungs. When a pathological right-to-left shunt is present a larger proportion of the labelled particles will bypass the lung bed and reach the systemic circulation. Calculation of the ratio of lung to whole body activity can then be used to estimate the magnitude of the shunt.

Invasive Studies

Invasive cardiac studies involving catheterization and angiographic techniques are used extensively for both diagnostic purposes and establishing the severity of conditions known to be present from clinical evidence and non-invasive investigations. Cardiac catheterization and angiocardiography are almost always necessary before cardiac surgery can be undertaken, though increasing sophistication of real-time ultrasound and Doppler studies has resulted in some conditions being amenable to non-invasive investigation only.

Right heart catheterization

A right heart catheterization can theoretically be performed from almost any vein in the arm or leg, though usually a median antecubital vein is used. Following a cutdown in the vein, the catheter is fed under fluoroscopic control to the right heart and usually also into the pulmonary artery. Pressure recordings are made in both right heart chambers and in the pulmonary artery. The catheter may be advanced until its tip becomes wedged in a peripheral pulmonary vessel, most commonly in the right lower lobe. The 'wedge' pressure, or 'pulmonary capillary' pressure measured in this site has been shown to be an acceptably accurate reflection of left atrial pressure. In cases where intracardiac left-to-right shunting is suspected, blood samples are obtained at various sites and estimated for oxygen saturation. Angiographic studies (angiocardiography) may be performed during right heart catheterization, though less frequently than in left heart studies. Angiocardiography is used to demonstrate the structure of the cardiac valves, cardiac chambers and the pulmonary vasculature. Ventricular function is more difficult to assess in the right heart than in the left because of the irregular shape of the cavity of right ventricle. Angiocardiography may also be valuable in demonstrating the site of right-to-left cardiac shunts.

Left heart catheterization

Left heart catheterization can be achieved in a number of ways. A direct arterial method involves the percutaneous puncture of the femoral artery or a cutdown into a brachial artery and feeding of the catheter retrogradely under fluoroscopic control through the great vessels to the aorta and left heart. These approaches may be unsuccessful in patients with aortic stenosis because of an inability to cross the valve and enter the left ventricular cavity. In such cases, an alternative is trans-septal catheterization in which a long catheter is introduced via the femoral vein and passed to the right atrium. A needle at the tip of the catheter is then used to puncture the interatrial septum and pass to the left side of the heart.

During left heart catheterization pressure recordings are obtained from the chambers and blood samples can be obtained if shunting is suspected. Contrast studies are frequently employed in the left heart, often in two planes, to provide anatomical information. In addition, left heart angiocardiography is generally accepted as the gold standard method of assessing left ventricular function (Fig. 3.5). Global ventricular function can be assessed by measuring ventricular volumes and by estimating the left ventricular ejection fraction. The ejection fraction is calculated by measuring certain longitudinal and transverse dimensions of the heart at end-systole and at end-diastole. Making certain geometric assumptions about the shape of the ventricle, formulae can then be employed to convert these dimensions into volumes and so measure the ejection fraction. Regional left ventricular function is determined by observing the motion of segments of the left ventricular wall during the cardiac cycle. The motion is usually graded from normal through various degrees of decreased motion (hypokinesia) to absent motion (akinesia). In patients with ventricular aneurysm — due to replacement of myocardium by fibrous tissue following myocardial infarction — dyskinetic or paradoxical ventricular wall motion is seen, that is the affected area will expand rather than contract during systole.

Coronary arteriography

A catheter is introduced via the femoral or brachial artery, passed retrogradely to the aortic root and the tip is then manipulated to enter the orifice of the right and left coronary arteries. Contrast is then injected down the vessels and films obtained, preferably in two planes. The films are assessed for the presence and locations of stenoses or occlusions of vessels. The presence of collateral flow is also evaluated.

Risks of cardiac catheterization

In skilled hands, the overall risks of cardiac catheterization are low. All of the procedures may result in haemorrhage from the punctured vessel. In

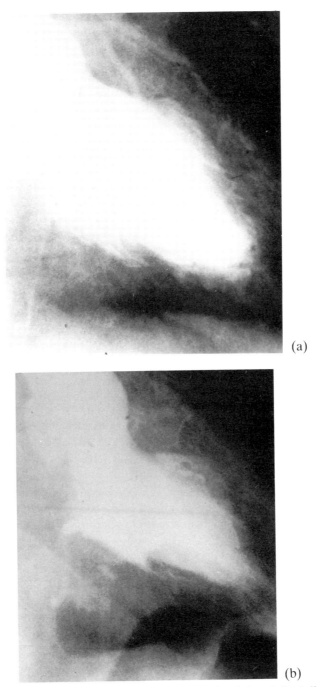

(a)

(b)

Fig. 3.5 Normal contrast left ventriculogram showing (a) end diastolic and end systolic frames. The end diastolic frame also shows contrast in coronary vessel from the preceding coronary arteriogram. (Courtesy Dr AR Lorimer.)

those which involve arterial puncture, thrombus formation at the site may cause vascular insufficiency, which occasionally requires surgical intervention. Rarely, distal embolism may occur. In patients with an intracardiac thrombus, catheterization may result in peripheral embolism. Vasovagal attacks may also occur during catheterization and patients are often given atropine intravenously before the study to reduce the likelihood of this occurring. Atrial and ventricular arrhythmias are common. They are usually minor, but heart block, ventricular tachycardia or asystole can occur. All patients undergoing cardiac catheterization or coronary arteriography must have continuous monitoring of the cardiac rhythm during the procedure, and full resuscitation equipment, including pacing facilities, must be available within the catheterization laboratory.

When contrast studies are being carried out, the osmotic load from the dye may precipitate left ventricular failure and the volume should always be kept to a minimum, especially in patients known to have a raised left atrial pressure. Hypersensitivity to the iodine within the contrast material may occur but is rare.

The risks of left heart catheterization and coronary arteriography are many times greater than those of right heart catheterization. In adults, the mortality rate for left heart catheterization should be less than 0·5 per cent overall and the rate of serious complications less than 2 per cent. In general terms, the mortality and morbidity rates rise in parallel to the severity of the cardiac disorder and the presence of other serious conditions. In ill neonates being investigated for congenital heart disease, the mortality rate may rise to 5 per cent.

Congenital Heart Disease

In congenital heart disease imaging may be used to elucidate the anatomy of the abnormality, to monitor the effects of the congenital abnormality on the lungs and to follow-up the response to the surgical correction or palliation of the lesion.

The chest X-ray may be used to assess changes in the lung vasculature. Increased lung vascular markings or lung plethora may be seen in any condition producing a left-to-right shunt. Increased lung vascular markings are common in atrial septal defect. They are less common in patients with patent ductus arteriosus or ventricular septal defects, being present only when the shunt is large in these conditions. Ventricular septal defect and patent ductus arteriosus are usually associated with a large shunt when they present in childhood. On fluoroscopy, patients with any cause of a left-to-right shunt may demonstrate 'hilar dance' — that is, increased pulsation of the main pulmonary arteries. This sign is seen most commonly with an atrial septal defect.

Oligaemia

Decreased lung vascularity (oligaemia) is seen in severe pulmonary stenosis and in many cases of Ebstein's anomaly in which there is downward displacement of the tricuspid valve and atrialization of part of the right ventricle, often with an associated atrial septal defect. In Fallot's tetralogy (right ventricular hypertrophy, large ventricular septal defect, overriding aorta and right ventricular outflow obstruction) the lung fields are oligaemic in severe cases and normal or even plethoric in less severe cases. In patients with left-to-right shunts who develop pulmonary hypertension and reversal of the shunt (Eisenmenger's syndrome or reaction), the main trunk and branches of the pulmonary vessels are large though the peripheral vessels are pruned. If the condition develops early in life large vessels may not be seen.

Cardiomegaly

Cardiomegaly is seen on the chest X-ray commonly in atrial septal defect and in most patients with Ebstein's anomaly. In coarctation of the aorta, left ventricular enlargement is commonly seen. Cardiomegaly is not seen until the lesion is severe in aortic stenosis, ventricular septal defect and patent ductus arteriosus. In pulmonary stenosis, the heart usually remains normal in size on chest X-ray until the onset of right ventricular failure. In Fallot's tetralogy, mild cases are associated with increased cardiac size, while in more severe cases, the heart is smaller.

The precise nature of the cardiac abnormality is often not obvious from the chest X-ray. However, a number of specific signs may be seen. In severe cases of Fallot's tetralogy, the small heart assumes a 'boot' shape (*coeur en sabot*) due to the combination of right ventricular hypertrophy and stenosis of the infundibulum of the right ventricle. In atrial septal defect, the pulmonary outflow trunk is prominent and the aortic knuckle small. Right atrial and ventricular enlargement are present. It is very difficult to distinguish this from left ventricular enlargement in ASD. In the rare condition of total anomalous pulmonary venous drainage, the chest X-ray may show the 'snowman heart' in which the upper circular shadow is formed by the anomalous venous pathway and the lower circle consists of the rest of the heart.

Assessment

In congenital heart disease, the visceral *situs* should be assessed. This can usually be done from the chest X-ray when the upper abdomen is visible. The right atrium is usually on the same side as the liver. Malposition of the heart and abnormal visceral *situs* are usually associated with complex heart defects. The side of the aortic arch should also be assessed. A right-sided aortic arch is seen in association with Fallot's tetralogy but may also occur in other less common complex abnormalities.

Changes in the vessels may indicate the nature of the cardiac abnormality. In aortic stenosis poststenotic dilatation of the aorta is sometimes seen especially when the obstruction is at the supravalvular level. In pulmonary stenosis, dilatation of the main pulmonary artery, and sometimes of the left pulmonary artery, is seen with valvular but not with subvalvular lesions. The degree of vessel dilatation is not a reliable guide to the severity of the valvular lesion.

In coarctation of the aorta the ascending aorta is often prominent. The '3' sign may also be seen in the region of the aortic knob, the bulges of the '3' being formed by the left subclavian artery above and poststenotic dilatation of the aorta below. Patients with coarctation may also show rib notching due to the erosion of the inferior surfaces of the ribs by enlarged intercostal vessels. Aneurysms may develop within these vessels and subsequently calcify.

In patent ductus arteriosus, a prominent pulmonary artery shadow running up towards a large aortic knuckle should lead to the diagnosis being suspected, but is not present in all patients. Later in life the ductus may calcify and become visible radiographically.

Echocardiography

Echocardiography has, in recent years, had an increasing place in the assessment of congenital heart disease. The echo study can be used to assess the presence of cardiac chambers, the size of the cavities and the presence of ventricular hypertrophy. Valvular abnormalities, such as pulmonary or aortic stenosis can be demonstrated and, by adding Doppler studies, the severity of valvular obstruction can be evaluated quite effectively. In valvular stenosis, echo studies can be valuable in demonstrating the site of obstruction and can, for example, differentiate valvular aortic stenosis from hypertrophic obstructive cardiomyopathy (HOCM) in which the abnormality lies in the left ventricular outflow tract. The presence of atrial or ventricular septal defects can also be demonstrated from ultrasound studies. In skilled hands, real-time and Doppler ultrasound can go a long way towards establishing the nature of complex congenital abnormalities and their effect on cardiac function, even in neonates.

A very important role of echocardiography is in excluding the presence of significant congenital heart disease without the need to resort to invasive studies. This is valuable in patients with a cardiac murmur detected on routine screening or in the cyanosed infant where it may be extremely difficult to distinguish cardiac from pulmonary disorders — either clinically or on plain X-rays.

Nuclear medicine procedures have little or no role in the investigation of congenital heart disease in infants. In older children or adults nuclear medicine may be used to follow the size of shunts or to assess ventricular function but even here they have only a very limited contribution to make. Occasionally, microsphere studies help in planning the surgery of patients with abnormal pulmonary arterial anatomy.

Previously, catheterization and angiographic studies were essential for the diagnosis of many cases of congenital heart disease and were also of invaluable use before surgical intervention. The increased use of echocardiography has greatly reduced the need for diagnostic invasive studies, though the latter are still frequently required in complex abnormalities and in small babies. The echo and angiographic studies are complementary, with the information obtained from the echo, making logical planning of the angiocardiography procedure more possible and aiding the interpretation of the results of the catheter studies. Catheterization studies are usually performed prior to surgery, but in some cases of ASD, VSD or pulmonary stenosis, the echocardiographic and other non-invasive procedures may be adequate. Echocardiography is also very valuable in the follow-up of disease progression or in assessing the results of surgical intervention and has dramatically reduced the need for repeated cardiac catheterization studies.

Valvular Heart Disease

As with congenital heart disease, valvular heart disease is increasingly investigated by non-invasive imaging methods with a consequent reduction in the need for cardiac catheterization.

Mitral Stenosis

In mitral stenosis, the chest X-ray may show characteristic changes resulting from enlargement of the left atrium. As already mentioned, this produces a dense double shadow in the right cardiac border on the PA film while prominence of the left atrial appendage causes loss of the normal concavity of the left heart border between the pulmonary artery and the left ventricular shadow. The overall cardiac size may be normal or increased. In older patients or those with long-standing mitral stenosis, the valve may calcify and become visible on the PA film just to the left of the spine above the left atrial shadow. The straight X-ray of chest is also helpful in screening for pulmonary hypertension or pulmonary oedema. The mitral valve is well assessed on echocardiography which can accurately demonstrate the severity of valvular stenosis, the degree of left atrial dilatation and the presence of thrombus in the left atrial appendage. Ultrasound is also useful in determining the presence of severe fibrosis or calcification of the valve which would necessitate valve replacement rather than valvotomy and is able to distinguish mitral stenosis from the much less common condition of atrial myxoma which has a very similar clinical presentation. Catheterization is not usually required in uncomplicated cases of mitral stenosis, even when surgery is being proposed. Angiocardiography may be necessary when additional valvular lesions are suspected and is essential if there is clinical evidence to suggest coincidental ischaemic heart disease.

In combined mitral stenosis and regurgitation, the chest X-ray appearances may be identical to those of mitral stenosis but as valvular regurgitation becomes more pronounced the left atrium may become extremely large (giant left atrium) and the overall cardiac size increases due to ventricular enlargement. Calcification may be seen within the wall of the left atrium. Echocardiography can be useful in showing a reduced diastolic closure rate of the valve and is valuable in assessing the size of the cardiac chambers and the presence of ventricular hypertrophy.

Mitral Regurgitation and Mitral Valve Prolapse

Whereas mitral stenosis is almost invariably due to rheumatic heart disease, isolated mitral regurgitation may be produced by many disease processes which either affect the valve cusps directly or act on the papillary muscles and chordae tendineae, or the valve ring. Any abnormality causing left ventricular dilatation may cause secondary mitral regurgitation by stretching the valve ring. Ischaemic heart disease produces mitral regurgitation either in this fashion or, more directly, due to the rupture or dysfunction of the chordae tendineae. Degenerative changes may occur in the mitral valve without any evidence of rheumatic heart disease and resulting in the so-called 'floppy mitral valve' which can produce significant mitral regurgitation. Mitral valve prolapse may also cause valvular incompetence and can arise from a variety of causes. Mitral valve prolapse is well demonstrated on echocardiography. When present as an isolated finding it is almost always of no pathological significance. The chest X-ray may be normal in many cases of mitral regurgitation, but with increasing degrees of regurgitant flow the signs of left atrial enlargement, ventricular enlargement, pulmonary hypertension and pulmonary oedema may appear. Echocardiography is useful in determining which part of the mitral valve apparatus is the cause of the dysfunction.

Aortic Stenosis and Aortic Regurgitation

Aortic stenosis usually results from congenital abnormalities, rheumatic heart disease or calcific degenerative changes. Similar clinical findings may be produced by hypertrophic obstructive cardiomyopathy (HOCM). The cardiac size on chest X-ray remains normal until a fairly late stage in pure aortic stenosis because of the late occurrence of ventricular dilatation. Early appearance of cardiac enlargement suggests the presence of other abnormalities such as unsuspected aortic regurgitation. Calcification of the aortic valve is common in older patients with aortic stenosis but is frequently not seen on the AP chest film because the valve then overlies the spine. The lateral view is better for demonstrating valve calcification. In patients with valvular aortic stenosis, the aortic root is usually dilated.

Echocardiography is used to demonstrate the mobility of the valve cusps, the presence of ventricular hypertrophy and dilatation and in assessing the severity of valve stenosis. The echocardiographic studies are particularly valuable in differentiating between valvular stenosis and abnormalities of the outflow tract such as HOCM. Doppler studies are being used in some centres to assess the severity of valvular stenosis but catheterization is usually required before valve replacement. In addition to assessing the gradient across the valve, angiography can measure left ventricular function and establish whether coronary artery disease is also present. In cases of severe aortic stenosis, the trans-septal approach may be required in order to enter the left ventricular cavity.

Aortic regurgitation may be produced by rheumatic heart disease, infective endocarditis, trauma, syphilis, various collagen disorders and dissecting aneurysms. Cardiac dilatation is an early feature of significant aortic regurgitation, and cardiac enlargement on chest X-ray appears early in the course of the disorder. Changes indicating pulmonary oedema or pulmonary hypertension are indicative of severe left ventricular disease. Valve calcification is frequently absent in pure aortic regurgitation. Echocardiography is useful in confirming the presence of aortic regurgitation and in demonstrating the impact on left ventricular function. Ultrasound can demonstrate vegetations or, less often, frank abscesses in cases of infective endocarditis, and can frequently visualize dissection of the ascending aorta; both pathologies can produce aortic regurgitation. Cardiac catheterization is used to confirm the diagnosis of aortic regurgitation by showing retrograde flow of dye from the ascending aorta to the left ventricle. Changes in left ventricular pressures occur late in the disease. Recently, it has been suggested that exercise radionuclide ventriculography is useful in timing the need for valve replacement in aortic regurgitation: operative intervention is required when the left ventricular ejection fraction fails to rise during stress.

Pulmonary and Tricuspid Valve Stenosis

Pulmonary stenosis is almost always congenital in aetiology and this has been discussed. Pulmonary regurgitation may develop secondary to long-standing and severe pulmonary hypertension producing dilatation of the valve ring. Contrast studies demonstrating reflux of dye from the pulmonary artery to the right ventricle are required to confirm the diagnosis.

Tricuspid stenosis usually results from rheumatic heart disease, almost always in association with mitral valve disease. The chest X-ray appearances are non-specific as the right atrial dilatation present may be seen with tricuspid regurgitation or may be mimicked by marked left atrial dilatation. Echocardiography can be used to confirm the presence of tricuspid stenosis as may contrast studies, though in some cases of tricuspid stenosis no pressure gradient is detected across the valve at catheterization because of the low cardiac output.

Tricuspid regurgitation is much more common than tricuspid stenosis and is usually functional, secondary to pulmonary hypertension. The chest X-ray appearances are non-specific and are largely dependent on the other cardiac abnormalities. Two-dimensional echocardiography and Doppler studies can be used to confirm the presence of the tricuspid regurgitation.

Echocardiography is frequently used to assess the success of valve replacement surgery, both in examining the adequacy of valve movement and in determining chamber size and ventricular function. Echocardiography is especially useful when a patient is suspected of having thrombosis around a prosthetic valve, resulting in impaired movement of the valve.

Infective Endocarditis

Infective endocarditis is due to infection of the endocardial surfaces of the heart or, more often, the valves. The condition occurs most commonly in patients with pre-existing heart disease — particularly rheumatic heart disease or some forms of congenital heart disease. Any heart lesion may be associated with infective endocarditis but the commonest site is the aortic valve, followed by the mitral valve. Prosthetic heart valves may also be the sites of infective endocarditis. Less commonly, normal hearts may harbour infective endocarditis, usually following a severe infective illness elsewhere in the body. Recently, the rise in the incidence of intravenous drug abuse has resulted in cases of infective endocarditis due to the use of unsterile syringes and needles. Unlike other forms of endocarditis, the lesions in these patients affect the valves on the right side of the heart predominantly.

The chest X-ray has no role in the diagnosis of infective endocarditis, although it may give some information on predisposing cardiac disease. The most important imaging investigation is two-dimensional echocardiography which can be used to demonstrate the presence of vegetations on the affected valves. It is also useful in serial measurements of valvular regurgitation, ventricular function and chamber sizes. It should be remembered, however, that bacterial endocarditis is essentially a clinical and microbiological diagnosis. False-negative ultrasound studies occur in endocarditis and failure to visualize vegetations does not exclude the diagnosis. Nuclear medicine imaging with indium-111-labelled leucocytes has been disappointing and usually is only positive when frank abscess formation has occurred. Emergency valve replacement, especially of the aortic valve, may become necessary because of rapidly progressive damage resulting from infection. In such circumstances contrast radiology is required prior to surgery.

Ischaemic Heart Disease

Diagnosis of Coronary Artery Disease

The firm diagnosis of coronary artery disease depends upon the demonstration of coronary artery stenoses by arteriography. Coronary artery disease results in areas of narrowing or even complete occlusion of the coronary vessels. In major coronary vessels, flow distal to an occlusion remains normal at rest until the stenosis is very severe — about 85 per cent of the intraluminal diameter. Under conditions of stress, such as exercise, flow is impaired distal to a stenosis of a lesser degree. For this reason, lesions occupying 50 or 70 per cent of the luminal diameter of a major vessel are taken to indicate significant coronary artery disease, the precise figure varying in different centres. The severity of coronary artery disease is classified by whether one, two or three of the major vessels (right, left anterior descending and left circumflex coronary arteries) are involved. Stenosis of the left main coronary artery is also looked for. Other information gathered from coronary arteriography is whether collateral flow is present to stenosed vessels and whether there is good distal run-off beyond the major stenoses. Left ventriculography is usually carried out at the same time as coronary arteriography and to establish global and regional left ventricular function. Non-invasive imaging techniques are very valuable in screening patients for coronary disease and as adjuncts to catheterization in patients known to have ischaemic heart disease.

In the patient with angina or chest pain thought to be of cardiac origin, the first non-invasive investigation is the electrocardiogram, often conducted during stress as well as at rest. A chest X-ray is also usually obtained early in the assessment of the patient, and is used to determine cardiac size, abnormalities of the aorta and the presence of pulmonary hypertension or pulmonary oedema. Specific signs of coronary artery disease are usually lacking. Coronary artery calcification can sometimes be seen, and in patients with prior myocardial infarction there may be localized bulging of the left ventricular contour due to a ventricular aneurysm.

In patients with classical angina and an abnormal stress electrocardiogram, non-invasive imaging techniques are rarely employed for the diagnosis of coronary artery disease. In this group of patients coronary arteriography is usually carried out when coronary artery bypass surgery is being contemplated, either because of continuing angina in spite of adequate medical therapy or because the patients are suspected of having high risk coronary disease (main left or triple vessel disease) which appears to have a better prognosis when treated surgically. Nuclear medicine procedures are sometimes valuable after coronary arteriography in demonstrating the functional effects of the coronary stenosis or myocardial blood flow and ventricular function during exercise. This is especially helpful when the coronary arteriogram shows lesions of only equivocal significance.

Non-invasive imaging techniques have a much greater role in the primary evaluation of patients with atypical chest pain. In conjunction with stress electrocardiography, exercise myocardial scintigraphy using thallium-201 and exercise radionuclide ventriculography are powerful tools for screening for the presence of coronary artery disease in this group of patients. A patient who exercises to a satisfactory level and has a negative stress ECG and a negative stress scintigram is very unlikely to have significant coronary artery disease and is unlikely to require arteriography. Patients with abnormal myocardial perfusion scintigrams are likely to have coronary artery disease and may need to proceed to coronary arteriography. An abnormal stress radionuclide ventriculogram in which there is an inadequate ejection fraction response may have coronary disease but similar results may be obtained in other forms of heart disease such as cardiomyopathy or in hypertension. The development of regional left ventricular wall motion abnormalities during exercise is more specific for coronary artery disease. While nuclear cardiology techniques are useful in the diagnosis of the presence of coronary artery disease, they are less valuable in predicting the extent of coronary disease. Echocardiography can be used to examine ventricular function in patients with possible ischaemic heart disease and can be useful in demonstrating other possible causes of chest pain such as mitral valve prolapse or aortic stenosis. Echocardiography at present, however, contributes little to the diagnosis of coronary artery disease.

Nuclear cardiology techniques should not be used for screening purposes in asymptomatic patients although they may be useful in clarifying abnormal or equivocal stress ECG results in asymptomatic subjects.

Myocardial Infarction

In established or suspected acute myocardial infarction, the chest X-ray is important in the early detection of pulmonary oedema. Assessment of heart size also provides some prognostic value. Myocardial scintigraphy with infarct-avid agents such as $^{99}Tc^m$-pyrophosphate 'hot spot imaging' has been advocated for the diagnosis and sizing of acute infarction especially when ECG and cardiac enzyme results are either equivocal or uninterpretable. Some centres use hot spot imaging but it has not been accepted routinely, partly because of the non-availability of suitable equipment in many hospitals and partly because the test only becomes positive at a relatively late stage in the course of the acute event.

Prognostic information may be gained from imaging techniques in acute infarction. Radionuclide ventriculography or echocardiography can be used to assess ventricular function and can provide valuable information on the efficacy of therapy being given. Because of their non-invasive nature they can be repeated easily.

Fluoroscopy or X-ray screening is an essential procedure in any coronary care unit: it is necessary for the insertion of transvenous (temporary) pacemakers in patients who develop heart block. Fluoroscopy is also used to monitor the position of catheters fed from a peripheral vein to the pulmonary circulation for pressure readings, especially the pulmonary capillary wedge pressure. These pressure recordings are valuable in defining and monitoring therapy for complications of acute infarction such as cardiac failure or cardiogenic shock.

Echocardiography is useful in the acute phase of infarction in confirming or excluding the complications of ventricular septal rupture and papillary muscle dysfunction or rupture.

In the convalescent phase, or after recovering from acute infarction, coronary arteriography is often employed to demonstrate whether patients have large areas of myocardium which are viable, i.e. not infarcted, but at risk because of stenosis of the coronary artery supplying them. Patients with large areas of residual jeopardized myocardium may merit coronary artery surgery. The nuclear cardiology methods can be used to screen patients in the early postinfarction period and they may be useful in identifying patients at risk because of large areas of ischaemic myocardium.

In the postinfarction period echocardiography can be useful in confirming a diagnosis of postmyocardial infarction syndrome (Dressler's syndrome). This is thought to be an autoimmune condition and typically occurs one to six weeks after the acute infarction. The patient develops pericarditis, often accompanied by fever. The demonstration of fluid in the pericardial sac by echocardiography confirms the diagnosis, though the absence of fluid does not exclude it. The chest X-ray may show pulmonary infiltrates and pleural effusions.

The presence of a ventricular aneurysm may be suspected from ECG changes and from bulging of the cardiac shadow on the chest X-ray. In many cases the diagnosis can be confirmed by echocardiography or by radionuclide ventriculography, though sometimes contrast ventriculography has to be employed. Sites of previous infarction, especially where there is aneurysm formation, may be associated with mural thrombus formation from which systemic embolization can ocur. Two-dimensional echocardiography can detect the presence of mural thrombus.

Ventricular aneurysms, if large, may cause congestive cardiac failure which may respond to surgical resection of the aneurysmal area. Radionuclide ventriculography or echocardiography can be employed to distinguish this cause of congestive failure from the diffuse myocardial impairment of widespread coronary artery disease (congestive cardiomyopathy) which is not amenable to surgery. Before proceeding to surgery, patients with ventricular aneurysms require cardiac catheter studies.

Assessment after Coronary Artery Surgery

Reassessment by imaging is not usually required if patients have good relief of angina after bypass surgery. Chest pain recurring after surgery may be due to myocardial ischaemia. This may result from failure of the graft to revascularize the myocardium, from progression of the disease to within ungrafted native vessels or grafted vessels distal to the graft site, or finally, it may be due to occlusion of the graft. Pain may also result from other causes such as musculoskeletal damage produced by the surgery or completely unrelated conditions such as an alimentary disorder or lung disease. Early in the postoperative period the patient may develop the postcardiomyotomy syndrome, which is autoimmune in nature and similar to the postinfarction syndrome.

If the pain suggests continuing ischaemia, stress electrocardiography should be carried out in the first instance. If this is equivocal, myocardial scintigraphy or radionuclide ventriculography can be used to demonstrate stress-induced abnormalities suggestive of myocardial ischaemia. Coronary arteriography is required for definitive determination of graft status and is always employed prior to any further surgery. In cases where the postcardiomyotomy syndrome is suspected, echocardiography is helpful in looking for pericardial fluid. When initial non-invasive tests show no evidence of myocardial ischaemia in patients with atypical or non-anginal chest pain, appropriate imaging of the lungs, upper gastrointestinal tract or musculoskeletal system should be performed to elucidate the diagnosis.

Percutaneous Transluminal Coronary Angioplasty (PTCA)

Coronary angioplasty utilizes specialized catheters which have an inflatable balloon at their tip. Using a transfemoral approach in most cases, the catheter is fed into the coronary arterial circulation and the balloon expanded at the site of a coronary stenosis. It is routine to carry out coronary arteriography immediately before and after dilatation of the stenosed vessel. Long-term follow-up of vessel patency requires a combination of coronary arteriography and non-invasive tests such as stress electrocardiography.

The major attraction of PTCA is its relatively non-traumatic nature when compared to coronary artery bypass grafting. The procedure is carried out under local anaesthesia and patients can often be discharged from hospital within 48 hours. Angioplasty does, however, carry some hazards. In addition to the complications previously listed for coronary arteriography and cardiac catheterization, PTCA can produce coronary occlusion either by dissection of the vessel wall, spasm or acute thrombosis. Should such occlusion occur and not be reversed immediately by vasodilators, then emergency coronary bypass surgery is required. Coronary bypass surgery should always be available at short notice if PTCA is being performed. The incidence of complications requiring emergency surgery was initially 5–10 per cent but is now falling to about 3 per cent in centres with wide experience. The myocardial infarction rate after PTCA is around 3 per cent and the overall mortality (including cases requiring emergency surgery) is around 1 per cent.

There is still debate over the precise role of coronary angioplasty. Some centres employ PTCA in acute infarction but this is controversial.

In patients with angina, the best results, both in terms of symptom relief and persisting patency of the vessel, are obtained when there is a short, discrete stenosis involving a single vessel. Some patients with multivessel disease may be suitable for complete or partial treatment with PTCA, but opinions differ on precise indications. It is generally agreed that left main coronary artery disease should not be treated by angioplasty in view of the very large mass of myocardium in jeopardy if acute occlusion of this vessel occurs.

It is possible that PTCA may bring about some reduction in the number of patients requiring coronary bypass grafting. It is also likely that it may offer effective therapy for patients who would not be considered for coronary artery surgery, for example, because of age or coexisting severe diseases.

Hypertension

In the hypertensive patient the chest X-ray is a standard initial investigation. The heart size is assessed. Concentric left ventricular hypertrophy is characteristic of hypertension but of itself does not produce significant heart enlargement. With the onset of left ventricular decompensation, the ventricle will dilate and the heart size increase. The chest X-ray should also be inspected for evidence of pulmonary venous congestion or frank pulmonary oedema. Some dilatation of the aorta is commonly seen and usually indicates coexistent aortic arteriosclerosis. Extreme aortic dilatation or a sudden increase in aortic shadow size may indicate aortic dissection, a recognized complication of hypertension.

The presence of left ventricular hypertrophy is usually assessed from the ECG but the echocardiogram probably provides a more accurate indication of left ventricular mass. Radionuclide ventriculography gives an accurate non-invasive measure of left ventricular function but, at present, nuclear cardiology methods are not used in the routine assessment of the hypertensive patient. Coronary arteriography may be required if angina or atypical chest pain develops.

Another important role of imaging investigations in hypertension is detecting the underlying cause of the hypertension — particularly renal parenchymal or renal vascular disease, aortic coarctation or endocrine diseases such as Cushing's syndrome, Conn's syndrome or phaeochromocytoma. The investigation of these conditions is considered elsewhere in this book.

Pericardial Disease

Pericardial effusion occurs in association with pericarditis, which can be due to infection, but is also seen in metabolic disorders, such as uraemia, autoimmune states such as collagen vascular diseases or the postmyocardial infarction and postcardiomyotomy syndromes and in malignant infiltration of the pericardium. Haemopericardium occurs after trauma as a complication of aortic dissection or in patients on anticoagulant therapy.

Pericardial effusion is primarily investigated by echocardiography. The heart size may appear normal on the chest X-ray if the effusion is small or loculated, but larger effusions will cause an increased cardiac size. The heart shadow is characteristically globular in pericardial effusion but it is usually impossible to determine from the chest X-ray whether an enlarged heart is due to pericardial effusion or cardiac chamber enlargement. Echocardiography enables definitive separation of these two possibilities. In pericardial effusion there is an increased echo-free space between the pericardial parietal lining and the free wall of the ventricles. Localized or loculated effusions can be shown by echocardiography and this can be useful in assisting aspiration of the fluid for diagnostic or therapeutic purposes. Cardiac catheterization is necessary only if echocardiography is not possible or if some underlying heart disease is suspected. Where the cause of a pericardial effusion is unclear, CT may be helpful in demonstrating invasion by mediastinal tumour. Nuclear medicine procedures have no specific role to play in suspected pericardial effusion.

Pericardial tamponade occurs when the pericardial effusion produces a high enough pressure to impair ventricular filling. In acute tamponade the cardiac size may appear small, but as any combination of tamponade and effusion may occur, the chest X-ray does not provide a firm diagnosis. The lung fields are characteristically clear with no evidence of pulmonary oedema. The diagnosis of pericardial tamponade is essentially a clinical one, but the demonstration of a pericardial effusion by echocardiography provides strong confirmatory evidence. In equivocal cases, cardiac catheterization with pressure measurements is required.

Pericardial constriction occurs as a long-term complication of pericardial inflammation. It may occur after any form of pericarditis though in the past it was frequently due to tuberculous infection of the pericardium. The main clinical feature is obstruction of right ventricular filling. The chest X-ray may show a normal-sized or an enlarged heart and the lung fields appear clear. The diagnostic finding is the presence of pericardial calcification which may be intermittent or continuous over large areas of the pericardial surface. In cases due to tuberculosis, lung changes of active or old disease may be seen. Echocardiography may be able to demonstrate pericardial fluid or pericardial thickening but is frequently unhelpful. Cardiac catheterization may show elevation of left and right atrial pressures.

Peripheral Arterial Disease

Imaging Techniques

Peripheral arterial disease is very common in developed countries and usually results from atherosclerosis. The abnormalities initially appear in high pressure arteries and, characteristically, are situated at points of turbulent flow such as major bifurcations of vessels or where there is kinking of vessels or pressure from external structures. It has been suggested that intimal damage caused by the disturbed flow at these sites initiates the atherosclerotic process. Chronic vascular insufficiency may also, less commonly, be due to arteritis, notably that associated with the collagen vascular and autoimmune disorders.

The non-invasive assessment of vascular disease is obtained from straight X-rays, looking for evidence of vascular calcification, and from ultrasound. Although Doppler will usually be required to study small- and medium-sized vessels, real-time ultrasound is employed in the detection and measurement of aneurysms of the abdominal aorta. Ultrasound may also be able to detect the presence of intraluminal clot. Sequential measurements of aortic diameter by ultrasound can detect progressive enlargement. Doppler techniques have been used increasingly in the last few years, especially in examining the carotid vessels and have yielded good results in certain centres when compared to angiographic or operative findings. Although a number of radionuclide methods such as thallium-201 scintigraphy of the legs, imaging with indium-111-labelled platelets, or regional clearance with xenon-133 have been employed in investigation of peripheral arterial disease, they still remain research techniques.

Angiography is the definitive method of assessment of peripheral arteries at present. In arteriography, contrast may be introduced by direct puncture of the vessel, as for example in translumbar aortography or, more commonly, by retrograde catheter techniques following femoral puncture. Angiography can demonstrate the presence of arterial stenosis or occlusion (Fig. 3.6), areas of aneurysm formation or intraluminal clot. The angiogram is also used to estimate the quality of vessels beyond the narrowed segment, an important factor in planning reconstructive vascular surgery. Angiography is associated with a definite morbidity. Complications include hypersensitivity to the contrast agent, haemorrhage or infection at the puncture site and arterial occlusion due to local thrombosis at the site of puncture or distal embolization either of blood clot or of fragments of an arteriosclerotic plaque.

Digital subtraction angiography has been introduced in recent years. The patient is given a bolus injection of dye via a peripheral venous catheter and computer techniques are employed to enhance pictures of the vessels under study and remove the effects of overlying tissues. This method has the great attraction of avoiding the hazards associated with traditional arteriography and is particularly useful in high risk patients such as those on anticoagulants or those with poor arterial access. Digital subtraction techniques can also be employed to improve the quality of standard angiographic images, and to reduce morbidity by using smaller arterial catheters and a smaller volume of contrast.

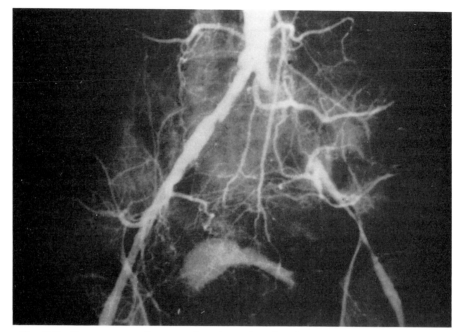

Fig. 3.6 Contrast arteriogram showing a localized complete occlusion of the left common iliac vessel. There is extensive collateral formation indicating that the occlusion is not acute.

Chronic Arterial Disease

The majority of patients with chronic arterial disease do not have detailed imaging investigations. These become necessary only if the symptoms become disabling or if there is evidence of a threat to limb viability. Under these circumstances angiography is performed to determine whether reconstructive surgery is possible.

In aortoiliac and lower limb arterial disease, endarterectomy (in which the lumen of the vessel is stripped to remove the arteriosclerotic lesion), prosthetic graft insertion to bypass stenosed or aneurysmal areas, and bypass grafts utilizing reversed autogenous saphenous veins are performed, sometimes in combination in the same patient. Disease extending above the renal arteries is difficult to deal with and graft occlusion is less likely if there is good flow above the graft and good distal run off beyond it. These factors can be assessed at arteriography which is essential in planning any vascular reconstructive surgery. A particular problem exists in diabetics who may have both microvascular and major arterial disease. Those diabetics with small vessel disease only and normal major vessels at arteriography will not respond to vascular surgery although reconstruction may be beneficial if there are major vessel abnormalities.

Angioplasty, in which the stenosed vessel is dilated by a balloon at the end of a catheter, may be an appropriate procedure in patients who are too ill for reconstructive surgery. It may allow preservation of a limb until the patient is fit for arterial surgery. In the future, intravascular laser therapy may also be helpful in this group.

Imaging is not routinely performed in the follow-up to vascular surgery but arteriography may be necessary if graft occlusion is suspected. In the immediate postoperative period, haemorrhage may occur from the anastomosed site requiring emergency surgery. CT or ultrasound can be useful in detecting leaks from an aortic anastomosis, and may indicate retroperitoneal leaks. Infection at the prosthetic site is a serious but, fortunately, rare complication usually necessitating removal of the graft. Ultrasound and indium-111-labelled leucocytes have had some success in detecting graft infection.

Aortic Aneurysm

When an abdominal aortic aneurysm is suspected it can usually be confirmed by either straight X-ray, which shows a dilated calcified vessel, or by ultrasound. The latter technique has the advantage of enabling more precise measurement of the dimensions of the aneurysm and also demonstrating the presence of clot within the aneurysm which may produce distal embolism. CT can demonstrate the extent of the aneurysm and whether the renal arteries are involved. It will also show any leak or periaortitis. Arteriography is required if reconstructive surgery is being contemplated because of the risk of aneurysmal rupture or distal arterial insufficiency.

A dissecting aortic aneurysm usually begins in the thoracic portion but may develop in the abdominal aorta. Extension of the aneurysm may occur for any length and may involve any of the branches of the aorta and the aortic root secondarily. Often the diagnosis can be confirmed by CT but definitive diagnosis may require arteriography, which in any case is necessary for surgery to be planned. When the ascending aorta is involved coronary arteriography is an essential part of the preoperative assessment.

Acute Vascular Insufficiency

Acute ischaemia of a limb may occur because of local thrombosis — usually superimposed on an atheromatous plaque — or due to an embolus occluding a major vessel. The embolus may originate from the surface of more proximal atheromatous plaques or from an intracardiac source. Clinical differentiation of thrombosis and embolism is not possible and emergency arteriography is required. Following the treatment of embolism, cardiac ultrasound should be performed to detect mural thrombosis, valvular heart disease or intra-atrial thrombus.

Following trauma, acute ischaemia may result from tearing or, more often, compression of major vessels. Typically, compression occurs due to displaced fractures but it may be secondary to massive oedema produced by tissue necrosis. Immediate arteriography is required if there is evidence of limb ischaemia after trauma.

Acute intestinal ischaemia is seen predominantly in elderly patients who present with features of an acute abdomen and incomplete intestinal obstruction. The diagnosis may be suspected from the presence of 'scalloping' or 'finger printing' in the large bowel on barium enema but selective arteriography is required to demonstrate the occlusion of a major vessel by thrombosis or embolism. A proportion of patients with acute intestinal ischaemia have no major arterial occlusion in which case the syndrome may be due to venous thrombosis or merely to a low flow state resulting from severe cardiac failure or shock.

Chronic intestinal ischaemia results in central abdominal pain — after meals, characteristically, ('abdominal angina') — and weight loss. The syndrome does not usually occur until only one of the coeliac, superior mesenteric or inferior mesenteric vessels remains patent. Angiography will show any occlusive lesions in these vessels but does not prove that this is the origin of the symptoms.

Vasospastic Disorders and Arteritis

Vascular insufficiency may arise because of spasm of the arterial wall rather than fixed stenosis. The condition is usually reversible and is most commonly seen in Raynaud's phenomenon in which the extremities become pale initially, then cyanosed, and finally, red as the circulation is restored. Classically, attacks are brought on by cold or by emotion. Raynaud's phenomenon is caused by a number of disorders including collagen diseases, especially scleroderma. If scleroderma is suspected, investigation should include a barium swallow as there is a high incidence of oesophageal abnormalities. A variant of scleroderma is the CRST syndrome (calcinosis, Raynaud's, sclerodactyly, telangiectasia) in which plain X-rays will reveal subcutaneous calcification, notably in the hands (Fig. 3.7). Unilateral Raynaud's phenomenon should raise suspicion of thoracic outlet compression. This isusually due to a cervical rib which will be demonstrated by a chest X-ray, though the syndrome can occur without any obvious anatomical anomaly. Subclavian artery compression is confirmed by arteriography which may also show poststenotic dilatation of the vessel. Raynaud's phenomenon can also be associated with atherosclerosis, though it is unclear whether the two are causally related in this situation. Atherosclerosis is relatively rare in the upper limb, and evidence of arterial insufficiency in the arms should always lead to a search for arterial compression, which is frequently reversible.

Vascular spasm may occur elsewhere in the body, notably in the intracranial vessels giving rise to migraine, and in the coronary vessels, causing myocardial ischaemia and angina. Coronary artery spasm may occur either in normal vessels or in those which are the site of atherosclerosis. It may be observed occurring spontaneously at coronary arteriography or it may be provoked in the catheter laboratory by the intravenous administration of

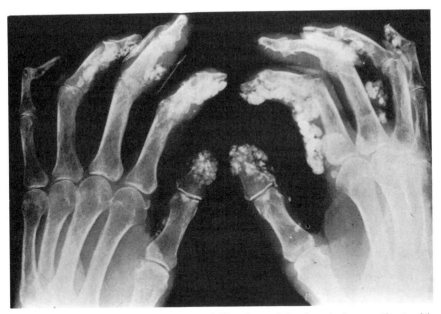

Fig. 3.7 Extensive subcutaneous calcification of the hands in a patient with CRST syndrome.

ergometrine. There is considerable debate on the need and wisdom of provoking coronary spasm in this fashion as it is not without some risk of inducing severe ischaemia or even infarction.

Arteritis may develop in association with various infections or haematological disorders but is a particular feature of the collagen vascular diseases. The investigation should be directed primarily towards establishing the nature of the underlying systemic disorder and is mainly dependent on serum tests for autoantibodies and histological examination of biopsied tissue. The role of imaging is often secondary. One exception to this is polyarteritis nodosa in which mesenteric or renal arteriography may demonstrate the characteristic findings of multiple aneurysms in medium and small calibre arteries. Angiography has also been applied in temporal arteritis in an attempt to establish the best site for biopsy but has proven disappointing.

Venous Disease

Deep vein thrombosis (DVT) is a frequent event in the postoperative period, especially after pelvic or lower limb procedures. An increased incidence of DVT is also seen following prolonged bed-rest (for any reason), in congestive cardiac failure, in women taking the oral contraceptive pill, and in a wide variety of malignant and inflammatory conditions. Clinical diagnosis is unreliable. Accurate diagnosis of deep vein thrombosis is essential in view of the potentially fatal effects of failing to give anticoagulants in major DVT and of the need to avoid these hazardous drugs in patients who do not have DVT.

The mainstay of confirmation is ascending venography (Fig. 3.8). Contrast dye is injected into veins on the dorsum of the foot and the flow of dye is monitored up the deep veins by fluoroscopy. Venography enables visualization of the deep veins of the leg and may also allow assessment of the external iliac veins and inferior vena cava. The presence of deep venous thrombosis is confirmed by the presence of intraluminal clot and is suggested by failure to visualize deep vessels. The presence of venous collaterals suggests a more long-standing thrombotic lesion and does not necessarily imply recent thrombosis. Major pulmonary embolism (PE) is unlikely to occur with thrombosis limited to the calf veins but becomes increasingly likely in the thigh and pelvis. The likelihood of PE is greatly increased when

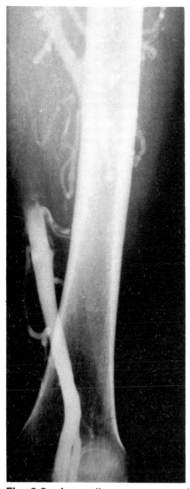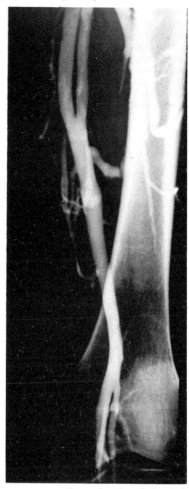

Fig. 3.8 Ascending venogram showing deep venous thrombosis of the left femoral vein. Patency is restored following fibrinolytic therapy one week later.

loosely adherent clot can be seen in the vessel lumen. Venography is unlikely to miss any significant venous thrombosis in the legs, but pelvic venous thrombosis may not be demonstrated. Venography is relatively contraindicated if there is a past history of reaction to contrast agents. A careful technique must be employed as subcutaneous extravasation of the medium may lead to considerable local tissue necrosis which on occasions has required skin grafting. Manipulation of the limbs during the procedure could potentially dislodge the thrombus resulting in pulmonary embolism, but, fortunately, this appears to be extremely rare. There is also some evidence that irritation of the venous endothelium by the contrast agent may cause extension of the DVT but again this is a minor problem and is outweighed by the value of the information obtained.

Scanning with ^{125}I-fibrinogen has been widely applied in the diagnosis of DVT. The tracer is injected and counts are then obtained over the next 7 days using a hand-held probe at predetermined points over the calf and thigh. A local increase in counts is taken to indicate the presence of DVT. The technique is unreliable in the upper thigh and pelvis because of the high blood pool levels in the background. False-positive results can result from any inflammatory lesion or from superficial thrombophlebitis. For these reasons, and because of the delay of at least some days in obtaining the diagnosis, ^{125}I-fibrinogen scanning is of very little value in individual patients with suspected DVT and is used more often in studies of the incidence of the condition in high risk populations.

A number of alternative radiopharmaceuticals suitable for use with gamma camera imaging, such as ^{99}Tc^{m}-labelled plasmin and urokinase, have been developed but have been disappointing in clinical practice. Radionuclide venography has been utilized in some centres. The technique is similar to that of ascending contrast venography with the tracer being injected into a dorsal foot vessel with the patient positioned under the gamma camera. It may be combined with lung scanning for suspected pulmonary embolism if labelled microspheres or macroaggregates are injected. This method may sometimes be useful in patients who are allergic to contrast material, and in detecting central venous occlusions which may not be detected on ascending venography.

Doppler techniques are of some help in demonstrating abnormalities of flow in the veins of the thigh and groin and thus, by inference, demonstrating the presence of thrombosis. It remains largely a research technique at present, as does the non-imaging investigation of impedance plethysmography. Standard ultrasound may be useful in confirming a muscle haematoma or a ruptured Baker's cyst, conditions which clinically may closely mimic venous thrombosis. A Baker's cyst is a synovial cyst of the knee joint and is seen particularly in association with chronic arthritis of the joint. Rupture releases irritant fluid into the popliteal fossa and calf with resultant inflammation. Rupture of a Baker's cyst can also be confirmed by arthrography. The presence of inferior vena caval thrombosis may be detected from ascending venography, but frequently ultrasound, contrast enhanced CT, radionuclide venography or venography through an IVC catheter are required to make this diagnosis.

Any patient with an unexplained deep venous thrombosis should be evaluated for the presence of underlying malignancy. Careful initial clinical examination may direct investigation in a particular line but in the absence of clinical clues a case can be made for a routine abdominal and pelvic ultrasound which is simple, not associated with any risk and may demonstrate a resectable lesion.

Lymphatic Disorders

Lymphatic obstruction may occur following trauma (notably, arm lymphoedema after mastectomy and axillary dissection), may occur due to acquired abnormalities of the lymph nodes or lymphatic vessels (e.g. filariasis), or be due to congenital hypoplasia of lymphatic channels. Severe lymphoedema is a considerable disability to the patient making the affected part difficult to use and liable to infection or trauma. Reconstructive surgery or excision of lesions causing lymphatic compression may produce marked improvement. Lymphangiography can be helpful in assessing patients for surgery. In the case of the lower limb, a coloured colloidal dye is injected into a toe web space and is taken up by a small lymphatic vessel on the dorsum of the foot. Contrast material is injected into the vessel and its passage up the limb imaged. The procedure usually requires a cutdown into the lymphatic vessel. An alternative technique, injecting dye into a regional lymph node, is less successful in demonstrating the lymph vessel structure. Recently, there has been considerable interest in the use of lymphoscintigraphy utilizing labelled colloids. The precise role of this technique in investigating lymphoedema has still to be established but it does have the major advantage of not requiring a cutdown into the lymphatic vessels (the tracer is injected into the interdigital space of the limb) and of allowing some semiquantitative estimation of lymphatic flow.

The investigation of lymphadenopathy and lymphoma is considered in Chapter 10.

Further Reading

Iskandrian, A.S. (1987). *Nuclear Cardiac Imaging — Principles and Applications*. F.R. Davis Co., Philadelphia.

Jefferson, K. and Rees, S. (1980). *Clinical Cardiac Radiology*. Butterworth Scientific Ltd, Guildford.

Miller, S.W. (Ed.). (1985). Advances in cardiac imaging. *Radiology Clinics of North America* Vol 23, **4**. W.B. Saunders & Co., Eastbourne & Philadelphia.

Weir, J. and Pridie, R.B. (1984). *Atlas of Clinical Echocardiography*. Pitman.

4

GASTROINTESTINAL DISORDERS

Imaging Techniques

Plain Radiographs

The plain abdominal film, obtained with the patient supine, provides a view of the whole abdomen which is inexpensive, quick and easy to obtain. The bony landmarks and soft-tissue features shown on the plain film are readily related to clinical findings. A technically adequate examination must include the hernial orifices inferiorly and the domes of the diaphragms superiorly. Two films will often be necessary to cover the full extent of the adult abdomen.

Gas is normally present within the bowel lumen but not elsewhere in the abdomen. The normal stomach and colon almost always contain some gas whereas the small bowel is usually gas-free in the healthy, fasting subject. Most patients with abdominal pain, nausea or vomiting will have gas in the small bowel as a result of air-swallowing. It is usually possible to distinguish the gas patterns of large and small bowel on the plain abdominal film and so identify any displacement or local dilatation or distention of the gut. Fluid levels within the bowel lumen are shown on films obtained with the patient erect or lying on the side (lateral decubitus position) using a horizontal X-ray beam. The length of air-fluid levels in the gut depends not only on the degree of distention but also on the relative proportions of air and fluid in the particular loop of bowel concerned. The diameter of the jejunum is usually a little wider than that of the ileum, and its mucosal folds are closer together so it is usually possible to distinguish the gas patterns of the proximal and distal small bowel. Where the lumen of the bowel is filled with gas and the outside of the bowel wall is surrounded by fat, it will be possible to see the thickness of the bowel wall — a useful diagnostic feature in patients suspected of having inflammatory bowel disease.

Gas shadows outside the bowel lumen indicate pathology or trauma. Air introduced into the peritoneal cavity at surgery takes up to a week to be absorbed. In the absence of surgery or trauma, intraperitoneal air indicates perforation of the gut, the most common causes being peptic ulcer, diverticulitis, appendicitis and colonic or gastric carcinomas. Intraperitoneal gas is usually visible on both supine and erect films; the smallest amounts of air are detected on horizontal beam films obtained after the patient has been lying on the left side for several minutes.

Retroperitoneal gas usually indicates abscess formation complicating local inflammatory disease but occasionally arises from posterior perforations of colonic tumours. Gas in the biliary tree (Fig. 4.1) is a common finding after

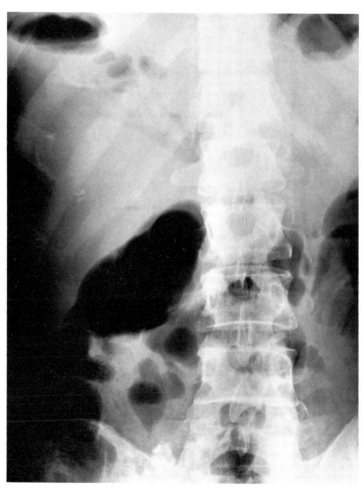

Fig. 4.1 Plain film of the upper abdomen showing gas in the biliary tree in addition to a cluster of gas pockets in the liver and subphrenic space. Diagnosis: liver abscess and cholangitis complicating pancreatic carcinoma. (Courtesy of Dr A Chapman).

sphincterotomy or surgical reimplantation of the bile duct and in these circumstances has no pathological import. However, gas appearing in the biliary tract of a patient who has not undergone such intervention suggests either cholangitis with gas-forming organisms or the passage of a large gallstone which has subsequently impacted in the distal ileum (gallstone ileus). Linear streaks of gas seen within the wall of any part of the gut or within the gall bladder wall indicate local tissue necrosis or infection with gas-forming organisms. In either case, this feature indicates severe pathology: often, impending infarction of bowel with subsequent perforation.

It is only possible to see soft-tissue outlines within the abdomen where the organs are surrounded by layers of fat. Even in thin subjects, the retroperitoneum usually contains enough fat to outline both kidneys and often the psoas edges will be visible. The free margin of the right lobe of the liver is usually separated from the chest wall by a little interstitial fat and a distinct seam of fat separates the left colon from the adjacent abdominal wall. The presence of fluid obliterates these soft-tissue edges and their absence may be taken as supporting evidence for the presence of abnormal fluid collections. However, since these edges are not visible in every subject, their absence cannot be regarded as unequivocal evidence of pathology. However, if, for example, one renal outline is clearly visible and the other is not visible at all, or if part of the fat seam separating the colon from the abdominal wall is clear and an adjacent section is not, this may be taken as good circumstantial evidence of inflammatory or neoplastic infiltration from the kidney or colon, respectively. Except in obese patients, there is no fat around the left lobe of the liver so this part of the liver edge is very rarely visible on the plain film. Enlargement of the liver or spleen is recognized more from the associated displacement of stomach and colon than from the visibility of the edges of these organs. Similarly, pathological soft-tissue masses within the abdomen are manifest by displacement of normal anatomical structures.

In the absence of radiographic contrast media or other foreign material, high density shadows on the abdominal film represent areas of calcification. Common and characteristic areas of calcification of little pathological significance are those seen in the costal cartilages, aorta, iliac and splenic arteries, mesenteric lymph nodes, and pelvic veins (phleboliths). Many other typical patterns of calcification may be recognized; some of these are benign and of little significance — e.g. uterine fibroids, prostatic calculi — others are also benign but clinically important — e.g. renal calculi, gallstones, calcification in chronic pancreatitis, calcified hydatids — while others still suggest malignancy — e.g. primary and secondary liver tumours, retroperitoneal masses.

A plain chest film should be obtained routinely in patients being investigated for gastrointestinal disorders. The chest film is simple, inexpensive and quickly taken; it provides a good deal of information for very little cost. In addition, it is important to remember that some types of pathology arising in the chest (e.g. lower lobe pneumonia) may present with abdominal symptoms and signs whereas some types of upper abdominal pathology (e.g. pancreatitis) often produce secondary effects in the chest.

Contrast Studies of the Gastrointestinal Tract

The demonstration of gastrointestinal pathology by abdominal radiography after ingestion of radio-opaque material was first achieved only a few years after the discovery of X-rays. The material used for most purposes currently is barium sulphate. Small particles of this insoluble compound are suspended in a watery solution with various additives which produce a smooth coating of the gastrointestinal mucosa, prevent precipitation of the particles (flocculation) and also prevent frothing if double contrast with air is used. Single-contrast barium studies produce a radiographic 'cast' of the lumen but the currently preferred techniques are those using double-contrast. This is achieved by inflating the lumen of the gut with air or, in the case of the small bowel, with a watery solution while maintaining a coating of barium suspension on the mucosa. The double-contrast technique gives an improved demonstration of the surface mucosal pattern (Fig. 4.2). Distending the gut wall also has the effect of rendering local pathology more easily visible and the elimination of large volumes of opaque material in the gut lumen allows overlapping loops of bowel to be distinguished.

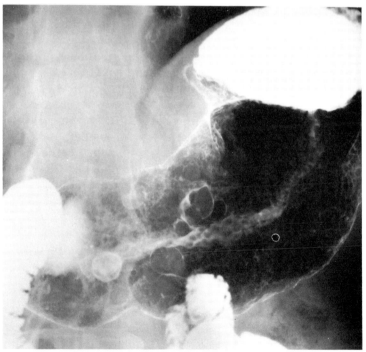

Fig. 4.2 Double-contrast barium meal showing multiple polyps arising from the gastric wall. (Courtesy of Dr A Chapman).

Barium meal

Barium sulphate is chemically stable and is non-toxic while it remains within the bowel. However, barium escaping into the peritoneal cavity together with faeces from the colon produces a severe peritonitis with high morbidity and mortality. Escape of barium from the oesophagus, stomach or duodenum produces much less morbidity but a significant inflammatory reaction is still likely to ensue. Barium inhaled into the bronchial tree is also likely to cause at least a degree of inflammatory reaction and may result in local bronchial plugging. For these reasons it is usual to avoid the use of barium sulphate for patients in whom perforation is suspected or where inhalation of barium is likely. Water-miscible media similar to those used for intravenous studies should be used in these circumstances. Double-contrast cannot be achieved as effectively as with barium and, in general, the level of anatomical detail achieved using these media is not as good as with barium sulphate.

Contrast examinations of the gastrointestinal tract are dynamic procedures and the series of films obtained represents a selection of brief moments from several minutes of fluoroscopic examination. The importance of the fluoroscopic data gained during the examination should not be underestimated. This is most significant in the upper GI tract and in many centres it is now common practice to retain a cine film or videotape record of a fluoroscopic examination, particularly in relation to oesophageal disorders.

The standard barium meal should examine the upper gastrointestinal tract from the cricoid cartilage to the duodenojejunal flexure. Even in patients with symptoms referable to the oesophagus, it is wise to include at least a limited examination of the stomach and duodenum since some patients with gastric outflow obstruction may present with oesophageal symptoms. Gaseous distention of the barium-coated stomach is obtained using a carbon dioxide releasing mixture in powder or tablet form (e.g. sodium bicarbonate and citric acid). Short-acting antispasmodics are usually used to allow maximal distention of the stomach and duodenum except where contraindicated by heart disease, glaucoma or urinary retention. Films are obtained erect, supine and in various degrees of obliquity. The quality of examination will be impaired if the patient is immobile or unable or unwilling to swallow the barium and the gas-forming mixture.

The small bowel

The small bowel can be examined by either a follow-through type of procedure, or by direct intubation of the jejunum. The simpler method is that of the small bowel meal or follow-through in which the patient drinks a barium mixture specially formulated for demonstration of the small bowel. Conventional straight abdominal radiographs are taken at intervals until the barium column reaches the caecum. At any time during the procedure the radiologist may examine any areas of particular interest by fluoroscopy and obtain spot films while doing so. The small bowel enema, on the other hand, requires the passage of a long nasogastric tube which is manipulated under fluoroscopic guidance into the proximal jejunum. A bolus of barium is injected through the tube and subsequently propelled through the small bowel by 1–2 litres of methyl cellulose solution which allows the barium to coat the small bowel mucosa whilst replacing the contrast in the lumen with clear fluid. Other versions of the small bowel enema use single-contrast with dilute barium. Spot films of each of the bowel loops are obtained with distention and compression during fluoroscopy. The small bowel meal is much easier for both the patient and the radiologist to perform and in many circumstances gives very satisfactory results. The small bowel enema gives a better demonstration of small bowel pathology (confirmed by comparative studies) but involves the additional trauma to the patient of nasogastric intubation; as a procedure it is more labour-intensive although no more time-consuming than the meal technique.

The colon
The double-contrast barium enema is the procedure of choice for examination of the colon. Since peristalsis in the colon is much slower than in the stomach and small bowel, fluoroscopy plays less of a role in diagnosis. The colon is partially filled with barium and sufficient air then introduced to distend the whole of the colonic lumen. Films are obtained from various projections after rotating the patient to promote coating of the whole of the colonic mucosa. An examination which does not show the caecum should be regarded as technically unsatisfactory. Reflux of barium may also outline the distal ileum but this cannot be relied upon as a method for demonstrating the terminal ileum. Spasm of the colon can usually be overcome by intravenous relaxants. Rectal incontinence is an occasional cause of difficulty in refluxing barium back to the proximal colon but the most common problems in performance and interpretation of barium enemas arise from failure to prepare the bowel adequately.

Bowel preparation
Cleansing the colon is the single most important factor influencing the quality of barium enema examinations. Methods of preparing the colon vary from one centre to another but most involve 48 hours or so of dietary restriction and purgation. In cases of acute inflammatory bowel disease or suspected colonic obstruction, an 'instant' enema using dilute barium or a water-miscible contrast medium may be enough for immediate diagnosis but the quality of detail obtained with this type of procedure is very limited. Small bowel examinations are also enhanced by prior evacuation of the colon and all contrast examinations of the alimentary tract should be preceded by a period of starvation to eliminate fluid and food residues.

Endoscopy

The development of fibre optic instruments has lead to the widespread use of endoscopy for the diagnosis of gastrointestinal disease. Compared with other imaging apparatus, endoscopes are portable and relatively inexpensive, and since, in most cases, no ionizing radiation is involved, the constraints on their use are few. Physicians, surgeons and radiologists may become proficient in their use. Most patients undergoing endoscopy will require some degree of sedation and topical anaesthesia so nursing support and access to a recovery area are needed.

The aim of endoscopy is for the operator to view the whole of the mucosa of the upper gastrointestinal tract or of the large bowel. Most mucosal pathology can be recognized visually and where there is doubt about the significance of the appearances, biopsy samples for histology should be easily obtained through the endoscope. Other therapeutic possibilities include photocoagulation for superficial bleeding lesions and the injection of sclerosants in or around oesophageal and gastric varices.

Although endoscopy provides a direct view of the accessible organs, its diagnostic use is restricted by several factors. (1) The operator can view only a small proportion of the mucosa at any one time so that it is sometimes difficult to be sure that all areas have been seen. (2) Some areas may be difficult to reach in patients with unusual configurations of the gastric fundus or colon. (3) Pathologies originating outside the mucosa (e.g. serosal tumours or extrinsic masses) may pass unrecognized. (4) Strictures of the oesophagus, gastric outlet, or colon may prevent the passage of the endoscope past these areas. (5) The presence of food residues in the stomach or faeces in the colon is always a major handicap to the endoscopist. Blood in the gut lumen may also prevent adequate viewing.

Ultrasound

Ultrasound scanning is of primary importance in imaging the liver and biliary tract. Examination of the pancreas is technically more difficult but often very rewarding. The applications of ultrasound to the gut are less developed but certainly worthy of attention. Patients are normally examined fasting or, if the pelvis is of interest, with a full bladder. Since ultrasound is scattered in air, the presence of gas-filled loops of distended bowel is a major handicap to ultrasound examination in patients with abdominal pathology. Obesity also degrades the image quality and the presence of surgical wounds, drains and dressings may interfere with the surface contact of the transducer and so some parts of the abdomen may not be accessible. Abdominal ultrasound is relatively simple for the patient but difficult for the operator. The introduction of real-time scanners has reduced the operator-dependence of the technique, but there is still a good deal of subjectivity in the interpretation of ultrasound images. Just as with fluoroscopy, it must be remembered that the recorded films are merely a selection of still frames from several minutes of continuous or intermittent observation.

Ultrasound is a tomographic technique and the plane of the scan is infinitely variable. This is an advantage to the operator who can vary the angle of the probe to produce the most informative image. It does, however, result in a degree of inconstancy of anatomical landmarks on the hard copy images which, to the uninitiated, can be disturbing. Because ultrasound is a tomographic technique it displays only part of the anatomy of interest at any one time and a picture of the whole organ must be built up from a series of parallel sections. The resolution of ultrasound images tends to diminish with increasing depth within tissue so that superficial structures are shown in better detail than deep structures, and thin patients are easier to examine than fat ones.

Although the application of ultrasound to detect gut disorders is less well developed than its use in liver and biliary tract disease, current areas of interest include the demonstration of colonic tumours, transrectal ultrasound for the staging of primary rectal tumours and the use of endoscopic ultrasound for improved staging of gastric and oesophageal tumours.

Computed Tomography

Because computed tomography (CT) displays a full range of tissue densities from gas through to bone, the anatomical information displayed is relatively easy to assimilate. Respiratory blur is avoided if the patients are able to stop breathing for the period of each scan (usually about 3–5 seconds). With this range of scan times bowel peristalsis does not cause troublesome artefacts. The presence of dense barium or metallic implants may, however, severely interfere with image quality. Mild obesity usually improves the anatomical differentiation of various structures since fat is easily distinguished from other soft tissues, but gross obesity has a degrading effect on the images. Slices through the torso are normally obtained only in the axial plane and although it is possible to obtain computer reconstructions of image data in sagittal, coronal and oblique planes from a stack of axial slices, the anatomical detail obtained by this method is relatively crude. In order to distinguish small bowel from lymph nodes and other soft-tissue structures, it is routine procedure to render the lumen of the bowel opaque by giving the patients a substantial volume (1000 ml or so) of dilute contrast medium by mouth. In some cases, rectal infusion of dilute contrast is also helpful and in many others intravenous contrast enhancement is used to improve the demonstration of abnormalities within the solid organs, to indicate the position and patency of the major vessels, and to give some indication of the vascularity of abnormal structures. Patients are normally examined fasting and bowel preparation may be helpful though not essential. CT examinations need to be tailored to the problem under investigation. In general, CT demonstration of small structures needs thinner slices to give better anatomical detail, whereas survey examination requires thicker slices to cover the area of interest more quickly. As with ultrasound and endoscopy, each image shows only a small part of the organ of interest and interpretation of the images requires a three-dimensional appreciation of sequential scans. The radiation dose from CT is comparable with that from two or three plain films of the abdomen and is less than would be expected from a barium examination.

Although demonstration of the gut mucosa by CT is by no means as detailed as that obtained by double contrast barium examination or by endoscopy, CT is much more effective in showing pathology outside the mucosal layer, e.g. bowel wall thickening, extrinsic tumours, and the spread of mucosal disease through the bowel wall.

Scintigraphy

The level of anatomical detail available from scintigraphic studies is relatively poor and the major aim of these tests is to allow a diagnosis of physiological abnormalities. As with other investigations of upper abdominal physiology, the patients are examined in the fasting state, but they need no other special preparation. Radiation doses from the procedures described below are no greater than those associated with conventional radiographic procedures, and in most cases are much less. The radionuclide tests usually involve a single intravenous injection or oral administration of the radiopharmaceutical and complications arising from these tests are virtually unknown.

Salivary gland function

When $^{99}Tc^m$-pertechnetate is given intravenously it is trapped and secreted by the salivary glands in the same way as iodine. Using the gamma camera, the normal parotid and submandibular glands can be identified. The trapped isotope can be discharged from the salivary glands following the oral administration of a secretory stimulus such as ascorbic acid or lemon juice. In addition to visual analysis, uptake and discharge can be measured by plotting activity/time curves if the camera is interfaced to a computer during acquisition.

The technique can be of value in the patient who presents with a dry mouth (xerostomia). In patients with Sjogren's syndrome, the salivary glands are not visualized on pertechnetate scanning, whereas in psychogenic xerostomia, both uptake into, and discharge of, tracer from the salivary glands are normal.

Isotope scanning of the salivary glands has also been used in patients producing excessive saliva. This problem is sometimes treated surgically by section of nerves supplying motor fibres to the salivary glands. Study of the discharge rate of tracer pre- and postoperatively gives a quantitative assessment of the efficacy of the surgery.

Salivary scanning has no role in patients with acute inflammation of the salivary glands, salivary gland tumours or salivary duct obstruction.

Oesophageal studies

Oesophageal transit is usually studied after an overnight fast. The patient is positioned supine under a gamma camera and asked to drink a small volume of radioisotope solution (such as 5–10 MBq ^{99}Tc^m-sulphur colloid in 15 ml of water) through a straw. Data is then acquired continuously for 30–60 seconds. Several swallows are usually performed. In normal subjects the oesophagus clears within 20 seconds, but patients with motor disorders of the oesophagus and some cases of reflux show prolonged transit. Radionuclide studies of oesophageal transit give a lower radiation dose than a barium swallow. They may also be more physiological as the density and viscosity of the barium influence its behaviour. The sensitivity of radionuclide studies for detecting motor abnormalities of the oesophagus appears to be higher than that of a barium swallow. Oesophageal manometry and studies of intraoesophageal pH have the disadvantage of requiring intubation which may in itself alter oesophageal function.

For the detection of gastroesophageal reflux, 10 MBq ^{99}Tc^m-sulphur colloid is given orally to the patient in either 300 ml of water or a mildly acidic mixture of orange juice and hydrochloric acid. After 15 minutes the patient is imaged in the erect position. Any activity in the oesophagus indicates either reflux or abnormal transit. If no oesophageal activity is seen in the initial image, an abdominal binder (put in place before the start of the study) can be progressively inflated at 20 mmHg increments to 100 mmHg. At each pressure increment an image is obtained to demonstrate any reflux. A similar technique may be used to evaluate the presence of gastropulmonary reflux, particularly in children. Last thing at night, a dose of ^{99}Tc^m-sulphur colloid is taken orally. Next morning the chest is imaged with the gamma camera and the presence of intrathoracic activity documented. The radionuclide method appears to be more sensitive than contrast radiology for demonstrating gastropulmonary reflux.

Gastric emptying studies

Abnormalities of gastric emptying may result from peptic ulcer disease primarily, from gastric surgery or from anticholinergic drugs. The investigation of abnormal gastric emptying (e.g dumping syndrome after gastroenterostomy) can be achieved using radionuclide tracers. If two isotopes are used, liquid and solid phases of emptying can be evaluated simultaneously. Solid markers include chicken liver labelled with ^{99}Tc^m-sulphur colloid, ^{99}Tc^m-labelled egg and small pieces of paper or bran labelled with ^{99}Tc^m. If a solid phase marker is being used, the liquid phase marker is usually a solution of indium-111-DTPA. If only the liquid phase is being studied, a solution of $^{99}Tc^m$-DTPA is employed. After an overnight fast the patient receives the radiolabelled meal and data is then collected for 1–3 hours. Images are usually obtained for 60 seconds at 5–15 minute intervals and stored on computer. A region of interest is drawn around the stomach and a time/activity curve plotted. The usual method of expressing results is to estimate the time after meal ingestion at which stomach activity reaches half the initial value. More complex analysis consists of plotting the entire gastric emptying curve against normal ranges. Because of the influence of test meal composition on speed of gastric emptying, each department must determine its own normal range. The curve derived for liquid emptying will usually be approximately monoexponential, while the solids show a zero-order curve. Normally, the liquid component of a mixed meal empties faster than the solid component. Delayed gastric emptying may result from mechanical obstruction, previous vagotomy, electrolyte imbalance, and various drugs including opiates and anticholinergic agents. Rapid gastric emptying results from pyloroplasty or partial gastrectomy, most often, and is an important element in the early dumping syndrome seen after peptic ulcer surgery. Occasionally, rapid gastric emptying may be seen with a gastrin-producing tumour of the pancreatic islets (Zollinger–Ellison syndrome) or in thyrotoxicosis.

Gastric reflux of bile

Reflux of bile from the duodenum into the stomach often follows peptic ulcer surgery. The irritant action of bile produces gastritis which can result in abdominal discomfort and vomiting. Radionuclide assessment of biliary reflux utilizes hepatobiliary imaging agents such as $^{99}Tc^m$-IDA given intravenously. After a 30 minute interval to allow hepatic uptake and gallbladder accumulation to occur, the patient drinks 200–300 ml of milk. Images are then obtained every 5 minutes for 1 hour. Using a computer, a region of interest for the stomach is outlined and the percentage of the injected dose of $^{99}Tc^m$-IDA entering the region is calculated. The results of the isotope test correlate well with more invasive methods of assessing biliary reflux.

Gastrointestinal bleeding

Three isotope techniques are available:
1. 400 MBq (10mCi) of $^{99}Tc^m$-sulphur colloid are given intravenously. Most of the injected activity is rapidly cleared by the liver and spleen. If active bleeding is taking place then the tracer will be seen to accumulate at the bleeding point due to continuing extrusion of residual blood activity.

2.   An intravascular marker such as $^{99}Tc^m$-albumin or $^{99}Tc^m$-autologous red
     blood cells is given. The technique for labelling autologous red cells is
     described in Chapter 3 (pp. 61–2). In patients with slow or intermittent
     bleeding, accumulation of the intravascular tracer at the bleeding site
     will be seen. To achieve maximum sensitivity images should be obtained
     at intervals up to 24 hours after injection. The sensitivity of detection of
     gastrointestinal bleeding is claimed to be better than that of arterio-
     graphy and the trauma to the patient is much less.
3.   Injection of $^{99}Tc^m$-pertechnetate localizes in gastric mucosa and
     so demonstrates the vast majority of symptomatic Meckel's diverti-
     cula.

SeHCAT

The selenium-labelled conjugated bile acid 23-selena-25-homotaurocholate
(SeHCAT) has recently been used in the assessment of terminal ileal
disease. The compound is administered orally and absorbed in the distal
small bowel. Following absorption it is excreted by the liver into the bile and
enters the enterohepatic circulation. Even in the presence of severe hepatic
disease, urinary excretion is minimal. In the case of terminal ileal disease,
the absorbed fraction is reduced and this is demonstrated by low whole body
retention values 4–7 days after giving the tracer. The precise place of
SeHCAT studies is still being defined but it may be useful as a method of
specifically studying terminal ileal function.

Abscess detection

Gallium-67 citrate or autologous white cells labelled with indium-111 can be
used to localize abscesses or areas of active inflammation. Gallium is
excreted by the liver so its appearance in the bowel is not particularly
diagnostic but labelled white cells may be used to demonstrate the extent
and activity of inflammatory bowel disease.

Because the data obtained from radionuclide techniques are essentially
physiological rather than anatomical, it is more difficult to assess the clinical
value of these tests than of the simpler anatomical imaging methods. In
addition to the techniques mentioned in the preceding paragraphs, many
other radionuclide techniques have been described for investigating different aspects of gastrointestinal pathophysiology. In many cases techniques
are appropriate for research purposes but have not been widely accepted yet
for routine clinical use. Some current areas of interest additional to those
described above are the use of labelled monoclonal antibodies for tumour
localization and new technetium-labelled agents for demonstrating inflammatory bowel disease and abscesses.

Diagnostic Angiography

The development of ultrasound and computed tomography has minimized the diagnostic applications of angiography in disorders of the solid organs. Angiography is usually undertaken now for diagnosis only in patients suspected of having one of the diseases in which anatomical abnormalities are characteristically seen in the vascular tree — e.g. ischaemic bowel disorders, polyarteritis, aneurysms and occlusions of the visceral arteries, vascular malformations and vascular tumours. Angiography also has a small but important role in the localization of sites of gastrointestinal bleeding. Angiography has further applications in the preoperative mapping of vascular anatomy before attempted resection of tumours of the liver and pancreas. The angiographic demonstration of portal venous anatomy plays an important part in the surgical management of portal hypertension. Therapeutic applications of angiography techniques are mentioned below.

Angiography requires a skilled operator and is costly in terms of consumables. There is a potential risk of local complications at the site of catheterization (usually the femoral artery). Patients require sedation and local anaesthesia but general anaesthesia is rarely necessary except in children. Residual barium in the bowel may obscure the field of view and usually renders visceral arteriography technically unsatisfactory. The development of digital angiography allows fine detail of vascular structures to be obtained with much smaller doses of contrast media which can be injected through narrower arterial catheters but the intravenous route is not usually satisfactory for visceral arterial studies.

Interventions

Percutaneous access to the vascular tree allows the radiologist to carry out therapeutic manoeuvres under fluoroscopic control. Embolization of either arteries (for gastrointestinal bleeding or vascular tumours) or veins (bleeding oesophageal varices) can be achieved using either absorbable or nonabsorbable materials. Chemotherapeutic agents may be delivered locally to the site of tumours by catheters placed percutaneously under radiological guidance.

Percutaneous needle biopsy of soft-tissue masses is facilitated by ultrasound or CT guidance. Almost any organ or soft-tissue mass within the abdomen can now be biopsied percutaneously. Either ultrasound or CT guidance can be used; ultrasound is generally quicker and simpler but CT is preferable for retroperitoneal structures and lesions close to the major vessels. With either technique, diagnostic scans are obtained to illustrate the position of the lesion and the required direction and depth of needling. Further scans are obtained with the needle *in situ* to ensure that the biopsy obtained is actually from the tissue of interest.

By similar methods, the percutaneous aspiration of fluid collections can be achieved in most patients. Where more prolonged drainage is required, placement of drainage catheters is greatly simplified by ultrasound or CT guidance. Again, either method can be used but as a general rule, ultrasound is preferable for fluid collections which are situated near the surface of the body, while CT is reserved for deep-seated abscesses, particularly retroperitoneal collections.

The negotiation of catheters through strictures of the oesophagus or bile ducts can be facilitated by imaging guidance which is also helpful in the subsequent dilatation of benign strictures using balloons or the insertion of tubes to maintain the lumen through malignant strictures.

Gastrointestinal Disease

The Oesophagus — Tumours, Strictures and Varices

Patients presenting with oesophageal symptoms should have a plain chest radiograph. An upper mediastinal mass containing an air-fluid level will suggest a pharyngeal pouch. A homogeneous mass in the lower part of the middle mediastinum could be caused by a duplication cyst; if this is suspected, radiographs of the spine should be obtained since such cysts are often associated with bony abnormalities. Where a mediastinal mass is discovered, a barium swallow is needed to show its relation to the oesophagus and computed tomography will demonstrate the relationship to other organs and often give indications of the likely pathology. The chest film may also show: evidence of aspiration into the lungs in patients with motility disorders, diffuse enlargement of the oesophagus with food and fluid retention (achalasia) or a dilated oesophagus filled with air (systemic sclerosis).

The choice between barium swallow and endoscopy for direct examination of the oesophagus is contentious, and may depend upon local expertise and availability. Barium usually allows both the upper and the lower limits of oesophageal lesions to be defined whereas endoscopy may be restricted to the upper extent of stricturing lesions. However, the ability to take biopsy samples is clearly an advantage of the latter technique.

Oesophageal varices can be demonstrated either by barium examination, obtaining films in the prone and supine positions (preferably with the use of smooth muscle relaxants), or by endoscopy which has the added advantage of enabling sclerosants to be injected at the same manoeuvre.

Endoscopy is the first-choice examination for haematemesis and should identify bleeding sources within the oesophagus, whether ulcers, varices or Mallory–Weiss tears. Contrast examinations will not usually show the Mallory–Weiss type of lesion, but remain essential for the detection of suspected oesophageal perforation or fistulae. Where the possibility of perforation exists, the safest contrast media to use are the low osmolality water-miscible compounds. The assessment of extraluminal spread of oesophageal tumours is best performed by computed tomography. This technique may give a good indication of the resectability of oesophageal tumours and should be considered before surgery.

Oesophagitis, Reflux and Motility Disorders

Good quality double-contrast examination of the oesophagus should give a clear indication of the state of the mucosal surface. Demonstration of deep ulcers and fissuring associated with severe oesophagitis should not be difficult. A fine granular pattern is seen with less severe oesophagitis but the earliest changes are probably best shown by endoscopy with biopsy. The altered mucosal surface of Barrett's oesophagus shows a characteristic double-contrast barium appearance in some but not all cases. Peptic strictures can usually (but not always) be differentiated from malignant strictures and in doubtful cases endoscopy and biopsy should be mandatory. The presence of hiatal hernia is readily shown by barium examination but the relationship between hiatal hernia, reflux and oesophagitis is doubtful. Gastroesophageal reflux may be observed spontaneously during the fluoro-scopic part of a barium examination, or additional manoeuvres — e.g. abdominal compression, head down position, water siphonage test, etc. — may be undergone to try to produce it. Current opinion tends towards the view that the demonstration of reflux *per se* is of very little importance. What is much more significant is the volume, frequency and duration of reflux episodes, and these aspects require quantitative estimation. This can be carried out by scintigraphic studies or by prolonged pH monitoring. Scintigraphic tests for reflux include the 'binder test' in which the stomach is filled with a technetium-labelled colloid in 500 ml of acidified orange juice. An inflatable abdominal binder is applied and the pressure increased in steps of 20 mmHg from 0–100 mmHg. Gamma camera images of the stomach and lower oesophagus are obtained at each level of pressure to determine whether any reflux has occurred. The validity of this test is controversial and a more physiological approach is simply to observe spontaneous episodes of reflux occurring during a period of 40 minutes or so after the ingestion of the labelled fluid meal. Episodes of reflux can be recorded individually, their volume measured, and the rate of oesophageal clearance estimated. This test is simple but time-consuming and has not yet found wide application.

   The diagnosis of motility disorders requires the dynamic observation of the oesophagus. This can be carried out by video-tape barium studies in which fluoroscopy of the oesophagus is recorded so that a qualitative assessment of oesophageal movement and peristalsis can be obtained. Characteristic overdistention of the oesophagus with retained fluid and food is seen in achalasia whereas systemic sclerosis produces a typical adynamic oesophagus of uniform width which empties only with the assistance of gravity.

Alternatively, a radionuclide method may be used, a bolus of technetium colloid in saline or fruit juice is swallowed by the patient and its progress monitored by rapid sequence gamma camera images. Time/activity curves are obtained for upper, middle and lower thirds and for the whole of the oesophagus. Different characteristic curve patterns are described for patients with adynamic and incoordinate movement. Gastroesophageal reflux and intraoesophageal reflux may also be recognized.

Peptic Ulcer, Gastritis and Abnormalities of Emptying

As with the oesophagus, most mucosal abnormalities of the stomach and duodenum can be demonstrated effectively by either endoscopy or a good quality double-contrast barium examination. Endoscopy always has the advantage of allowing multiple biopsies to be obtained, but the histology of endoscopic biopsies of the stomach is often, surprisingly, non-contributory. In the presence of duodenal scarring, it is often difficult to decide on barium examination whether an ulcer is active, whereas the presence of local inflammation and stigmata of recent bleeding are better recognized by endoscopy. Both techniques are greatly handicapped by the presence of retained fluid and food, so once pyloric obstruction has been confirmed by the presence of residues in the fasting stomach, gastric aspiration is required before diagnostic efforts are renewed. Distinction between benign and malignant pyloric obstruction can often be made on barium examination while endoscopy allows biopsies to be taken. It will usually be impossible to pass the endoscope through an obstructed pylorus whereas it is usually feasible to get some barium past the obstruction and thus demonstrate the distal extent of the lesion. With gastric masses of substantial size, percu-taneous needle biopsy can safely be performed at the time of the barium examination.

Double-contrast examination of the stomach relies on gaseous distention and temporary paralysis of the organ with a coating of the mucosa by high density barium. Normal mucosal pattern (areae gastricae) is disturbed by tumour, local inflammation or mucosal infiltration. It is sometimes possible to visualize the thickness of the stomach wall which may be of diagnostic significance in the differentiation of gastritis from diffuse infiltrations of the gastric wall, e.g. in lymphoma or Crohn's disease.

The demonstration of recurrent ulceration in the postoperative stomach is probably best achieved by endoscopy, but it may be impossible to demons-trate the afferent loop by this method. Contrast examinations often allow filling of the afferent loop but, again, this is not always possible. In patients where afferent loop obstruction is suspected clinically, a scintigraphic study using technetium-labelled iminodiacetic acid (IDA) should demonstrate the flow of bile into the afferent loop and any stasis there should be readily detected.

Measurement of the rates of gastric emptying of both liquid and solid meals is most easily assessed by isotopic methods. One such technique uses indium–111-labelled fruit juice as a liquid component and technetium-colloid in scrambled egg as a solid component. Scintigraphic demonstration of the rate of gastric emptying allows calculation of half-emptying times for liquid and solid phases separately. Typically, vagotomy and drainage procedures produce more rapid emptying of the fluid phase but often delay in solid phase emptying. Patients with rapid solid phase emptying after gastric operations may have symptoms of dumping, while complaints of upper abdominal distention and nausea after eating are often associated with delayed emptying.

Gastric Tumours

Either endoscopy or barium examination may be sufficient for diagnosis. Endoscopy, if easily available, is probably the first choice since confirmatory biopsies can be obtained easily. Barium examination is probably better at showing the extent of the tumour preoperatively, particularly the relation of lesser curve tumours to the pylorus and to the cardia. Enlarged glands around the pylorus or around the cardia may produce characteristic deformities. A more precise demonstration of the extramural extent of gastric tumours is obtained by computed tomography. Recent work suggests that preoperative CT is helpful in showing invasion posteriorly into the pancreas, coeliac axis, and anterior invasion to the left lobe of the liver — findings which will render the tumour unresectable. Demonstration of enlarged lymph nodes around the coeliac axis or along the lesser curve of the stomach, and liver metastases are also important preoperative findings which may aid the surgical management. Further discussion of screening for liver metastases follows in Chapter 5.

Non-Tumorous Bowel Disorders

Patients with inflammatory bowel disease will require barium enema examination of the colon and, if Crohn's disease is suspected, some type of barium examination of the small bowel will be needed. The barium examinations should give information about the mucosal surface, i.e. the presence of ulceration, nodularity and fistulae, and also about the thickness of the bowel wall, the retention of normal landmarks (haustration in the colon, concentric folds in the small bowel) and the presence of strictures. In assessing the extent of small bowel disease, it is particularly important to be able to show which segments of the bowel are normal as well as demonstrating the areas of abnormality. Confirming anatomical normality of the small bowel is also a useful role for the barium examination in patients with malabsorption so that many of the causes of this syndrome can be excluded, e.g. small bowel diverticulosis, blind loop syndromes, lymphoma, parasitic infestations and other conditions producing typical abnormalities on barium examination. Differentiation of Crohn's disease from other diffuse or patchy disorders of the small bowel is not always possible from the barium procedure, but in many cases characteristic appearances will be found (Fig. 4.3).

Differentiation between ulcerative colitis and Crohn's disease of the large bowel is sometimes clear cut on the barium enema findings, but it is becoming increasingly apparent that an area of overlap exists where not only radiological but also pathological differentiation is impossible. Ischaemic and infective colitis may also produce similar appearances.

Preservation of the haustral pattern, asymmetric distribution of disease across the bowel lumen, 'skip lesions', sparing of the rectum, and deep ulceration are all features favouring Crohn's disease rather than ulcerative colitis. In the latter condition, the ulceration is usually less deep and more uniform, the haustral pattern tends to be lost in the affected areas, the bowel wall is not unduly thickened, and the distribution differs in that the rectum is

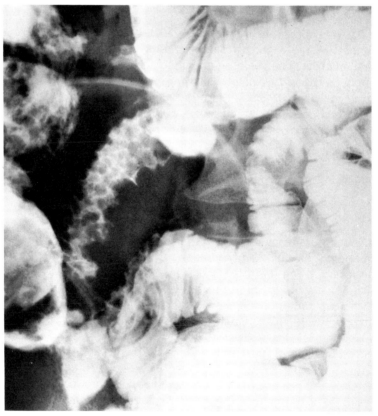

Fig. 4.3 Spot view of the terminal ileum during double-contrast small bowel enema showing the typical 'cobblestone' pattern and spicular ulceration of Crohn's disease.

almost invariably involved and the disease is usually continuous in extent with the rectum being the most severely affected area.

Diverticular disease of the colon and the polyposis syndromes require demonstration of the colon by double-contrast barium examination. Careful inspection will usually allow differentiation between isolated polyps and diverticula. In patients with diverticular disease, the complications of stricture formation, inflammatory mass, local perforation or pericolic abscess are much more important than recognition of the presence of diverticula. Where most of the pathology is outside the lumen of the bowel, barium examination may give quite an incomplete picture and needs to be supplemented by ultrasound or CT scanning of the abdomen and pelvis. These methods should show pericolic fluid collections and, particularly with CT, local bowel wall thickening can easily be shown.

Infective foci in and around the bowel can also be elegantly demonstrated by scintigraphic studies using radiolabelled white cells or gallium. Gallium is excreted through the bile so its appearance in the bowel is a normal feature but areas of uptake which persist after bowel clearance usually indicate extrinsic inflammatory lesions. Labelled white cells are more precise in demonstrating not only focal abscesses and inflammatory masses but also in indicating the extent of activity in patients with inflammatory bowel disease. Results with labelled white cells are generally better in acute than in chronic infection.

The technique requires meticulous harvesting and labelling of patients' own white cells and so is labour-intensive but, if the necessary facilities are available, gives a simple, non-invasive method of assessing disease activity which patients may find easier to tolerate than barium procedures or colonoscopy.

Tumours of the Large and Small Bowel

Most bowel tumours present either as bleeding or obstructing lesions, or as a change in bowel habit. Unless there is a classic history of small bowel colic, double-contrast barium examination of the large bowel will usually be performed first (except in a patient presenting with an acute abdomen — see below). Carcinomas in the left side of the colon usually appear as irregular strictures with well-defined shouldered ends. A complete demonstration of the colon including the caecum should be obtained even if a sigmoid lesion is found during the early part of the examination, since multiple synchronous tumours are present in roughly 5 per cent of cases. Carcinomas arising in the right colon are more often large polypoid masses producing local filling defects within the lumen of the bowel (Fig. 4.4). The barium enema will include films of the rectum down to the anal verge but rectal tumours should be diagnosed by proctoscopy or sigmoidoscopy which should always precede the barium enema. In cases of equivocal or unexpected findings on the barium enema, colonoscopy and biopsy are often decisive.

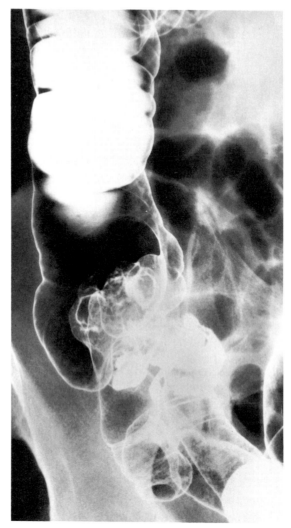

Fig. 4.4 A polypoid carcinoma of the caecum shown by double-contrast barium enema. (Courtesy of Dr A Chapman).

Small bowel tumours are shown as localized filling defects or strictures on barium examination. In some cases, the particular features are characteristic enough to indicate the histological diagnosis, but often the appearances are non-specific. This does not matter in most cases since virtually all patients will need laparotomy for relief of mechanical symptoms.

Computed tomography has a supplementary role in the staging of neoplasms of the bowel. CT shows the extent of tumours through the bowel wall and will usually indicate whether or not the tumour is attached to surrounding structures. At the same time, a search for lymph node and liver metastases can be made. Some success in staging has also been achieved with ultrasound of the abdomen and with per-rectal ultrasound for rectal carcinoma.

Imaging for Specific Clinical Problems

Dysphagia

Patients presenting with dysphagia require either endoscopic or barium examination of the oesophagus. The current view is that rigid oesophagoscopy should not be carried out before the site and likely nature of the obstructing lesion have been shown by barium swallow. However, the flexible fibreoptic instruments, being much safer, may be used more readily. Endoscopy has the advantages of allowing direct biopsy of abnormal areas, and also of providing a route for transluminal dilatation of strictures. Barium swallow is more likely to show the full extent of the lesion, outlining both its upper and lower margins, and showing the anatomic relation of the lesion to surrounding mediastinal structures, particularly the aortic arch, left main bronchus, inferior pulmonary veins, and gastric cardia. If oesophageal perforation is suspected, a contrast swallow using low osmolality aqueous media or an oily contrast medium is indicated. The same technique should be used if there is clinical suspicion of aspiration of swallowed material into the lungs. A chest radiograph is often helpful and should not be omitted. If dysphagia is caused by extrinsic compression of the oesophagus, imaging of mediastinal structures will be needed; CT is the preferred method.

If no anatomic obstruction is shown, the cause of the symptoms may lie in disturbed neuromuscular function. Motility studies using either radionuclide techniques or video-taped barium swallow can be used. Infants and small children with dysphagia will be investigated in the same way as adults, but the possibilities of congenital anomalies such as duplication cysts, tracheoesophageal fistulae, vascular rings, and a diaphragmatic hernia, should be borne in mind.

Dyspepsia

The choice between endoscopy and barium meal examination is arguable. Either could be the first-choice test for patients with dyspepsia, both tests being used if the results of the initial procedure are indecisive or unexpected. If the oesophagus, stomach and duodenum appear normal, attention should then turn to the right upper quadrant. An ultrasound examination here should detect gallstones if any are present and also allow the size and consistency of the liver to be assessed. If still no abnormality is found, it may be worth pursuing the pancreas with ultrasound or CT, and endoscopic retrograde cholangiopancreatography (ERCP) to rule out minor degrees of chronic pancreatitis.

Patients with dyspeptic symptoms after gastric surgery may be suffering from delayed gastric emptying, dumping, or bile reflux gastritis. The first two of these conditions can be investigated usefully by scintigraphic studies of gastric emptying, particularly when liquid phase and solid phase emptying can be measured separately. The presence of bile reflux can be detected non-invasively by IDA scintigraphy, a quantitative estimate of the amount of bile being refluxed being feasible with this technique. Recurrent ulceration can be often recognized on double contrast barium examination, but endoscopy is probably better for this particular application. Persistently recurring or multiple ulcers should raise the question of the Zollinger-Ellison syndrome and, if serum gastrin levels are raised, a search for a pancreatic islet cell tumour will be needed; ultrasound and CT scanning are firstly indicated, but if they are inconclusive, careful pancreatic arteriography is suggested; if this also fails to locate the tumour, transhepatic sampling of pancreatic veins is the final option.

Embarrassing mistakes will be made if plain chest and abdominal radiographs are not obtained at some early stage of the patients' investigation.

Altered Bowel Habit

If malabsorption is suspected clinically, then attention should be focused first on the small bowel and pancreas. Some type of small bowel barium examination should be used, preferably a double-contrast small bowel enema. If no anatomical lesion is found but biochemical and clinical evidence points to small bowel disease, duodenal or jejunal biopsies may be the next useful step. Although not strictly speaking an imaging technique, the SeHCAT retention test is proving to be most useful in distinguishing patients with diarrhoea due to the malabsorption of bile acids from other causes of this symptom.

Biochemical evidence of pancreatic insufficiency should stimulate examination of the pancreas, either by ultrasound or CT or, if these techniques are equivocal or negative, ERCP. In some patients, positive findings on small bowel barium examination may be sufficient for diagnosis, e.g. demonstration of small bowel diverticulosis or patients with ischaemic ileitis following therapeutic irradiation. Most focal small bowel lesions (tumours, strictures, lymphoma etc.) will require surgical intervention for mechanical reasons and histological evidence can then be obtained. If the small bowel series shows inflammatory bowel disease, then examination of the colon should follow.

In most patients with altered bowel habit, the colon will be the first area looked at. Normally, the distal sigmoid and rectum will be examined endoscopically first, then a barium examination of the whole of the large bowel carried out. The use of ultrasound and CT in the preoperative staging of colorectal tumours, and scintigraphic studies in inflammatory disease have been mentioned earlier.

Intestinal Obstruction

Chest and supine abdominal films must be obtained. Whether an erect or horizontal decubitus film of the abdomen is also essential is a current controversy at the time of writing. In most cases, the plain film findings together with clinical assessment will indicate not only the presence of obstruction but the approximate level of the lesion. If there is doubt about the presence of obstruction, a large bowel enema using water-soluble contrast will usually be decisive. This is particularly helpful in elderly patients where 'pseudoobstruction' may present similar clinical and plain film findings. Small bowel obstruction is usually recognizable from the plain film which, in some cases, will also give clues to the cause, e.g. gallstone ileus, incarcerated hernia, appendix mass. With intermittent or dubious obstruction of the small bowel, particularly in patients who have undergone previous surgery, it may be helpful to define more precisely both the level and the anatomical appearance of the obstructing lesion. This is best achieved by a small bowel enema examination obtained after aspirating as much fluid as possible from the stomach, duodenum and proximal jejunum. There is little point in giving these patients a cupful of contrast medium orally and hoping for the best.

The Acute Abdomen, Ischaemia and Trauma

After obtaining plain chest and abdominal films, indications for further imaging depend largely on the clinical localization of the likely site of pathology. If a gynaecological problem is thought likely (acute salpingitis, ectopic pregnancy), pelvic ultrasound is indicated. Symptoms or signs pointing to the right upper quadrant also warrant ultrasound as the next investigation. If the likely pathology is retroperitoneal (acute pancreatitis, leaking aortic aneurysm), CT scanning is most likely to be contributory. Contrast studies of the stomach or intestine are valuable in patients with suspected perforation or penetrating injuries (Fig. 4.5). Ultrasound is most useful for screening the parenchymal organs for evidence of injury after abdominal trauma, but may be technically difficult once an ileus has developed. Where there is clinical doubt about soft-tissue injury, CT is often helpful. In cases of severe trauma, some information on the viability of tissue fragments can be obtained by observing the degree of enhancement seen after intravenous contrast injection during CT scanning. Iminodiacetic acid (IDA) scintigraphy may be useful in patients with bile leakage; doubt about the integrity of the pancreatic duct after severe trauma may require ERCP for clarification. Abdominal angiography is occasionally used for defining vascular injuries (traumatic fistulae, false aneurysms) and in showing the extent of ischaemia in patients with traumatic vascular occlusion. Visceral arteriography may also be used to confirm a clinically suspected diagnosis of mesenteric venous thrombosis, arterial thrombosis or

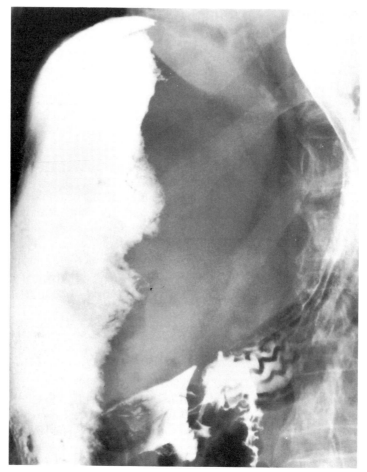

Fig. 4.5 Film taken during upper GI contrast series in a patient with a perforated duodenal ulcer. Contrast fills the right subphrenic recess of the peritoneal cavity. (Courtesy of Dr A Chapman).

embolism, but few patients with acute intestinal ischaemia are suitable for surgical treatment. The significance of chronic arterial insufficiency, particularly compression of the coeliac axis, is, to say the least, dubious. The current consensus is that the combination of a narrow or compressed coeliac axis with upper abdominal pain is not an adequate indication for major reconstructive surgery.

Abdominal Mass

As always, a chest film should be obtained. Plain abdominal films will sometimes illustrate the relation of a soft-tissue mass to adjacent viscera, and disturbances in the contours of the liver, spleen or kidneys may indicate the site of origin. Some types of calcification suggest a specific pathology, e.g. dermoid cyst, hydatid cyst, some hepatomas.

Ultrasound scanning should be the next examination. If a mass is palpable there should be no technical difficulty in obtaining images of the lesion. In most cases ultrasound will show not only the structure of the mass (whether cystic, solid or mixed) but also its relation to surrounding anatomy. As long as normal landmarks are not grossly distorted, it should be possible to ascertain the site of origin of the mass, although this may be difficult with tumours of the bowel, mesentery or omentum. With very large masses anatomy is often so severely disturbed that it may be impossible to decide from which organ the mass originates. Percutaneous aspiration of cysts or needle biopsy of solid masses may then be carried out under ultrasound guidance.

When the results of ultrasound are equivocal or unexpected, CT is suggested as the next step (Fig. 4.6). CT will also be more informative than ultrasound with most retroperitoneal masses so it could be regarded as routine in these cases. In particular, tumours of the pancreas, adrenals and lymph nodes are shown best by CT; the absence of artefact from bowel gas enables the relation of masses to the gut to be readily appreciated. The

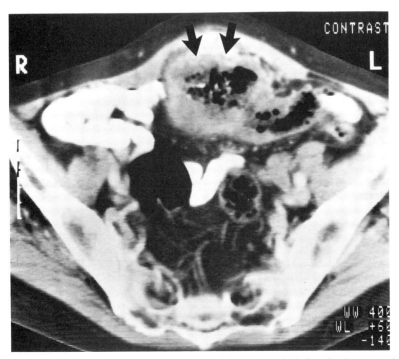

Fig. 4.6 CT scan of a patient presenting with a central abdominal mass which was difficult to interpret on ultrasound. CT showed the presence of gas within the mass (arrows) and illustrated its continuity with the large bowel. Diagnosis: colonic carcinoma.

effect of intravenous contrast on the lesion as seen by CT will give an indication of its vascularity, an important consideration if percutaneous biopsy is to be performed with a cutting needle.

If the clinical pointers or the initial scans suggest a bowel origin for a mass, barium examination or endoscopy is indicated. Even in patients with tumours of the gut, however, demonstration of the extraluminal extent of the mass by CT or ultrasound is helpful in planning treatment. If, after ultrasound and CT examination, the site of origin of a mass is still not clear, arteriography may be helpful even though most such patients will undergo surgical exploration.

In addition to its role in the diagnosis of abdominal masses, imaging plays a part in the preoperative assessment and staging of tumours. The operability of tumours arising in any of the abdominal organs can often be predicted from imaging procedures, particularly CT and arteriography. In some cases, for example liver and pancreatic tumours, the angiographic demonstration of vascular involvement is of crucial importance in deciding on resectability. The demonstration of liver metastases by scintigraphy, ultrasound or CT may decisively affect the management of a primary tumour in patients with gastric or colorectal cancer.

Patients with a renal mass should have either intravenous urography or a scintigraphic renal study. In addition to adding further information about the anatomical and functional disturbance caused in the affected kidney, it is most important to establish the presence and the level of function in the unaffected kidney prior to surgery.

The discovery of enlarged lymph nodes will initiate a search for abnormal nodes elsewhere and, if nodal histology shows metastatic disease, for a primary tumour site.

Gastrointestinal Bleeding

Difficulties in the localization of GI bleeding sites arise from the fact that most bleeding is intermittent and that, in some patients, multiple potential bleeding sites are present. In most patients with acute bleeding, upper GI endoscopy will demonstrate the source. Difficulties arise if the stomach is full of blood or retained food and gastric juice, or if a potential bleeding site is found but no actual bleeding seen. For example, oesophagitis, varices or a peptic ulcer may be apparent but if there are no stigmata of recent bleeding, it will be difficult to decide whether the lesion seen is the one responsible for the haemorrhage. The possibility of further bleeding sites elsewhere has to be considered. If endoscopy is equivocal, double-contrast barium meal may help to resolve doubtful lesions or show areas of the stomach which were not well seen by the initial examination, and also the distal parts of the duodenum.

If upper GI endoscopy is negative, a search for a bleeding site elsewhere must be made. Sigmoidoscopy followed by double-contrast barium enema will usually be the next examinations, except in patients with massive rectal blood loss. Colonoscopy is a reasonable alternative, but the risk of failing to visualize the caecum is probably higher with this technique. Blood in the colon often obscures the view. As with the upper GI series, lesions may be shown which could be responsible for bleeding although active haemorrhage is not seen — for example, diverticular disease or small colonic polyps.

In patients with massive rectal bleeding, it may be impossible to clear the contents of the colon sufficiently to obtain an adequate view at colonoscopy, and these patients, together with those in whom colonic examination is negative, need further imaging with radionuclide localization and followed, in some cases, by arteriography. The site of extravasation of blood within the bowel can be shown by colloid scintigraphy with bleeding rates less than 1 ml/minute as long as the patient is actively bleeding during the first few minutes after injection of the labelled material. If the colloid study is positive, an indication of the approximate level of bleeding site in the large or small bowel will be obtained within a few minutes (Fig. 4.7). Knowledge of the level of bleeding can then guide surgery or arteriography (Fig. 4.8).

Demonstration of extravasation of contrast after selective catheterization of the mesenteric or gastroduodenal vessels gives a more definitive localization. In appropriate cases, bleeding can be controlled angiographically

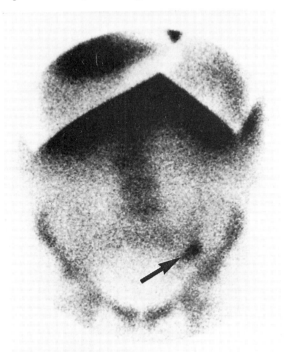

Fig. 4.7 Scintigram of the abdomen and pelvis 15 min after injection of labelled colloid. A focus of activity (arrow) indicates localized extravasation in the left iliac fossa; diverticular bleeding was confirmed at surgery.

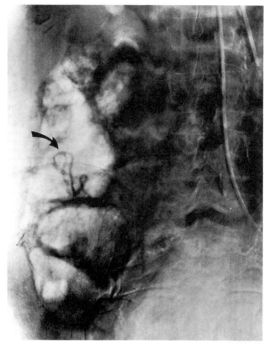

Fig. 4.8 Venous phase of selective superior mesenteric arteriogram showing a small vascular malformation in the ascending colon (arrow).

either by the infusion of vasopressin or by therapeutic embolization of the feeding vessels.

If the colloid study fails to show a bleeding site, labelled autologous red cells can be given. Extravasation which is intermittent or very slow can then be detected by regular scintigraphic images over several hours. Negative red cell scintigraphy indicates that the patient has stopped bleeding. A positive result indicates continuing bleeding and should give an indication of the approximate site. Once the level of bleeding is known, either angiographic or surgical control can be attempted.

Patients with rapid bleeding who need urgent endoscopic, angiographic or surgical therapy should not be given barium since this will make angiography impossible, and also interfere with surgery. In order to show extravasation by angiography, the rate of bleeding must be 1–2 ml/minute or more. Bleeding which is not fast enough to produce a melaena stool will not produce positive results with angiography or the scintigraphic tests. In those patients who present with chronic anaemia in whom a positive chemical test for faecal occult blood is the only evidence of gastrointestinal bleeding, a small bowel barium is suggested if the upper GI endoscopy and barium enema are normal. If no structural lesion is shown in the small bowel, arteriography is then indicated since, even in the absence of active bleeding, it may show angiodysplasia of the colon or other structural vascular lesions which could represent bleeding sites.

Fluid Collections and Abscesses

Imaging procedures should readily demonstrate the presence and distribution of ascites; simple collections of blood, bile, urine, lymph or pancreatic exudate should also be readily recognized. Abscesses, with their complex structure, are a little more difficult, but even so, the introduction of newer imaging techniques has substantially improved the management of abdominal abscesses.

Gross ascites produces a number of changes on the plain abdominal film, most of which are also seen in abdominal obesity. The only specific signs are those caused by fluid accumulating along the paracolic gutters and around the edges of the liver separating the viscera from the adjacent abdominal and chest wall. In patients with unequivocal evidence of ascites on the plain film, the presence of fluid will usually be obvious clinically. Plain films cannot be relied upon to show small amounts of ascites but both ultrasound and CT are much more sensitive.

Small amounts of free fluid are seen by ultrasound or CT in the upper abdomen as a thin layer surrounding the right lobe of the liver or in the pelvis as a fluid-filled space in the rectovesical or rectovaginal pouch.

Loculated fluid collections may give the appearance on the plain film of a soft-tissue mass displacing surrounding organs. Abscesses show the same feature together with the presence of gas, seen either as streaks or small bubbles within the contour of the lesion, or as an air-fluid level on films taken with a horizontal beam. The location of a fluid collection or abscess can be best demonstrated by ultrasound or CT. Either technique should show the position of the lesion within the abdomen or pelvis, and its relation to surrounding structures. Distinction should also be made between clear fluid (such as urine, lymph or ascitic transudate) and fluid collections with a solid component such as fresh haematoma, abscess or necrotic tumour. With CT, measurement of the average attenuation value within the fluid collection will give some indication of its nature. Lymph and urine will have attenuation values close to that of water, active abscesses, fresh haematomas and necrotic tumours will have attenuation values about 20–30 Hounsfield units, whereas collections of old blood or non-infected bile show a density which is usually intermediate. Gas bubbles are easily recognized on CT and also produce characteristic densely echogenic foci on ultrasound.

Although either ultrasound or CT may be used, ultrasound has the advantage of being more readily available and equipment can be transported to the bedside of patients who are severely unwell. Ultrasound is likely to be successful in patients with collections near to the surface of the abdomen and is generally reliable in the pelvis and the right upper quadrant (Fig. 4.9). The presence of gas in the splenic flexure and stomach makes the left upper

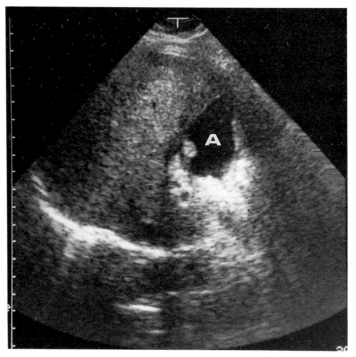

Fig. 4.9   Sagittal ultrasound scan through right lobe of liver in a patient with pyrexia and pain áfter cholecystectomy. There is a small fluid collection (A) in the gall bladder bed.

quadrant a more difficult area for ultrasound; the retroperitoneal area, particularly of the lower abdomen, is also difficult to access, so lesions in these areas are usually better shown by CT. Percutaneous aspiration or the introduction of drainage catheters can be guided by either method. Ultrasound guidance is technically simpler since the beam can be directed along the line of approach of the needle or catheter in virtually any direction. With CT, it is only possible to obtain direct scans in the axial plane, so that if an oblique approach is necessary, a transverse scan will only show part of the needle. Nevertheless, CT provides the clearest possible visualization of the retroperitoneal anatomy, and, for example, can be used to guide the placement of catheters extraperitoneally for the drainage of pancreatic collections or psoas abscesses (Fig. 4.10).

Most patients with abdominal abscesses, particularly following surgery, will have clinical signs suggesting the location of the lesion. Some such patients will have no localizing signs and others may present with pyrexia of uncertain origin (PUO). In these cases, scintigraphy using labelled white cells or gallium is often the most helpful procedure. Scintigraphic techniques have the advantage that a focal abnormality will be shown if there is an abnormal accumulation of white cells so that not only mature abscesses but

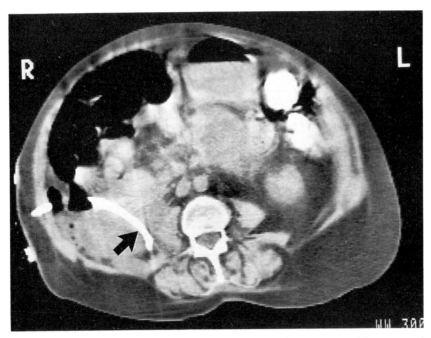

Fig. 4.10 CT scan through the upper abdomen of a patient with retroperitoneal abscesses complicating acute pancreatitis. A sump drain has been placed percutaneously under CT guidance (arrow).

also focal inflammatory masses and 'pre-abscesses' will also be detected. Bowel infarcts or fat necrosis may also produce positive results, however. A further application of these tests is in distinguishing whether or not infection exists in a fluid collection which has been demonstrated by anatomical methods. A negative white cell or gallium scan will suggest that the fluid collection is sterile and thus unlikely to be contributing to symptoms and signs of active infection.

Abdominal Trauma

Some patients with abdominal trauma require immediate operation while others will be investigated before treatment. Plain abdominal radiographs can be helpful before more complex investigations. The bony integrity of the spine and pelvis, the bowel gas pattern, the soft-tissue outlines and the presence of free intraperitoneal gas can be assessed. Generalized dilatation of large and small bowel with gas seen down to the rectum usually results from an ileus secondary to the trauma. A localized loop of dilated bowel may indicate trauma to a particular organ, e.g. a dilated transverse colon with pancreatic trauma. The outlines of the inferior margins of the liver and spleen, the kidney, the psoas muscles, and the properitoneal fat planes may be effaced by trauma or by the presence of effusions. Obliteration of a psoas outline, particularly if the ipsilateral renal outline is obscured, points to pathology in the retroperitoneum. A careful examination of the lumbar spine should follow so as not to overlook a forced flexion injury or Chance fracture.

If injury to a solid organ is thought likely, either ultrasound or CT should be performed. The presence or absence of free intraperitoneal fluid will be shown and, after IV contrast, CT gives an assessment of renal function and vascular damage. An ileus will hamper the ultrasound examination of solid organs, particularly the pancreas, and no comment can be made concerning renal function or skeletal damage. Ultrasound examinations may also be hindered by wounds, drains and dressings. CT scans in patients with abdominal trauma are usually performed after intravenous contrast to maximize the differentiation between enhancing parenchyma and non-enhancing haematoma. It is important to scan down into the pelvis in order to see collections of fluid or blood in the pouch of Douglas. The appearance of blood on CT varies with its age and site. Free peritoneal blood tends to be less dense than fresh clotted blood and the density will diminish with time. After 48–72 hours, free blood may have a similar density to ascites.

Real-time ultrasound can provide a bedside or emergency room service to assess the abdominal organs, pelvis and chest wall with minimal disturbance to an ill patient. The requirement for patient co-operation is less with ultrasound than for CT.

Further Reading

Cosgrove, D.O. and McCready, V.R. (1982). *Ultrasound Imaging of the Liver, Spleen and Pancreas.* John Wiley, Chichester.

McCort, J. (1981). *Abdominal Radiology.* Williams and Wilkins, Baltimore & London.

Megibow, A.J. and Balthazar, E.J. (1986). *Computed Tomography of the Gastrointestinal Tract.* C.V. Mosby, Ontario.

Robinson, P.J. (Ed.). (1986). *Nuclear Gastroenterology.* Churchill Livingstone, Edinburgh.

Simeone, J.F. (Ed.). (1984). Co-ordinated diagnostic imaging. *Clinics in Diagnostic Ultrasound* **14**, Churchill Livingstone, Edinburgh.

5

THE LIVER, BILIARY SYSTEM AND PANCREAS

Imaging Techniques

Conventional Radiology

Abdominal radiograph

An abdominal radiograph is generally requested to identify gallstones. As less than one third of gallstones contain sufficient calcium to appear on a plain X-ray, this investigation represents rather poor diagnostic value. When opaque, calculi may be laminated, facetted or occasionally gas-containing. Although 15–20 per cent of patients with multiple gallstones will have duct stones shown at surgery, in practice, only 1–2 per cent of these can be identified on the plain X-rays. Traditionally, a plain radiograph is obtained prior to oral cholecystography to detect opaque calculi before gall bladder opacification. However, as the number of calculi missed by omitting the control film is so small, most centres no longer require this as a matter of routine.

Other gall bladder pathology can be shown on a plain X-ray. 'Limey' bile occurs in association with chronic partial obstruction of the cystic duct usually by a calculus. A porcelain gall bladder, an uncommon sequel to chronic inflammation, can be identified on a plain radiograph and is of clinical importance because of the association with gall bladder malignancy. Approximately, 20 per cent of porcelain gall bladders removed at surgery will have evidence of malignancy on histological inspection.

Liver parenchymal calcification can be produced by a number of pathologies. Past infection such as tuberculosis and hydatid disease or calcified metastases from a primary in the gut or ovary may give single or multiple calcified foci on an abdominal radiograph. The investigation of a patient with chronic pancreatitis — be it due to alcohol abuse or cystic fibrosis — may include an abdominal radiograph to detect pancreatic calcification. CT is much more sensitive at detecting low levels of calcification which cannot be appreciated on conventional X-rays.

The abdominal radiograph is of somewhat greater value to the clinician when a patient presents with an acute abdomen. Two examples of biliary disease can be diagnosed from this investigation. In the presence of small bowel obstruction, a review of the right upper quadrant may reveal gas in the biliary tree. This combination of X-ray findings should indicate gallstone ileus. The actual calculus causing the obstruction is infrequently visible as most are cholesterol-based and therefore non-opaque. The typical site of obstruction is in the distal ileum. It should be remembered that one of the commonest reasons for biliary tree gas on an abdominal radiograph is following sphincterotomy. Any clinical doubt concerning the diagnosis of gallstone ileus may be resolved by giving oral contrast which often reveals the fistulous connection between the bowel and the biliary tree. The second acute situation where the abdominal radiograph may be of value is in the diagnosis of emphysematous cholecystitis. Gas may be identified in the gall bladder lumen, gall bladder wall and, occasionally, within the bile ducts. Emphysematous cholecystitis can also be diagnosed by ultrasound, (see p. 147).

Oral cholecystography (OCG)

Contrast examination of the gall bladder was introduced by Graham and Cole in 1924 and, until the advent of ultrasound, oral cholecystography remained the mainstay of gall bladder investigation. As with all iodinated contrast agents which require oral ingestion, side-effects include nausea, vomiting and diarrhoea. The need for oral ingestion, absorption from the GI tract, and hepatic conjugation before gall bladder opacification will occur, are limiting factors in patients who are acutely unwell and in those with hyperbilirubinaemia. When the bilirubin level exceeds 30 mmol/litre, the gall bladder will not opacify.

Currently, a patient attending for oral cholecystography is required to take two doses of contrast media, the first on the preceding night, and the second some 2 hours before the examination. As modern contrast agents are concentrated both in the liver and in the gall bladder, some opacification of the biliary tree will occur. Opacification of the common duct is generally aided by gall bladder contraction following a fatty meal. Where duct contrast is shown in the absence of gall bladder opacification, this may reflect cystic duct obstruction. However, the gall bladder will fail to opacify in a substantial minority of patients with chronic cholecystitis and a patent cystic duct. It is also well recognized that the gall bladder may fail to opacify in patients with acute pancreatitis, peritonitis, after surgery, and after major trauma.

Intravenous cholangiography (IVC)

For many years, IVC remained the only non-invasive radiological investigation available to assess the common bile duct both before and after cholecystectomy. Following the introduction of ultrasound, scintigraphy and ERCP, the IVC has become a rarity in most X-ray departments. Contra-indications to the investigation include a bilirubin of greater than 45 mmol/litre, combined severe liver and renal disease, an oral cholecystogram up to 48 hours previously, paraproteinaemia and thyrotoxicosis. The older in-travenous biliary contrast agents were associated with significant morbidity and a mortality rate of 1 in 5000. Modern agents are better but remain among the more toxic of current radiological contrast agents. Older texts state that the indications for an IVC include postcholecystectomy symptoms, a failed oral cholecystogram, and the diagnosis of acute cholecystitis. The availability of newer techniques largely negates these indications. Apart from the associated morbidity, the IVC has been shown to be unacceptably inaccurate both in detecting duct calculi and in assessing cystic duct patency.

Infusion tomography of the gall bladder

Gall bladder tomography following infusion of intravenous contrast has been advocated as a means of assessing gall bladder disease. Although initial reports concluded that the examination was of value in acute cholecystitis when gall bladder wall thickening could be detected, disputes concerning the normal range of gall bladder wall thickness and much observer variation in assessing the findings, have resulted in little enthusiasm for the technique.

Direct Cholangiography

T-tube cholangiography

During surgical exploration of the common duct, the insertion of a T-tube provides a convenient route for subsequent opacification of the biliary tree to ensure that no duct stones remain. Traditionally, this is performed some 7–10 days after surgery.

More recently, the technique has been extended to allow the percu-taneous removal of retained duct calculi. Instead of removing the T-tube, it remains *in situ* for approximately 5 weeks to allow the establishment of a fibrous tract. After T-tube removal, a guide wire and steerable catheter system can be passed down the T-tube tract and directed under screening control into the duct containing the calculus. Stones can then be extracted using a Dormia basket. The success rate of the procedure depends on the expertise of the operator. A collected series from 20 British centres showed an overall success rate of 70 per cent. The mortality rate reported in the literature is less than 0·1 per cent which contrasts with the reported mortality rate of endoscopic sphincterotomy of approximately 1 per cent.

Percutaneous transhepatic cholangiography (PTC)

Before the mid-70s, percutaneous opacification of the biliary tree involved the use of relatively large calibre, rigid needles. This resulted in strict limitations on the number of punctures attempted and when an obstructed duct system was shown, the patient subsequently required surgery to prevent the development of biliary peritonitis. The development of a flexible, narrow calibre (23 gauge) needle allowed multiple punctures of the liver to be performed with minimal risk to the patient. Then, PTC became widely used both in the diagnosis of obstructive jaundice and in the assessment of which patients with intrahepatic cholestasis were suitable for surgery. The success of PTC is directly related to the presence of duct dilatation. Successful puncture of a dilated system occurs in 99 per cent of cases. With normal calibre ducts, the success rate falls to approximately 80 per cent.

The more widespread availability of endoscopic retrograde cholangiopancreatography (ERCP) has resulted in some reduction in the demand for PTC. However, PTC does not require the use of expensive equipment or particular operator expertise as is the case with ERCP. PTC is considered of particular value to assess the intrahepatic ducts in high obstruction of the biliary tree when surgery is being planned.

There are few absolute contraindications to PTC apart from bleeding and coagulation disorders. In the presence of active cholangitis, care must be taken to avoid precipitating septic shock. Generally, antibiotic cover and careful use of small quantities of contrast will suffice.

Endoscopic retrograde cholangiopancreatography (ERCP)

ERCP allows direct opacification of the biliary tree and pancreatic ductal system. As the procedure requires a number of trained personnel and relatively expensive equipment, this investigation is generally limited to the larger centres. Indications with respect to the biliary system include the investigation of persistent or recurrent jaundice, and in the assessment of biliary cirrhosis due to conditions such as sclerosing cholangitis. ERCP is playing an ever-increasing role in the management of choledocholithiasis. Although ultrasound is excellent for demonstrating gallstones, its performance in diagnosing duct stones is less impressive. ERCP is used when ultrasound is inconclusive, where ultrasound shows normal calibre ducts in the presence of biochemical and clinical evidence of obstruction, and finally when duct stones have been demonstrated and sphincterotomy is indicated. ERCP has therefore an important role to play in the management of postcholecystectomy patients who have recurrent biliary symptoms.

Retrograde opacification of the pancreatic duct will give a better demonstration of the ductal anatomy than any other available technique. Main duct and side branch morphology will be demonstrated and connections between the ductal system and any pancreatic or peripancreatic fluid collections will be outlined. Pancreatography is also a good technique for displaying anatomical variants such as the unfused ventral pancreas. As contrast injection into the pancreatic duct may cause or exacerbate acute pancreatitis, retrograde cannulation is contraindicated during an attack of acute pancreatic inflammation.

Contraindications to ERCP are a recent episode of pancreatitis or the presence of Australian antigenaemia.

Ultrasound

Ultrasound has revolutionized the approach to biliary tract disease. The gall bladder and the biliary tree can be examined without the necessity of contrast media or ionizing radiation. Real-time ultrasound is quickly performed, is inexpensive, and harmless to the patient. The lack of ionizing radiation allows repeat examinations as required and portable ultrasound machines permit bedside examinations in acutely unwell patients. A major advantage of ultrasound over other imaging techniques is its survey potential. This means the ability of ultrasound to rapidly assess adjacent anatomy and demonstrate pathology in other organs in the upper abdomen as well as the biliary tract. While CT shares this survey potential, it is both less widely available and significantly more expensive.

There are no specific or relative contraindications to ultrasound, therefore it should be the initial radiological approach to any gall bladder or biliary tree problem. Further investigations are largely determined by the ultrasound findings. Multiple sections in a variety of spatial planes can be obtained by altering the direction of the ultrasound beam without moving the patient. This has obvious advantages in the acutely unwell or traumatized patient. As the ultrasound beam will not penetrate gas, this may be a problem in an acutely unwell patient with an ileus. The examination of a dilated common bile duct may be compromised, when the lower end and pancreatic head are obscured by intestinal gas. In this situation, further studies using CT or direct cholangiography will be required.

The liver

Ultrasound scanning of the liver, as well as demonstrating the biliary tree, will give information concerning the size and patency of the portal and hepatic venous system. Most mass lesions are distinguishable by echo patterns which differ from those of normal surrounding liver. Both focal and diffuse hepatic abnormalities may result in an alteration to the liver contour which can be appreciated during the ultrasound examination. Examples include the irregular contour of the cirrhotic liver and enlargement with anterior bulging when the liver is diffusely involved with metastatic disease. Pathologies such as fatty infiltration and fibrotic liver disease should be recognized by altered echo patterns. These are generally appreciated by comparing the echo patterns of the liver and the adjacent right renal cortex and by observing changes in the prominence of portal vein wall echoes.

The pancreas

Ultrasound examination of the pancreas is technically more difficult than liver or biliary scanning. Intestinal gas is the major determinant of success or failure and despite attempts at bowel preparation and fasting prior to scanning, gas may completely obscure the pancreas. A successful study may be achieved by scanning with the patient in the erect position or after a water load when the fluid filled stomach is used as an acoustic window. Ultrasound will assess the overall size and shape of the pancreas, identify focal areas of altered echo pattern including calcification, demonstrate dilatation of the ductal system, and assess the effects of acute inflammation on the pancreas itself and the peripancreatic tissues. Ultrasound cannot differentiate between a mass due to focal pancreatitis and one due to a pancreatic malignancy. Both may appear as focal areas of reduced reflectivity and percutaneous fine-needle biopsy is often indicated to achieve a diagnosis.

Ultrasound guided percutaneous puncture of a dilated pancreatic duct can be helpful when endoscopic retrograde cannulation fails. Contrast can be injected percutaneously and the ductal system and any extra pancreatic connections demonstrated.

Computed Tomography (CT)

In many parts of the UK, access to a CT scanner remains very limited. When available, CT provides an excellent means of liver and biliary tract investigation and, as with ultrasound, adjacent organs can be assessed at the same examination. Because of its expense and limited availability, CT is used infrequently as the primary approach to a biliary tract problem but is generally used when ultrasound or other radiological investigations prove inconclusive. This applies particularly when gas has impaired an ultrasound examination. Anatomical detail of the liver with CT is approximately equivalent to that obtained by ultrasound. Technical failures with CT are very rare so demonstration of the pancreas is more reliable than when using ultrasound and the level of anatomical detail obtained is again superior. Demonstration of the gall bladder with CT is less effective than with ultrasound but the lower half of the common bile duct is shown more reliably by CT. A technically satisfactory CT examination requires patient co-operation. This may prove problematic in the acutely ill or traumatized patient and in children. Where appropriate, intravenous sedation or general anaesthetic may be required. CT is particularly sensitive in detecting concentrations of calcium far less than that required for demonstration by conventional radiography. Therefore, gallstones, duct calculi, and pancreatic calcification which are not identified on a plain radiograph, can be demonstrated easily by CT. As with ultrasound, CT cannot differentiate between focal pancreatitis and focal malignancy. CT is the better technique for demonstrating the spread of both malignancy and inflammation to the peripancreatic structures.

Scintigraphy

Labelled colloid is selectively trapped by reticuloendothelial cells to give an image of the liver and spleen. Characteristic abnormalities are seen in patients with diffuse liver disease and with mass lesions in the liver or spleen. Unlike ultrasound and CT, scintigraphic images include information from all parts of the organ being studied so it is much easier to obtain an overall impression of its size, shape and position. The relative contributions of portal venous and hepatic arterial blood supply to the liver can be estimated by a simple 'first-pass' type of technique; this ratio is valuable in the diagnosis of some types of liver disease.

Hepatobiliary scintigraphy

Hepatobiliary scintigraphy involves the use of radiopharmaceuticals which are extracted from the blood by hepatocytes. These agents are subsequently secreted into either plasma or the bile cannaliculi resulting in both biliary and urinary excretion. One of the earliest agents used to image the biliary system was iodine-131 Rose Bengal. Although successful in demonstrating the biliary tree, the use of ^{131}I was less than ideal because of its long half-life and poor physical characteristics for imaging. In 1975, a new generation of hepatobiliary agents were developed using ^{99}Tc^m-labelled iminodiacetic acid (IDA). Because of its short half-life (6 hours) and pure gamma emission, ^{99}Tc^m is an ideal isotope for imaging.

Since 1975, a number of N-substituted IDA derivatives have been developed each with its own particular excretion dynamics. One of the most commonly used agents is dimethyl IDA, or 'HIDA'. Hepatocyte uptake of a radiopharmaceutical is dependent on active transport mechanisms using specific carriers. These mechanisms can be blocked by the competitive binding of inhibitors, the most important of these being bilirubin. HIDA has the least hepatobiliary specificity and therefore has the greatest renal excretion. This renal excretion increases with increasing hyperbilirubinaemia, making HIDA an unsuitable agent in the presence of jaundice. Other IDA derivatives, for example, di-isopropyl IDA or DISIDA are more appropriate when hyperbilirubinaemia is present. This agent can still give effective biliary tract images when the bilirubin level reaches 500 mmol/litre.

Biliary scintigraphy requires the patient to fast for 4 hours. The patient is then placed under a gamma camera and following an intravenous injection of the chosen radiopharmaceutical, images are obtained at set time intervals appropriate to the requirements of the examination. The data is stored on computer. On a normal scan, activity will be shown in the liver, major intrahepatic ducts, common bile duct, gall bladder and small intestine within the first hour.

Apart from high levels of bilirubin, there are no specific contraindications to biliary scintigraphy. An isotope study demonstrates biliary function as well as anatomy. It should be remembered that the resolution of scintigraphy is inferior to other forms of imaging — for example, gallstones cannot be imaged. Indications for scintigraphy include the diagnosis of acute cholecystitis by assessing the patency of the cystic duct, the detection of developmental anomalies of the biliary tree, assessing biliary dynamics, and in the investigation of complications including bile leaks following biliary tract surgery.

Interventions

Percutaneous transhepatic techniques can be taken a stage further to allow the percutaneous drainage of an obstructed system. This may take the form of a temporary external drainage arrangement which will allow the patient time to improve physically and biochemically before planned surgery or other therapy. Alternatively, a more permanent palliative procedure can be performed. Following opacification of the biliary system percutaneously, an endoprosthesis can be advanced into the biliary tree, through the obstruction, and into the duodenum to allow internal biliary drainage.

Liver Disease

Diffuse Liver Disease

Imaging is supplementary to the major clinical and biochemical criteria for detecting diffuse liver disease. Some types of pathology produce distinctive appearances so that imaging procedures contribute to the specific diagnosis. Imaging is also useful in recognizing complications of diffuse liver disease. Scintigraphy, ultrasound and CT all have a contribution to make.

The size and particularly the shape of the normal liver are so variable that it is difficult to define a normal range. Conventional scintigraphy using labelled colloid probably gives the best overall impression of the size and shape of the liver and spleen as well as indicating the distribution of function, not only within the liver but also between the liver and the extrahepatic sites of reticuloendothelial cell activity (Fig. 5.1). Ultrasound and CT show the relationship of the liver to surrounding anatomical structures and by adding together the measured cross-sectional areas of adjacent tomographic slices, a precise measurement of liver volume can be

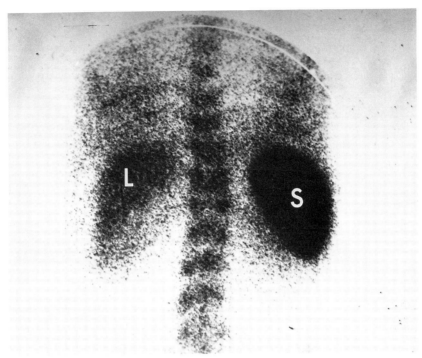

Fig. 5.1 Colloid scintigraphy in cirrhosis: the spleen (S) is enlarged and shows increased activity compared with the liver (L). There is also increased colloid shift into the bone marrow and the lungs.

made. The clinical value of detailed liver volume measurement has yet to be established. Clinical doubt about the size and shape of the liver can be resolved by imaging. For example, in patients with emphysema, the low position of the diaphragm may result in the edge of a normal liver being felt well below the costal margin, a Riedl's lobe may present as a dubious mass in the right flank or in the iliac fossa, and imaging should show whether a clinically suspected mass in the epigastrium is in fact the left lobe of the liver or a separate structure. Characteristic alterations in the shape of the liver are seen in patients with cirrhosis (relative enlargement of the left lobe) and in hepatic vein thrombosis (hypertrophy of the caudate lobe) whereas in uncomplicated hepatitis or fatty infiltration the relative size of the different lobes is usually normal.

Fatty liver produces a characteristic pattern of increased echogenicity on ultrasound. The patchy fibrosis seen in many patients with cirrhosis gives rise to similar ultrasound appearances so, in patients with fatty liver, distinction must rely on the preservation of normal lobar and vascular architecture. More specific recognition of fatty infiltration of the liver is possible with CT where the altered X-ray attenuation caused by fat deposition is unmistakable (Fig. 5.2). Changes in attenuation of the liver can

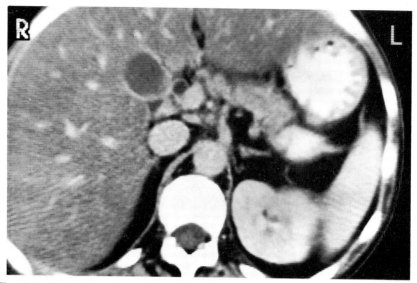

Fig. 5.2 CT of a patient with diffuse fatty infiltration of the liver. The liver substance shows grossly reduced attenuation (appears darker) compared with the spleen, aorta and vena cava.

also result from excessive iron deposition in haemochromatosis but recognition of specific attenuation changes in Wilson's disease and glycogen storage disorders has not proved reliable. The role of ultrasound or CT in patients with acute hepatitis is firstly to confirm the absence of biliary obstruction in jaundiced patients, and secondly to assess the size and, particularly, the progressive changes in the size of the liver — a small liver with severe functional impairment being associated with a poor prognosis. Patchy loss of contrast enhancement on CT is suggestive of infarction, a complication usually associated with trauma, surgery or occlusion of the abdominal aorta or visceral arteries.

Bone-seeking radionuclides accumulate in amyloid tissue, so, if this condition is suspected, scintigraphy using one of the bone agents will be useful. Colloid scintigraphy will show large volumes of ascites by displacement of the liver image from the adjacent rib uptake, but smaller volumes of ascites can be shown either by ultrasound or CT.

Portal Hypertension

In most patients, portal hypertension is secondary to liver disease, so, in these cases, imaging of the liver will be helpful in providing confirmatory evidence of a diffuse liver lesion. Where the cause of portal hypertension is uncertain, and in patients with variceal bleeding, imaging of the portal venous anatomy is crucial. Ultrasound of the liver should include an appraisal of the size and patency of the portal vein and its main branches.

Similar information can be obtained by CT with contrast enhancement. CT is also particularly effective in demonstrating the varices, not just around the splenic hilum and pancreatic bed, but in many cases, it will show enlargement of the gastric coronary vein, umbilical vein, and retroperitoneal collaterals. Both ultrasound and CT also normally show patency of hepatic veins draining into the vena cava.

However, if hepatic veins are not visualized, it is often impossible to be sure whether this is the result of hepatic venous thrombosis or of oedema of the liver compressing the veins to a size much smaller than normal. Where hepatic vein obstruction is clinically suspected, colloid scintigraphy shows characteristic predominance of the caudate lobe in about half of the cases. In other patients, the scintigraphic features are indistinguishable from those of cirrhosis and hepatic vein catheterization is necessary to confirm or refute the diagnosis.

Preoperative demonstration of portal venous anatomy is best achieved by opacifying the vascular tree. The traditional technique of splenoportography, with its associated risk of bleeding at the splenic puncture site, has now largely been discarded in favour of arterial portography which is safer and gives additional information about the arterial tree. Selective catheterization of the splenic, hepatic and superior mesenteric arteries gives a full demonstration of the arterial supply to the liver as well as showing the portal venous anatomy. The position and patency of the mesenteric, splenic and portal veins are shown to help the surgeon decide on which type of shunt procedure to perform. Demonstration of the patency and position of the inferior vena cava and left renal vein form part of this procedure. Where variceal bleeding cannot be controlled by endoscopic injection of sclerosants, percutaneous transhepatic injection is an alternative. This technique requires transhepatic catheterization of the portal vein with subsequent selective injection of embolic material into the main variceal channels. It is usually possible to occlude the gastric coronary vein and large varices arising from the splenic vein, but the short gastric veins are difficult to get at by this method. Using gelfoam emboli or absolute alcohol, bleeding from the varices can usually be stopped, but the natural history of this condition is such that recanalization or the development of new varices is relatively common. At the time of writing, transhepatic sclerosis of varices is used mainly as a holding procedure in patients in whom endoscopic sclerosis fails so that the patients can undergo a period of resuscitation to prepare for a more definitive surgical decompression.

The arteriographic pattern within the liver is often characteristic in patients with cirrhosis and a normal hepatic arteriogram in a patient with portal hypertension should initiate a search for an extrahepatic cause. Pancreatic tumours or chronic pancreatitis can be recognized by characteristic arteriographic changes in the pancreatic circulation, but in some cases the cause of splenic, mesenteric or portal venous thrombosis will not be found. If there is doubt about the normality of the liver in patients with portal hypertension, wedged hepatic venous pressure can be measured at the time of angiography.

Metastases in the Liver

The search for liver metastases is of key importance in patients with primary malignancies of the gastrointestinal tract. For example, in patients with colorectal cancer, the presence of liver metastases appears to be linked more closely with the prognosis than does the local staging of the primary tumour. The presence of liver metastases also has an influence on the management of other primary tumours, particularly carcinomas of the bronchus and of the breast. With these and other tumours spreading to the liver by the systemic route, the presence of liver metastases indicates a poor prognosis and may militate against major surgery for resection of primaries.

In addition to patients with known primary malignancy, liver metastases must be included in the differential diagnosis of patients presenting with pain in the right upper quadrant, hepatomegaly, deranged liver function or jaundice, and unexplained pyrexia.

Ultrasound

In patients with symptoms or signs of right upper quadrant disease, ultrasound should be the first examination since not only does it give information about the size and consistency of the liver, but will also allow inspection of the gall bladder and bile ducts and of the adjacent upper abdominal structures. With extensive metastatic disease the liver is enlarged, but in many early cases, the organ is of normal size and shape. Metastases appear on ultrasound as areas of differing echogenicity, commonly less echogenic than normal liver tissue, occasionally more echogenic and, in some cases, containing areas of both increased and reduced echoes. Occasionally, metastases may be indistinguishable in their echo patterns from the surrounding liver, but fortunately this is seen only rarely.

Computed tomography (CT)

Metastases appear on CT as areas of reduced attenuation (very rarely, increased attenuation) within the liver parenchyma. Most metastases are less vascular than the surrounding liver tissue and so become more easily visible when CT scans are obtained with intravenous contrast enhancement. However, since — albeit rarely — liver lesions may become less conspicuous after contrast enhancement it may be argued that scans should be obtained both before and after IV contrast. Nevertheless, the practicalities of the availability of scanning time necessitate a single series in most patients and this should be obtained by rapid sequence scanning after IV contrast injection. Occasional metastases show marked contrast enhancement on CT and appear as 'blushing' lesions on arteriography. These include metastases from melanoma, islet cell tumours of the pancreas, and renal cell carcinomas. Calcification in liver metastases, which may be either subtle (seen only on CT or ultrasound) or gross enough to be visible on plain films, is usually associated with colloid carcinomas of the colon, ovary or breast, but successful chemotherapy may lead to calcification of liver metastases from tumours elsewhere, e.g. testicular teratoma and apudoma.

Colloid scintigraphy shows metastases as simple areas of absent function within the contour of the liver. This appearance is also produced by polycystic disease, whereas either ultrasound or CT will readily differentiate clear-cut fluid-containing cysts from tumours which are largely solid even though in some cases central necrosis leads to a degree of liquefaction of the lesion. Additionally, polycystic disease of the liver is almost invariably associated with similar abnormalities in both kidneys, and occasionally in the pancreas, findings which are easily confirmed by ultrasound or CT.

Early identification

The early identification of liver metastases in patients with known primary tumours presents a different problem. The difficulty is not to characterize the liver lesion, but to determine whether the liver is normal or not. The physical constraints of imaging techniques mean that scintigraphy is unlikely to show lesions at the surface of the liver any smaller than about 2 cm across, whereas the resolution at depth is nearer to 4 cm. Ultrasound also shows decreasing resolution with increasing depth, but should be able to show most lesions in the 1–2 cm size range and some smaller than this, provided that the echo pattern is sufficiently distinct from surrounding tissue. The resolution of CT is not affected by depth but is influenced by the relative attenuation of the lesions compared with normal liver. Smaller lesions can be shown by obtaining thinner CT sections but time constraints and the difficulty of reproducing the liver position for each scan due to variations in respiratory excursion are additional limiting factors. Another problem is that livers which otherwise appear normal often contain small imperfections, either cysts or benign tumours, which are difficult to distinguish from metastases in the same size range (under 1 cm). In general it may be argued that current imaging methods are unlikely to show many metastases smaller than 1 cm in diameter, but lesions larger than this should be detected by one or other of the methods.

Imaging methods

Differences in resolution between scintigraphy, ultrasound and CT do not result in significant differences in the performance of these methods when compared with each other. This is largely explained by the observation that patients with liver metastases almost always have lesions of various sizes so that even if the less sensitive methods show only the larger lesions, the distinction between normal and abnormal is not affected. A second major difficulty in comparing the sensitivity of different methods lies in the choice of 'gold standard'. Surgical inspection of the liver at laparotomy has usually been chosen as the final arbiter, but it is clear from postmortem studies on patients dying within a few weeks of operation and also from longitudinal studies of postoperative patients, that surgical palpation and inspection of the liver fails to detect metastases which are actually present in a substantial proportion of cases, possibly as many as 50 per cent. The only imaging technique which is, apparently, more sensitive than surgical examination of the liver is perfusion scintigraphy. This technique requires the acquisition of time/activity data over the liver using a gamma camera after a rapid bolus injection of labelled colloid. It can be performed simply as part of conventional liver scintigraphy. Data from the time/activity curves allows the calculation of the relative contributions of hepatic arterial and portal venous inflow to the liver. This ratio has a normal range which is exceeded in patients with overt or occult metastases. At the time of writing, this technique has not yet been adopted widely, but there is good evidence that, at least in patients with colorectal cancer, the method provides the earliest possible indication of the presence of liver metastases.

Since the introduction of ultrasound and CT, arteriography of the liver has a very limited role in diagnosis. However, therapeutic manoeuvres can be mediated through a percutaneously placed catheter in the hepatic artery. This method has been used most effectively in patients with unresectable metastatic carcinoid tumours. Considerable relief of pain and systemic symptoms can be achieved by therapeutic embolism of the supplying vessels. Since primary and secondary liver tumours receive their blood supply almost totally from the arterial side of the hepatic circulation, arterial embolization using gelfoam or absolute alcohol results in considerable tumour necrosis. Angiographically placed catheters can also be used for local infusion chemotherapy.

Primary Liver Tumours

Imaging has a role in detecting primary liver tumours, in characterizing different types of tumour, and in showing the extent and blood supply of tumours in patients undergoing surgery. As with metastases, primary liver tumours are shown as areas of non-function (scintigraphy), altered echogenicity (ultrasound) or altered attenuation (CT). Plain abdominal films may show liver enlargement or a localized bulge of the liver contour; calcification is sometimes a feature of hepatocellular carcinoma and phlebo-liths may be seen in hepatic haemangiomas. Ultrasound evidence of dilated bile ducts points towards cholangiocarcinoma whereas hepatocellular carcin-omas are usually large heterogeneous masses with areas of cystic and solid elements. The CT appearance of hepatocellular carcinoma varies from well-defined, almost encapsulated, masses with large fluid components (Fig. 5.3) to ill-defined areas of subtle attenuation changes which may be no clearer after intravenous contrast enhancement. The latter type of appearance is associated with well-differentiated hepatoma.

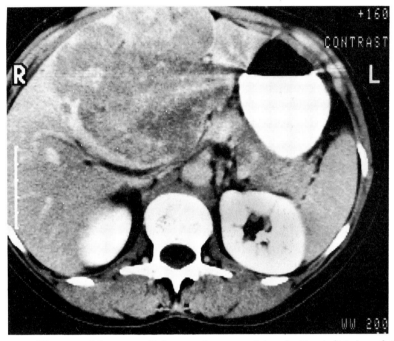

Fig. 5.3 CT scan of hepatocellular carcinoma arising in the left lobe of the liver and compressing main portal vein. The right lobe is free from disease.

Diagnosis

In most cases, the diagnosis of primary liver tumours can be achieved by demonstration of the mass by ultrasound or CT, followed by percutaneous needle biopsy, under imaging guidance if necessary. In patients in whom percutaneous biopsy is undesirable (e.g. those with a prolonged prothrombin time or a large amount of ascites), gallium scintigraphy is often useful. Gallium is concentrated in the vast majority of hepatocellular carcinomas, but to a much lesser extent in most metastases and not at all in fibrotic areas of cirrhotic liver. Comparing the gallium image of the liver with that obtained by sulphur-colloid, matched defects are likely to be the result of cirrhotic fibrosis, whereas a mismatched area of high gallium uptake, but with reduced or absent colloid concentration, is likely to be the site of a hepatoma.

Cholangiocarcinoma is an infiltrating tumour producing only subtle attenuation changes on CT although contrast enhancement usually allows the edge of the tumour to be defined (Fig. 5.5). With the exception of haemangioma, benign tumours of the liver are uncommon and present a variety of appearances on imaging. Histology is essential for diagnosis. Haemangioma is probably the commonest of all liver tumours. Small lesions are frequently found in patients undergoing CT or ultrasound scanning of the liver. A typical haemangioma produces a well-defined area of marked hyperechogenicity on ultrasound (Fig. 5.4), while on CT, the appearance of delayed central enhancement with IV contrast and the demonstration of large vascular channels around the periphery of the lesion in the period immediately after injection of contrast is characteristic. If there is diagnostic doubt about the nature of the lesion found on ultrasound or CT, arteriography is sometimes helpful. Haemangiomas show a characteristic pattern of dilated vessels around the periphery of the mass. Many of the lesions found incidentally are, however, less than 1 cm in diameter and do not show the characteristic features of the larger haemangiomas. In these cases, the incidental finding of a small liver lesion might be the only evidence of residual disease in a patient successfully treated for malignancy elsewhere. Management will then depend on the level of clinical suspicion, alternative possibilities being percutaneous or open biopsy of the lesion, or careful follow-up with repeated imaging to assess any change in its size.

Preoperative assessment

Patients who are candidates for surgical removal of liver tumours will require preoperative assessment of the extent of the tumour and its relation to the major vessels. CT or ultrasound should show intrahepatic metastases (a not uncommon feature of hepatoma) and extension of tumour into the portahepatis. In particular, if there is invasion of the portal vein or of the inferior vena cava, or if the tumour has spread across both sides of the falciform ligament this will often be enough to rule out successful resection. Arteriography is helpful in showing not only the vascularity of the tumour but also the source of its major feeding vessels (Fig. 5.6), and may sometimes show small vascular metastases which have been missed by other scanning techniques. The relationship of the tumour to the portal vein is most important, and the surgeon will also be aided by knowledge of the anatomy of the arterial supply to the liver, which is very variable.

The plain chest radiograph must not be overlooked in patients with liver tumours. The presence of lung metastases or invasion of the tumour through the diaphragm will simplify therapeutic decisions considerably.

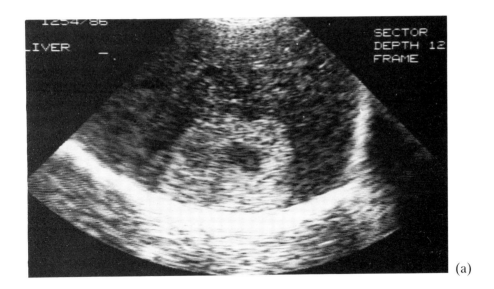

(a)

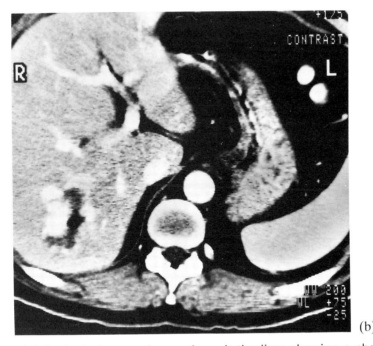

(b)

Fig. 5.4 (a) Sagittal ultrasound scan through the liver showing a sharply demarcated echogenic mass adjacent to the diaphragmatic surface. Diagnosis: benign haemangioma. (b) CT scan through the liver immediately after contrast injection showing intense enhancement around the rim of a mass lesion with relative ischaemia of the centre. Diagnosis: benign haemangioma.

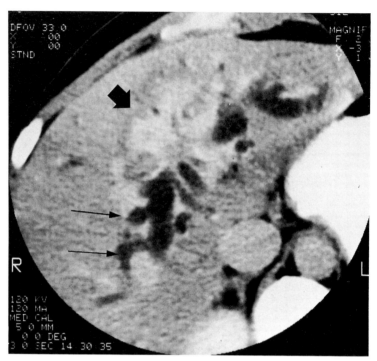

Fig. 5.5 Contrast-enhanced CT scan showing grossly dilated bile ducts (thin arrows) obstructed by an ill-defined tumour at the porta hepatis (large arrow). Diagnosis: cholangiocarcinoma.

Non-Tumourous Masses in and around the Liver

Polycystic disease, as mentioned above, produces characteristic appearances on both ultrasound and CT. Infection or haemorrhage within a cyst is often recognizable by the presence of echogenic material on ultrasound or increased attenuation on CT. If there is clinical doubt, guided aspiration can be carried out. Solitary cysts of the liver, unless large enough to cause mechanical effects, are usually found incidentally on ultrasound or CT scanning. Aspiration is more likely to show old blood and a few inflammatory cells than the presence of bile, so the origin of these cysts remains uncertain. Geographic areas of reduced function, altered echogenicity or abnormal attenuation can be recognized by scintigraphy, ultrasound and CT, respectively. The cause of these appearances is often obscure. Patchy infarction should be considered in patients with occlusion of or recent surgery involving the upper abdominal aorta. Patchy fatty infiltration is an occasional feature of diabetic patients. However, most of these unusual geographic lesions will require histology for adequate diagnosis.

   Blind percutaneous needle biopsy of the liver occasionally produces small subcapsular haematomas recognizable on CT and ultrasound and, rarely, a large haematoma may result. With substantial bleeding into an enclosed

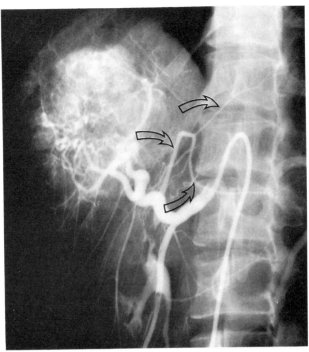

Fig. 5.6 Selective hepatic arteriogram showing a very vascular tumour in the right lobe of the liver. Note the relatively normal branches to the left lobe (curved arrows) which was uninvolved. Diagnosis: hepatocellular carcinoma.

space, a haematocrit-like effect can be seen on either ultrasound or CT, with the formed elements of high echogenicity and high attenuation lying in a horizontal band below a supernatant layer of liquid consistency. A solid haematoma is more difficult to distinguish from adjacent liver tissue, but its total lack of contrast enhancement on CT should allow its recognition — for example, in patients with trauma to the upper abdomen. Amoebic and mature pyogenic abscesses in the liver produce similar findings of a well-defined region of reduced attenuation on CT, reduced echogenicity on ultrasound, and non-function on colloid scintigraphy. The contents of the lesion are of the attenuation of turbid fluid on CT and produce a few low level echoes on ultrasound. A rim of contrast enhancement may be shown on CT. Percutaneous aspiration under imaging guidance is useful to establish the causative pathogen and may also have a therapeutic effect by evacuating the contents of a unilocular collection. With multilocular abscesses, more prolonged drainage is usually necessary, but again, this can often be achieved by percutaneous placement of drainage catheters. In the early stages of their formation, pyogenic abscesses may not be well-defined or show a fluid centre. If there is doubt about the infectivity of lesions producing only subtle changes on CT or ultrasound, scintigraphy with gallium or labelled white cells will usually be decisive. Concentration of these agents over and above the normal liver uptake strongly suggests an active focus of infection or tumour. Abscesses in the right subphrenic and subhepatic spaces are usually well shown by ultrasound, but CT can be used if there are technical difficulties with the former method. Either method can be used to place drainage catheters in or around the liver.

Hydatid disease of the liver usually appears as unilocular or multilocular cystic masses within the parenchyma. The cyst contents may be of turbid fluid density and contain some echogenic material, so the differentiation from abscess is occasionally difficult. Hydatids usually have a more clear-cut edge with a thinner wall and evidence of septation or daughter cysts within the main lesion. Current practice suggests that hydatids should not be punctured percutaneously. Masses arising in the gall bladder and bile ducts are discussed later (see pp. 149, 154).

Liver Trauma

Plain radiographs provide a useful initial assessment of the patient with liver trauma. Right-sided rib fractures, a local ileus in the right upper quadrant, effacement of the inferior liver outline, and elevation of the right hemi-diaphragm, perhaps associated with a pleural effusion, may be shown. Haemorrhage in or around the liver may result in displacement of adjacent bowel loops and ptosis of the right kidney.

Trauma to the hepatic parenchyma is best assessed by ultrasound or CT. Both techniques will help differentiate between patients who need surgical intervention and those for whom conservative management would be more appropriate. A subcapsular haematoma occurring as an isolated finding is relatively rare in liver trauma as most hepatic injuries result in parenchymal laceration. When a parenchymal tear is associated with intraperitoneal bleeding, surgery is usually required. The extent of free intraperitoneal haemorrhage appears to be the most significant finding when evaluating the need for surgical intervention. CT or ultrasound should demonstrate free fluid in the abdomen or pelvis.

The Biliary System

The Gall Bladder

Gallstones

Oral cholecystography is the traditional approach to gallstone diagnosis and over the years its accuracy has been established. During the late 70s, real-time ultrasound became widely available and this has gradually replaced the oral cholecystogram as the investigation of choice in gallstone diagnosis. High resolution real-time ultrasound has an accuracy of greater than 95 per cent. Both techniques have a false-negative rate of 2–9 per cent and studies with operative confirmation have shown that most calculi missed are less than 2 mm. These are frequently impacted near the gall bladder neck. This is a difficult site for ultrasound diagnosis as the region of the valve of Heister characteristically appears strongly echogenic and may mimic a gallstone.

Recent studies have suggested that the accuracy of oral cholecystography is nearer 90 per cent as 2–3 mm stones can be missed. False-positive diagnoses are unusual with both ultrasound and oral cholecystography therefore the techniques are very highly specific. It has been shown that some stones missed on oral cholecystography can be seen on ultrasound, therefore ultrasound appears slightly more sensitive for stone diagnosis.

The ultrasound diagnosis for gallstones (Fig. 5.7) depends on all of the following criteria.

1. One or more echogenic foci demonstrated within a bile filled gall bladder lumen
2. Clean acoustic shadowing arising from the echogenic focus
3. The echogenic focus changes position with change in patient position.

When these strict diagnostic criteria are used, the accuracy of ultrasound approaches 100 per cent. The patient should always be fasted to achieve a distended, bile-filled gall bladder. The detection of mobile, dependent,

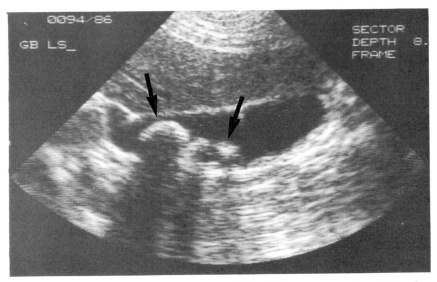

Fig. 5.7 Sagittal ultrasound scan through the right upper quadrant showing stones (arrows) within the gall bladder. Note the acoustic shadows cast by the stones.

echogenic foci associated with acoustic shadowing results in a confident diagnosis of gallstones.

Failure to identify the gall bladder in fasting patients has been shown to be associated with primary gall bladder pathology in over 90 per cent of cases. In patients where the gall bladder is fibrotic and scarred, the gall bladder itself may not be identified, but strong acoustic shadowing may be demonstrated arising from the gall bladder fossa. If this shadowing is shown to be constant in appearance in a variety of patient positions over a period of time, then it can be differentiated with confidence from other causes of shadowing around the gall bladder fossa, e.g. duodenal gas. In this situation, the shadowing most often represents a contracted gall bladder containing calculi.

Occasionally, echogenic foci are demonstrated within the gall bladder lumen which are freely mobile but are not associated with acoustic shadowing. The previously described criteria for gallstone diagnosis are therefore not fulfilled. Acoustic shadowing is due to a combination of reflection of the incident beam and absorption of sound by the stone, and has been shown to be independent of calculus composition, shape and surface characteristics. What is important for the production of acoustic shadowing is transducer frequency, depth of focus of the ultrasound beam, depth and size of the calculus and centering of the calculus to the beam. The greater the proportion of beam stopped by the stone, the greater the chance of producing a shadow. Studies *in vitro* have shown that all gallstones will cast an acoustic shadow. When non-shadowing, intraluminal foci are demonstrated during an ultrasound examination, a number of these will be gallstones. By altering the transducer frequency and focal length, acoustic shadowing may be produced and gallstones can be confirmed. Most stones which do not cast shadows on routine examination are less than 5 mm.

Although the majority of non-shadowing, discrete, intraluminal echoes whether mobile or not will be gallstones, other diagnoses require consideration. If a non-shadowing, echogenic focus bears a constant relationship to the gall bladder wall, this may represent a polyp (see p. 149). Gall bladder 'sludge' which is due to a mixture of calcium bilirubinate and cholesterol, produces homogenous low or mid-level echoes which form a dependent layer and slowly change position with gravity. Sludge can be demonstrated in the normal gall bladder following a fast but it is commonly seen in patients on parenteral feeding and hyperalimentation.

The oral cholecystogram has been shown to be superior to ultrasound in the follow-up of patients undergoing gallstone dissolution therapy. Ultrasound measurements of strongly reflective foci are generally inaccurate because of the physical characteristics of the ultrasound beam. As response to treatment tends to be assessed by detecting a reduction in gallstone size, oral cholecystography is the preferred approach.

CT is capable of detecting faintly calcified calculi which will not be seen on conventional radiographs. Gallstones are therefore commonly found during the course of an abdominal examination for some other cause. However, CT is not indicated for the diagnosis of gallstones *per se*. Scintigraphy is of no value in the demonstration of gallstones because of its inherent poor anatomical resolution.

Future developments for gallstone diagnosis may include intraoperative ultrasound. Early experience has shown that scanning the gall bladder during surgery is more accurate in the detection of gall bladder calculi than preoperative scanning.

Acute cholecystitis

It has been estimated that 15–25 per cent of patients with gallstones will develop acute cholecystitis. Conversely, almost all patients with acute cholecystitis will have gall bladder calculi. The consequence of cystic duct obstruction, most commonly by calculi, is a combination of vascular compromise, chemical injury due to lithogenic bile, and secondary bacterial infection.

Although still contentious among surgeons in the UK, the trend world-wide has been away from conservative management of acute cholecystitis to early surgery. Clinical assessment has been shown to be particularly unreliable when deciding which patient presenting with right upper quadrant pain has acute cholecystitis and which has some other cause. The option of early surgery is dependent on an accurate preoperative diagnosis and therefore relies heavily on diagnostic imaging for its success or failure. Ultrasound and scintigraphy are of most value to the clinician. Oral cholecystography cannot be advised because of problems with oral ingestion of contrast in the acutely ill patient, and the gall bladder's failure to opacify in a substantial minority of patients with chronic cholecystitis and a patent cystic duct. Intravenous cholangiography has been used over the years to diagnose acute cholecystitis, but this examination is notoriously unreliable in diagnosing cystic duct obstruction while being associated with significant morbidity.

Ultrasound of acute cholecystitis
The ultrasound features of acute cholecystitis (Fig. 5.8) can be divided into primary and secondary signs and these are shown in Table 5.1.

It is important to emphasize that the detection of gallstones *per se* in any

Table 5.1   Ultrasound features in acute cholecystitis

Primary signs	Gallstones
Secondary signs	Focal gall bladder thickening
	Gall bladder wall thickening
	Sonolucent layer in gall bladder wall
	Gall bladder distention

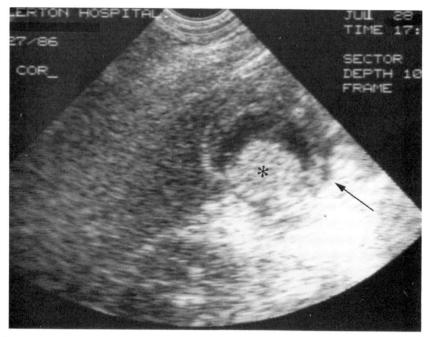

Fig. 5.8 Sagittal ultrasound scan through the right upper quadrant in a patient with acute cholecystitis showing a thickened gall bladder wall (arrow) and debris within the gall bladder lumen (*).

investigation of possible acute cholecystitis is not diagnostic of that condition. Gallstones are commonly found in the general population and therefore may be present in a patient with right upper quadrant pain due to pyelonephritis, perforated duodenal ulcer, retrocaecal appendix, etc. Other criteria must be demonstrated to make the diagnosis. If a gallstone can be displayed impacted in the region of the gall bladder neck along with other ultrasound criteria for acute cholecystitis, then the sensitivity and specificity of the technique is high. It must be remembered that with the exception of focal gall bladder tenderness, there are many other causes for the secondary ultrasound signs of acute cholecystitis and that these ultrasound features are not pathognomic but only consistent with the diagnosis. However, recent studies have shown that the combination of gallstones and either focal gall bladder tenderness (positive sonographic Murphy's sign), or a thickening of the gall bladder wall gives a very high positive predictive value for acute cholecystitis in patients presenting with clinical suspicion of the condition.

Scintigraphy in acute cholecystitis
By definition, acute cholecystitis is due to obstruction of the cystic duct — usually by stone but occasionally by debris, lymph nodes, or tumour. The

only imaging technique which demonstrates cystic duct obstruction reliably is scintigraphy. The time taken to visualize the gall bladder is variable but all normal gall bladders will be demonstrated by 60 minutes. There are, basically, three abnormal patterns:

1. Absence of gall bladder activity
2. Delayed gall bladder visualization
3. Obstructive pattern. No activity is demonstrated in the gall bladder, common bile duct or small intestine despite satisfactory liver uptake.

When the cystic duct is patent, the gall bladder will be visualized by 60 minutes postinjection. This effectively excludes the diagnosis of acute cholecystitis. When the cystic duct is obstructed, no gall bladder activity will be demonstrated (Fig. 5.9). It is important to continue imaging up to 4 hours postinjection, because patients with partial cystic duct obstruction — for example in chronic cholecystitis — can have absent gall bladder activity at 60 minutes, but continued scanning may demonstrate delayed visualization on the later images. A small proportion (less than 5 per cent) of patients with acute cholecystitis also show this pattern of delayed visualization. Therefore, a diagnosis of acute cholecystitis cannot be excluded when this pattern occurs. HIDA scanning is most accurate (almost 100 per cent) when a normal result is obtained; if activity is shown in the gall bladder, the cystic duct is unequivocally patent and acute cholecystitis cannot be present. This is of obvious importance to the clinician who is contemplating early surgery. An 'obstructive pattern' is found in obstruction of the common bile duct from whatever cause. Other investigations are required to determine the aetiology.

Complications of acute cholecystitis
A variety of complications may occur in the patient with acute cholecystitis. These include empyema of the gall bladder, gangrenous cholecystitis, perforation and pericholecystic collections. The most useful radiological technique to assess these complications is ultrasound. As ultrasound is totally non-invasive and avoids the use of ionizing radiation, repeat studies (as required) can detect improvement or deterioration in the gall bladder appearances and allow appropriate clinical action to be taken.

CT is also useful for demonstrating the complications of cholecystitis but because of its limited availability in many parts of the UK, ultrasound should be the primary approach.

While ultrasound is undoubtedly an excellent method of assessing the patient with acute right upper quadrant pain, there are limitations in acutely ill patients. Emergency ultrasound tends to be performed in the non-fasting state, which can lead to difficulty in assessing such features such as gall bladder size and wall thickness. Gallstones are more difficult to diagnose in the acute abdomen as often a local ileus may obscure the gall bladder fossa. A major advantage of ultrasound is its ability to show the anatomy

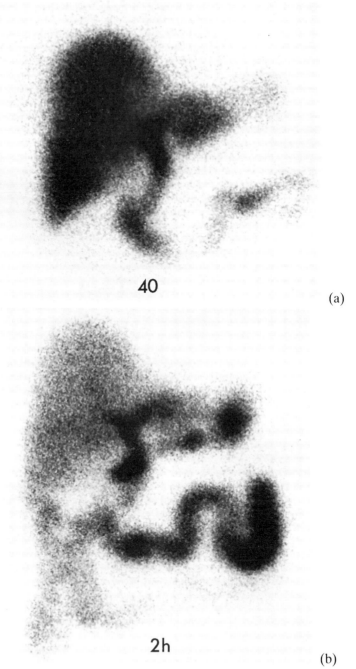

40

(a)

2 h

(b)

Fig. 5.9 Scintigrams obtained (a) 40 min and (b) 2 h after injection of technetium-labelled iminodiacetic acid (IDA). (a) shows diffuse liver activity with excretion into the common bile duct and duodenum. In (b) the liver activity has faded and most of the compound has now been excreted into the small bowel although some still remains in the biliary tree. At no stage has the gall bladder filled. Diagnosis: acute cholecystitis.

surrounding the gall bladder. It is accepted that of all patients presenting with right upper quadrant pain, only one third will have acute cholecystitis as the cause of their symptoms. An ultrasound examination will allow assessment of the liver, right kidney and pancreas.

Two other forms of acute cholecystitis require consideration.

Acalculous cholecystitis

Acalculous cholecystitis is a condition characterized by acute or chronic gall bladder wall inflammation occurring in the absence of calculi. Clinically and pathologically, acalculous cholecystitis is indistinguishable from calculous cholecystitis, but if left untreated, the incidence of progression to gangrenous change and perforation may be as high as 50 per cent. The mortality rate is at least double that of calculous cholecystitis. Many conditions have been reported in association with acalculous cholecystitis, but most cases occur as a complication of serious illness, particularly sepsis, major surgery, severe burns or multiple injury.

In acalculous cholecystitis both the oral cholecystogram and ultrasound findings may be entirely normal. Because of the associated increased morbidity and mortality, an early diagnosis is of great importance, but radiological investigation has proved frustrating. It has been shown that the sensitivity of both ultrasound and scintigraphy falls significantly in acalculous and cholecystitis when compared with calculous cholecystitis. However, the majority of patients will give a similar scintigraphy picture to acute calculous cholecystitis. Acalculous cholecystitis should always be considered in a patient presenting with clinical evidence of acute cholecystitis, a normal gall bladder ultrasound, and no other apparent cause to explain the clinical findings.

Emphysematous cholecystitis

Emphysematous cholecystitis is a condition affecting males four times as frequently as females. Some 25–33 per cent of patients are diabetic. Obliterative vascular disease compromising the cystic artery is considered a likely precipitating event. Morbidity is high, with some 75 per cent of patients developing gangrenous change leading to perforation in 20 per cent. At surgery, calculi are absent in 25 per cent and bile culture is nearly always positive. *Clostridium* is responsible in 46 per cent.

As previously described, the plain abdominal radiograph may show characteristic features of emphysematous cholecystitis. Ultrasound can diagnose emphysematous cholecystitis by demonstrating gas in the gall bladder wall or lumen. Sonographically, gas appears strongly echogenic and is associated with distal shadowing. Shadowing from gas tends to be poorly defined as opposed to the sharp clean acoustic shadows from calculi. If an ultrasound scan suggests gas in the biliary system, then an abdominal radiograph generally confirms the diagnosis.

Chronic cholecystitis

Oral cholecystography may demonstrate gallstones and/or a non-functioning or poorly functioning gall bladder. The value of scintigraphy is limited because a spectrum of appearances may be demonstrated. These include a delayed visualization pattern and, rarely, a similar pattern to acute cholecystitis. Because of the non-specificity of the technique, scintigraphy is not indicated to diagnose chronic cholecystitis. Ultrasound can also show a range of appearances. Gallstones associated with gall bladder wall thickening without evidence of focal tenderness, gall bladder distention, or wall sonolucency are the typical appearances. As previously described, failure to identify a gall bladder in a fasting patient is associated with a high incidence of primary gall bladder disease. CT may show a shrunken, thick-walled gall bladder with calculi.

Functional disorders of the gall bladder

Biliary dyskinesia remains a poorly understood subject. The oral cholecystogram was the initial radiological method used to assess gall bladder dynamics and, when combined with a fatty meal, or an intravenous injection of cholecystokinin, poor gall bladder contraction or a reproduction of the patient's symptoms was considered significant. Cholecystectomy has been successful in some patients with typical biliary symptoms and altered gall bladder dynamics. The subject is controversial partly due to the difficulties experienced in assessing the histology of gall bladders removed at operation, and partly because of problems in interpreting the cholecystographic findings.

Scintigraphy can also be used to assess gall bladder dynamics. The use of computer technology during HIDA scintigraphy can be used to give a more accurate assessment of gall bladder function. Regions of interest drawn around the gall bladder during HIDA scintigraphy can be used to estimate an 'ejection fraction'. Imaging these regions of interest before and after either a fatty meal or an injection of cholecystokinin, will allow an analysis of any change in the levels of gall bladder activity.

Abnormalities of the gall bladder wall

Diffuse gall bladder wall thickening occurs in many conditions (Table 5.2), not all of them directly related to the biliary tree. Focal or asymmetrical gall bladder wall thickening is often secondary to intrinsic gall bladder disease.

Table 5.2   Conditions causing thickening of the gall bladder wall

Contracted normal gall bladder
Cholecystitis: acute and chronic
Hypoalbuminaemia
Ascites
Hepatitis
Alcoholic liver disease
Heart failure
Renal failure
Adenomyomatosis
Carcinoma of the gall bladder
Multiple myeloma

Polypoid cholesterolosis
In cholesterolosis, abnormal deposition of triglycerides and cholesterol are found in macrophages in the gall bladder wall. Cholesterol deposits can be focal or diffuse. While diffuse disease cannot be diagnosed radiologically, focal cholesterol deposits exceeding 2 mm, may be detected on an oral cholecystogram where a fixed mural defect may be shown. These can also be demonstrated on ultrasound where, typically, single or multiple echogenic foci fixed to the gall bladder wall and not associated with acoustic shadowing are demonstrated. Cholesterol polyps and inflammatory polyps found in chronic cholecystitis have no malignant potential but cannot be distinguished by oral cholecystography or ultrasound from adenomatous polyps which are premalignant and of similar appearance.

Gall bladder malignancy

Metastatic disease to the gall bladder is uncommon, with melanoma and pancreatic tumours said to be the commonest primary malignancies to metastasize to the gall bladder.

Primary gall bladder malignancy is not common. It occurs more frequently in females and is a disease of the sixth and seventh decades. Gallstones are present in 80 per cent of cases and, typically, symptoms occur at a late stage of the disease.

The clinical presentation of gall bladder carcinoma is often non-specific and radiological differentiation from chronic cholecystitis can be very difficult. Oral cholecystography typically results in a non-functioning gall bladder. Where function is maintained, a filling defect may be demonstrated. A variety of ultrasound appearances have been reported. The most common is a poorly echogenic mass arising from the gall bladder fossa which can contain a highly echogenic focus which represents an engulfed gallstone. The gall bladder wall may be diffusely or focally thickened and, occasionally, an irregular hypoechoic mass may be demonstrated protruding from the gall bladder wall into the gall bladder lumen. Occasionally, carcinoma of the gall bladder may present as acute cholecystitis. A diagnosis of carcinoma is rarely made on ultrasound alone but occasionally nodes may be detected around the porta hepatis which should suggest a diagnosis of malignancy rather than simple inflammation.

CT may be useful in the diagnosis of gall bladder carcinoma. Focal or diffuse gall bladder wall thickening or a gall bladder mass can be shown. Liver invasion occurs in 75 per cent of cases of carcinoma of the gall bladder and CT may demonstrate extension of a mass from the gall bladder fossa merging into the liver parenchyma. CT, like ultrasound, can demonstrate liver metastases, dilatation of the biliary tree, and involvement of the draining lymph node chains.

As the gall bladder fails to opacify in over 90 per cent of patients with carcinoma of the gall bladder, oral cholecystography has no role to play in its diagnosis. CT would appear to have some advantages over ultrasound but, typically, the diagnosis is more often made at operation than preoperatively.

The Bile Ducts

Radiological investigation of the biliary tree has significantly changed with the advent of ultrasound and CT. Before these techniques were available, the choice was limited to intravenous or direct cholangiography. Demand for intravenous cholangiography has fallen considerably — partly because of the morbidity associated with cholangiographic contrast agents, partly also because of its poor diagnostic accuracy, but mainly because of the success of ultrasound diagnosis.

Traditionally, investigation of the jaundiced patient has been aimed at determining whether the cause is 'medical' or 'surgical', and radiology has been used to establish whether the biliary system is dilated or non-dilated. However, as with reliance on biochemical indices in jaundice, the simple division of patients into those with dilated or normal calibre ducts is not always indicative of obstructive or non-obstructive jaundice. There is no direct relationship between the calibre of the biliary tree and the presence or absence of clinically significant obstruction. This is particularly true in early, low grade or intermittent obstruction when a normal calibre common bile duct may be present. Often a combination of imaging techniques are required to establish, firstly, if obstruction is present and, secondly, to determine the level of obstruction.

Ultrasound

Normal calibre intrahepatic ducts are not usually identified by conventional ultrasound machines. Recent technological advances in the field of computed sonography has resulted in normal calibre ducts being identified during routine scanning. Dilated ducts are shown running parallel to portal vein branches. Both generalized and localized intrahepatic duct dilatation can be demonstrated by ultrasound easily. The upper limit of normal for the common bile duct as assessed by ultrasound varies slightly from department to department. The commonly accepted figure is 6 mm and this relates to the proximal extrahepatic common duct just as it emerges from the liver. This section of duct is chosen for specific measurement as it can be identified in nearly every patient. Moreover, the distal common duct may be bulbous and variable in size and therefore should not be used in any assessment of biliary obstruction. Although 6 mm is accepted as the upper limit of normal for duct calibre, the common duct can occasionally be larger with no evidence of obstruction. Further information may be gained by the use of a fatty meal similar to that following oral cholecystography. Generally, the common duct reduces in calibre following a fatty meal. If the calibre increases, then this is an abnormal response and may be due to calculi or tumour in the duct, or spasm at the sphincter of Oddi. Further investigation is then required.

False-negative ultrasound studies occur when there is obstruction but no duct dilatation. Causes ranging from small calculi to malignancy have been well documented. Typically, these patients have liver function tests which show an obstructive pattern with elevation of bilirubin and alkaline phosphatase. Although ultrasound is an excellent technique for diagnosing biliary tract obstruction, the actual level of obstruction will not be demonstrated in up to one third of patients, so other techniques are required to establish the diagnosis.

Computed tomography

The role of CT in biliary imaging is largely complementary to ultrasound. Because of its lack of ionizing radiation, easy availability, low cost and speed of examination, ultrasound should always be the initial approach to biliary disease. CT can be very helpful in determining the level of biliary obstruction especially when intestinal gas persistently obscures the lower common duct and head of the pancreas. CT can demonstrate dilated intra- and extrahepatic ducts, establish whether the gall bladder appears normal or abnormal, give the level of obstruction, and, by assessing the characteristics of the dilated system, suggest a likely diagnosis.

CT cannot demonstrate normal calibre intrahepatic ducts. The common bile duct measures some 2–6 mm in diameter and is identified in approximately one third of patients. Definite dilatation is present when the duct measures more than 10 mm in diameter. As has been previously stated, there is no direct relationship between the calibre of the biliary tree and the presence or absence of clinically significant obstruction.

Two further CT appearances may give diagnostic information as to the aetiology of biliary tract obstruction. Firstly, the degree of duct dilatation may be helpful because marked dilatation is most often associated with a malignant aetiology. Mild to moderate dilatation is a non-specific sign and may be due to malignant or benign disease. Secondly, the shape of the common duct is important. An abrupt transition in duct calibre is most often associated with malignancy. A longer, more tapered, change in calibre is more typical of benign disease.

Detection of choledocholithiasis

As 15 per cent of patients with gallstones have duct stones, their detection is of considerable importance to the clinician investigating gall bladder disease. The use of intravenous cholangiography to detect duct stones is no longer acceptable as various studies have shown the technique to be very inaccurate. In a review which compared the findings at intravenous cholangiography with either surgical or direct cholangiographic follow-up, an overall error rate of 45 per cent was found. Technically satisfactory opacification of the common duct which allowed confident interpretation occurred in approximately 50 per cent of cases. Even when technically adequate studies were considered, a 40 per cent error rate in interpretation was demonstrated. The majority of these errors were missed calculi, the very reason for most of the examination requests.

The sensitivity of ultrasound at detecting choledocholithiasis is poor. Recent prospective studies have shown an accuracy of 20–30 per cent in detecting duct calculi when compared with ERCP. Approximately one third of patients with duct stones have normal calibre ducts which partly accounts for this low rate of detection. Furthermore, stones often become impacted in the distal common duct which is often obscured by intestinal gas. Early experience with intraoperative ultrasound scanning has claimed an improved accuracy for the detection of common duct stones. This may provide an alternative to operative cholangiography, but the technique as yet is not widely used in the UK. If duct stones are thought likely clinically and ultrasound is not confirmatory, then direct cholangiography using ERCP or PTC is required.

CT is very sensitive in detecting low levels of calcification and therefore provides an alternative technique for the detection of choledocholithiasis. When calculus obstruction is present, the biliary tree may be only mildly dilated, the common duct tends to taper, and a high density calculus will be demonstrated in 80–90 per cent of cases.

Duct stones are usually easily detected with PTC. Occasionally, diagnostic difficulty may occur when biliary stagnation is present. Bile plugs can form and these may mimic duct calculi. Similarly, blood clot and, occasionally, intraluminal tumour can mimic calculus obstruction. In an obstructed biliary system, infection is present in 80 per cent of cases. Care must therefore be taken during direct cholangiography to avoid overinjection with contrast which may precipitate septicaemia.

In the presence of calculus obstruction, ERCP may show the common duct to be dilated or to be of normal calibre. It is an accepted rule of thumb that when complete obstruction to the common duct is shown at ERCP, this is never due to calculi. The only exception to this rule is when a calculus becomes impacted at the level of the ampulla. A major advantage of ERCP in the investigation of choledocholithiasis is the potential for endoscopic sphincterotomy (Fig. 5.10). Most centres quote success rates of 80 per cent, a complication rate of 8 per cent and a reported mortality rate of 1 per cent. The most common complications are perforation, cholangitis, haemorrhage, and pancreatitis. Sphincterotomy is playing an ever increasing role in the management of patients with duct stones and, particularly in the elderly, provides a means of stone removal without resorting to surgery.

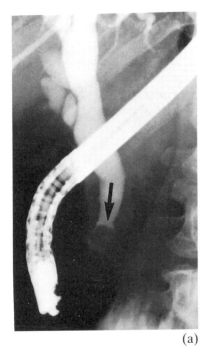 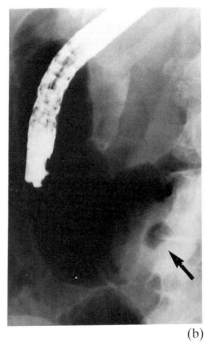

(a)   (b)

Fig. 5.10   (a) Endoscopic retrograde cholangiogram showing a retained stone in the common bile duct (arrow). (b) After sphincterotomy the stone has passed through into the duodenum (arrow).

Imaging of the postcholecystectomy patient

Both ultrasound and CT can be helpful in the management of the post-cholecystectomy patient during the immediate postoperative period. Both techniques can demonstrate a gall bladder bed haematoma or abscess formation, their progression or resolution can be monitored by repeat examination, and percutaneous drainage under CT or ultrasound control can be performed if required.

Postcholecystectomy symptoms are common and represent a difficult management problem. A British study has reported that 43 per cent of patients who have undergone cholecystectomy claim dissatisfaction with the outcome after one year. Early ultrasound reports suggest that the common duct dilates following cholecystectomy. Recent prospective ultrasound studies have demonstrated that if the proximal extrahepatic portion of the common duct is of normal calibre before surgery, then it will remain so unless some other cause of obstruction develops. The distal common duct may increase in calibre following surgery but because of the distensibility of this part of the duct, measurements here are of limited value. No relationship between common bile duct dilatation and intraoperative cholangiography has been demonstrated.

As has already been described, ERCP plays a major role in the investigation of the postcholecystectomy patient with persistent biliary symptoms, allowing both demonstration of bile duct calculi and the option of therapeutic sphincterotomy. Biliary leaks occur most commonly following surgery or trauma. Scintigraphy can detect a bile leak and show its location and extent. Because scintigraphy is a dynamic investigation, the preferential flow of bile from the leak can be demonstrated. This has important management consequences because, if the predominant flow of bile is shown to be through a biliary-enteric anastomosis or down the common duct and into the duodenum, conservative management may suffice.

Biliary tract tumours

While some authorities use cholangiocarcinoma to describe malignant tumours of the intrahepatic ducts and, correspondingly, primary bile duct carcinoma for the extrahepatic ducts, for the purposes of the following discussion, cholangiocarcinoma will refer to both the intra- and the extrahepatic biliary system.

Non-invasive imaging, i.e. ultrasound and CT only infrequently make a specific diagnosis of biliary tract malignancy. Although a solid intraluminal mass can occasionally be demonstrated, ultrasound may only show a stricture with dilated ducts proximally, while the commonest finding on CT is a focal or generalized dilatation of the biliary tree. Neither technique can confidently differentiate between a peripheral, i.e. intrahepatic cholangiocarcinoma, and a primary or secondary tumour of the liver itself.

Direct cholangiography may also produce non-specific appearances and, in particular, cholangiographic differentiation between malignant and benign strictures can be very difficult. Cholangiocarcinomas can produce short or long strictures, polypoid filling defects within the duct, or a shouldered stenosis occasionally with a rat tail appearance. ERCP can provide additional information by visualizing the pancreatic duct. While cholangiocarcinoma, characteristically, only blocks the common bile duct, pancreatic carcinoma may block both the common bile duct and the pancreatic duct. However, it is important to note that cholangiocarcinoma, carcinoma of the pancreas, metastatic carcinoma, and papillary tumours can all produce similar appearances on ERCP. Complete obstruction is almost always due to malignant disease.

PTC has a major role to play in the preoperative assessment of cholangiocarcinoma where surgery is being contemplated. Extension of the tumour into second order branches of both lobes, or beyond, means that it is unresectable. If second order branches of a single lobe are involved, then that lobe cannot be salvaged.

Benign strictures

The difficulty in radiological differentiation between benign and malignant strictures of the biliary tree should be emphasized. The commonest causes of benign bile duct strictures are chronic pancreatitis and surgery.

A history of previous surgery may be of value in deciding the underlying aetiology. A mid common bile duct stricture in a patient who has had a cholecystectomy is most likely to be benign. Any stricture in the mid common duct which occurs in a patient who has not had biliary tract surgery must be considered malignant until proved otherwise. Anastomotic strictures are said to occur in 30 per cent of patients who have undergone biliary-enteric bypass. Chronic pancreatitis characteristically gives a long smooth stricture of the pancreatic portion of the common bile duct. Typically, obstruction is incomplete. CT will, generally, demonstrate a moderately dilated biliary system with gradual tapering of the common bile duct on successive caudal slices. CT is particularly useful in detecting foci of calcification within the pancreatic tissue. ERCP has the advantage of being able to assess the pancreatic ductal system.

Sclerosing cholangitis
Primary sclerosing cholangitis is a condition of unknown aetiology which results in diffuse periductal fibrosis. Although the extrahepatic ducts are most commonly involved, the intrahepatic ducts may also show characteristic changes. The gall bladder and cystic duct are usually spared from the fibrotic process. There is a strong association with ulcerative colitis, and patients with sclerosing cholangitis are at risk of developing biliary tract malignancy. Direct cholangiography shows irregular, nodular, beaded narrowing of the extra- and intrahepatic ductal systems. Patients with sclerosing cholangitis require antibiotic cover before direct cholangiography is performed because of the risk of stasis and sepsis.

CT will show mild duct dilatation with a distorted irregular branching pattern. Characteristically, in sclerosing cholangitis, the ducts do not taper towards the periphery of the liver which is the typical pattern found in other causes of duct obstruction. Rarely, thickening of the common bile duct wall may be demonstrated. As periductal fibrosis restricts the degree of duct dilatation, PTC may be technically difficult and the diagnosis is best made on ERCP. Although the cholangiographic findings are fairly characteristic, it should be remembered that the rare multifocal cholangiocarcinoma can give identical appearances.

Biliary tract obstruction

As has already been stated, ultrasound may be unable to demonstrate the level of obstruction in up to one third of patients with dilated ducts. CT has proved rather more successful when compared with ultrasound in establishing the exact level of obstruction.

For diagnostic purposes, the biliary tree is generally considered in five sections: intrahepatic, junctional, suprapancreatic, pancreatic and ampullary. With intrahepatic dilatation, the ducts are dilated in only one segment or lobe of the liver. Junctional obstruction will show dilatation of both the right and left main hepatic ducts with an obliterated or normal calibre common duct. When a dilated common duct is identified on one or more slices with a duct of normal size seen in the pancreatic head, then a suprapancreatic obstruction is present. When the dilated duct extends down into the pancreas, an intrinsic pancreatic lesion is the cause. Finally, with ampullary obstruction, the dilated duct is identified into the pancreatic head and it may be associated with a dilated pancreatic duct. The appearance of the gall bladder may give further information. When obstruction occurs below the cystic duct/common bile duct junction, the gall bladder is generally distended. However, the gall bladder may not be able to distend if there has been prior gall bladder disease. The commonest causes of obstruction at these levels are given in Table 5.3.

Table 5.3 Causes

Intrahepatic	Primary or secondary liver tumours Sclerosing cholangitis Cholangiocarcinoma
Junctional	Surgical complications Carcinoma gall bladder Cholangiocarcinoma
Suprapancreatic	Porta hepatis lymphadenopathy Cholangiocarcinoma Pancreatic carcinoma extension
Pancreatic	Malignant — carcinoma Benign — stricture, pancreatitis, calculi
Ampullary	Calculi Carcinoma

Long-standing biliary obstruction may lead to hepatic atrophy. This generally occurs as a sequel to cholangiocarcinoma, or surgical complications affecting the region of the porta hepatis. Duct crowding, dilatation, and tortuousity may result. If an obstruction is restricted to a single lobe or segment, the remaining liver parenchyma may show compensatory hypertrophy.

Developmental abnormalities of the biliary tree

Recent imaging techniques have allowed a non-invasive approach to the diagnosis of congenital disorders of the biliary tree.

Choledochal cyst
One of the more common developmental abnormalities is a choledochal cyst. The classical triad of abdominal pain, a palpable mass in the right upper quadrant and jaundice occurs in 50 per cent of patients. The most common type of choledochal cyst results in a localized dilatation of the common bile duct, frequently associated with distal narrowing. This dilatation usually occurs below the cystic duct but may, on occasion, involve both the intra- and extrahepatic system. Both oral cholecystography and intravenous cholangiography may be unsuccessful in demonstrating a choledochal cyst. Ultrasound, CT, and biliary scintigraphy have proved successful without the need to resort to direct cholangiography. CT may give a characteristic appearance. When intrahepatic duct dilatation is present, this is, typically, limited to the central portions of the right and left main intrahepatic ducts, as opposed to acquired obstruction where generalized dilatation occurs. Scintigraphy will show uptake of activity by the mass, and its communication with the biliary tree will be confirmed.

Choledochocele
This developmental dilatation of the intramural portion of the distal common bile duct generally presents with pain, jaundice and vomiting. The condition may be suspected first during a barium series where a negative filling defect can be identified at the ampulla on the medial wall of the duodenum. Occasionally, the choledochocele may appear as a mass projecting into the duodenum. Direct cholangiography is the best method of diagnosis.

Biliary atresia
In the jaundiced neonate, it is important to differentiate neonatal hepatitis with a patent biliary tree from biliary atresia which can be surgically corrected. Ultrasound may be of value if a normal common bile duct is demonstrated. Biliary scintigraphy will exclude a diagnosis of biliary atresia when intestinal activity is demonstrated. The role of scintigraphy remains controversial because of difficulty in differentiating biliary tract obstruction

from severe hepatic parenchymal disorders. In the presence of severe hepatic dysfunction, hepatocyte uptake of isotope will be compromised. In this situation, stool collections to detect the presence of activity may be of some value, but care must be taken to avoid contamination with urine because of renal excretion of the radiopharmaceutical.

Pancreatic Disease

Acute Pancreatitis

The diagnosis of acute pancreatitis is made primarily on clinical and biochemical grounds. The role of imaging procedures is largely in the detection and monitoring (and sometimes in the treatment) of its complications, but simple confirmatory procedures can aid in initial diagnosis. A chest film and plain abdominal film should be obtained in all patients with acute abdominal symptoms. These films may show features which support the diagnosis of acute pancreatitis, such as basal pleural effusions, upper abdominal ileus — particularly isolated duodenal dilatation ('sentinel loop') — broadening of the duodenal loop indicating pancreatic enlargement, or absence of gas from the middle part of the transverse colon ('colon cut-off sign'). Gallstones are present in roughly half the patients with acute pancreatitis and in a minority of cases the stones will be sufficiently opaque to be visible on the plain abdominal film. In other cases, ultrasound of the upper abdomen will show stones in the gall bladder or, less frequently, stones in the common duct.

Imaging of the pancreas itself is useful if there is doubt about the diagnosis by clinical and biochemical criteria. The pancreas is usually enlarged and oedematous with inflammatory changes spreading into the surrounding tissues. These changes are reflected in the ultrasound appearances of pancreatic enlargement with reduced echogenicity and lowering of the normal retroperitoneal landmarks. Ultrasound is often technically difficult in these patients because of bowel gas associated with local or generalized ileus, and CT scanning is generally more informative. Typical features on CT are diffuse enlargement of the gland with blurring of its outline, heterogeneous attenuation (density) of the parenchyma, a streaky increase in the attenuation of mesenteric and retroperitoneal fat indicating inflammatory changes, thickening of the fascial planes around the pancreas, and fluid exudates, seen most commonly in the left anterior pararenal space (Fig. 5.11).

Most cases of acute pancreatitis are associated with gallstones or with alcohol abuse but if neither of these causes can be established, then a careful search for an underlying carcinoma must be made. In pancreatitis secondary to carcinoma, obstruction of the pancreatic duct can generally be detected by evidence of duct dilatation seen on ultrasound or CT. The tumour itself

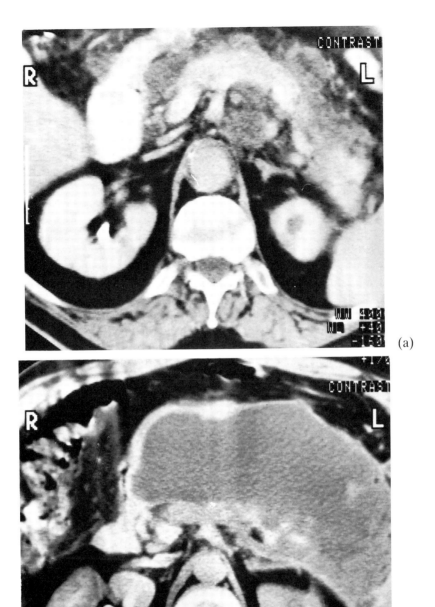

(a)

(b)

Fig. 5.11 (a) Contrast-enhanced CT scan of a patient with severe acute pancreatitis. There is exudate both anterior and posterior to the pancreas but the substance of the gland remains intact. (b) CT scan of another patient with severe acute pancreatitis. The glandular tissue has virtually disappeared and been replaced by a large collection of turbid fluid. Pancreatic necrosis was confirmed surgically.

may be quite small and a careful examination of the pancreatic head must be made by ultrasound, CT, or both. Pancreatitis resulting from trauma is often associated with other soft-tissue injuries in the upper abdomen, and CT gives the most complete demonstration of the injury. The viability of fragments of pancreatic tissue can be assessed by their enhancement after contrast infusion with rapid sequence CT.

Complications

The complications of acute pancreatitis require careful monitoring using ultrasound, CT, or both techniques. Pseudocysts appear on ultrasound as areas of fluid density in relation to the pancreas sometimes containing solid debris (Fig. 5.12). CT shows the more subtle changes with early development of exudates in the root of the mesentery and around all surfaces of the pancreas, as well as collections within the gland itself. Only a minority of pseudocysts actually occupy the lesser sac, fluid collections within the gland itself or on the anterior surface being more common. In some cases, fluid may extend down into the iliac fossa or even into the pelvis, and collections in the right subphrenic space and the subhepatic space are not infrequent. Simple fluid collections can be aspirated percutaneously under either ultrasound or CT guidance. Ultrasound is usually preferred for anteriorly placed collections, but CT is preferable for aspiration of retroperitoneal collections. Recurrent or persisting fluid collections may require drainage, and again, this can be carried out percutaneously with imaging guidance.

The rare complication of pancreatic abscess is suggested by the presence of gas within a pancreatic fluid collection, although this may sometimes result from the spontaneous formation of a fistula into stomach or bowel. Multiple tiny bubbles of gas seen within a heterogeneous pancreatic phlegmon may indicate aseptic necrosis. It is most important to recognize pancreatic necrosis should this complication occur since it is almost invariably lethal unless necrotic tissue is removed surgically. Pancreatic necrosis can be recognized on CT by demonstrating fragments of the parenchyma which fail to show enhancement after intravenous contrast in patients with other changes of severe pancreatitis.

Indium-labelled leucocyte scintigrams must be interpreted with caution in patients with acute pancreatitis since focal 'hot spots' may indicate areas of fat necrosis rather than abscesses.

Chronic Pancreatitis

The recognition of morphological abnormalities plays a major role in the diagnosis of chronic pancreatitis, so imaging procedures are important here. As is often the case, the most sensitive techniques are also the most invasive so the recommended approach is to start with the simplest tests and continue until the diagnosis has been confirmed.

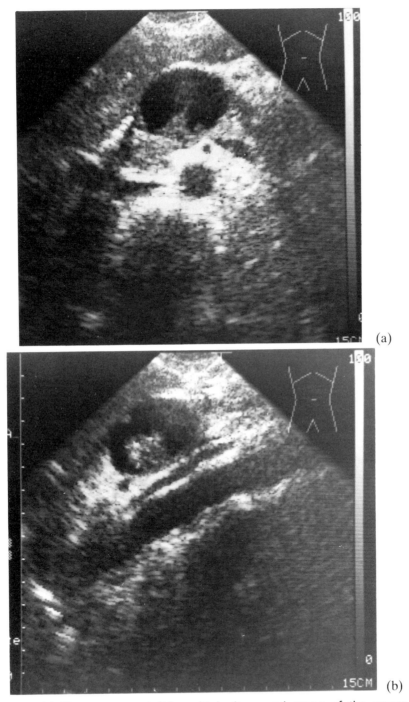

(a)

(b)

Fig. 5.12 (a) Transverse and (b) sagittal ultrasound scans of the upper abdomen in a patient with pancreatitis. Both scans show the superior mesenteric artery anterior to the aorta. Anterior to the SMA lies a rounded fluid collection containing echogenic debris. Diagnosis: pancreatic pseudocyst.

A plain film of the abdomen showing extensive pancreatic calcification is enough to establish the diagnosis of chronic pancreatitis. If the plain film is normal, ultrasound or CT should be the next step. In chronic pancreatitis the gland may be small or normal in size but localized areas of enlargement are quite common. Calcification which is not dense enough to show on the plain film may often be detected by CT and sometimes also by ultrasound. Both techniques show diffuse or patchy dilatation of the duct system and small cysts or pseudocysts representing areas of branch duct obstruction. The texture of the parenchyma appears heterogeneous on CT and on ultrasound usually shows increased echogenicity which is again often heterogeneous. Focal masses of fibrotic and inflammatory tissue may produce appearances similar to those of carcinoma and, although distinguishing features are present in most cases, it is sometimes impossible to differentiate the two conditions without histology, and percutaneous biopsy using a fine needle can be readily carried out under imaging guidance.

Where the ultrasound and CT findings are inconclusive or unexpected, ERCP may resolve doubt. This is the only widely available technique for direct opacification of the pancreatic ducts, and comparative studies have shown it to be the most sensitive detector of chronic pancreatitis. The significance of minor irregularities of the side branches ('minimal-change pancreatitis') is contentious, but even if these cases of dubious pathology are excluded, ERCP findings will usually be regarded as the final arbiter. This technique does, however, carry a risk of provoking acute pancreatitis particularly in patients with pre-existing pancreatic disease and it should be used with care. Of the non-invasive tests, CT has been shown to be a little more accurate than ultrasound but the accuracy of both of these tests can be enhanced by combining it with either a Lundh test or its modern variant, the tubeless test meal.

Angiographic changes are well described in chronic pancreatitis. The small arteries within the pancreas show irregularity of calibre and pruning. The splenic vein may be narrowed or even occluded by focal inflammatory disease and in the latter case left-sided varices are likely. These may also be recognized on CT and ultrasound by the presence of splenic enlargement, failure of the splenic vein to show contrast enhancement (on CT), and the presence of abnormal vessels around the splenic hilum and tail of the pancreas. Arteriography is also the best method for showing the relatively rare complication of false aneurysm in chronic pancreatitis which may be suspected if the patient in whom the diagnosis is already established develops a sudden severe pain in the upper abdomen or back.

Patients with chronic pancreatitis are usually thin with very little intrabdominal fat — just the opposite of patients with acute pancreatitis. Because of the absence of fat, interpretation of CT is sometimes difficult and ultrasound is technically easier than in the obese patient. In children, particularly, ultrasound is at least as informative as CT, and is the preferred technique in, for instance, children with cystic fibrosis.

Pancreatic Tumours

The early diagnosis of pancreatic carcinoma remains an unsolved problem. By the time they present clinically, many tumours are unresectable. Although the development of improved imaging methods, particularly ultrasound and CT, has led to more accurate and probably also earlier diagnosis, it has not yet been possible to show an improvement in the natural history of the disease through earlier treatment. Imaging has an important role both in the initial diagnosis of pancreatic tumours and in the preoperative assessment of their extent.

Plain film and barium examinations are of little value, showing changes in only the most advanced cases. Ultrasound and CT provide the mainstay of morphological diagnosis. Carcinomas which are large enough to deform the contour of the gland should be recognized by either of these methods. The tumours themselves are sometimes indistinguishable from surrounding pancreatic parenchyma but in many cases they show reduced echogenicity on ultrasound and reduced attenuation on CT (Fig. 5.13). The tumours are usually hypovascular and show reduced contrast enhancement compared with normal pancreatic tissue. Secondary effects of the tumours, such as dilatation of the main pancreatic duct or of the biliary tree in patients presenting with jaundice, are also readily detected by ultrasound or CT. If chronic pancreatitis is an alternative diagnosis, distinction between inflammatory and malignant pancreatic masses may be dependent on histology. Percutaneous biopsy with imaging guidance is considered safe in these circumstances if a skinny needle is used.

Patients who are candidates for resection of pancreatic tumours need a careful preoperative survey of their upper abdominal anatomy. Successful resection is likely to be ruled out if CT shows posterior extension of a pancreatic carcinoma to involve the coeliac axis, superior mesenteric artery or the aorta, or causing portal vein occlusion, metastases in the liver, or enlarged lymph nodes in the paraaortic areas or in the porta hepatis. If CT findings suggest that the tumour may be resectable, arteriography should be the next step. In addition to demonstrating characteristic patterns of malignant vessels in the area of the tumour, the vascular studies should confirm whether or not the splenic and mesenteric veins are encased or occluded, should demonstrate the presence of anomalous arterial supply to the liver, and should also allow a further view of the liver parenchyma to look again for metastatic lesions.

Functioning islet cell tumours of the pancreas present clinically with syndromes associated with hypersecretion of insulin, glucagon, gastrin, somatostatin, or VIP (vasoactive intestinal polypeptide). Although the diagnosis rests on clinical and biochemical grounds, the demonstration of an appropriate pancreatic tumour provides strong supportive evidence for the diagnosis, and in any case surgical treatment will usually be considered so that anatomical localization is of great importance. The relationship be-

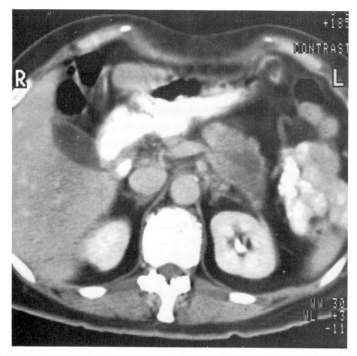

Fig. 5.13 CT scan of the upper abdomen in a patient with adult onset diabetes, weight loss and back pain. A small irregular tumour lies within the tail of the pancreas. Diagnosis: adenocarcinoma.

tween the size of the mass and the endocrine effects varies with the different tumour types. Most of the rare tumours producing somatostatin, glucagon and VIP are of substantial size at the time of clinical presentation, and so can be recognized by ultrasound or CT scanning as a visible mass enlarging the pancreas. Insulinomas are usually 1–3 cm in diameter at the time of presentation, so that although the majority of these tumours will be recognized by CT and ultrasound, a substantial minority will be undetectable. Gastrinomas are often less than 1 cm in diameter so that their detection by scanning methods is relatively infrequent. The demonstration of a single pancreatic tumour in these patients should not be regarded as the end of the story since multiple tumours are quite common, and malignancy with metastasis to the liver is also relatively frequent. Where scanning methods fail to show a pancreatic tumour in patients with strong clinical and biochemical suspicion of the diagnosis, pancreatic arteriography should be performed. This gives a more detailed view of the vascular anatomy of the pancreas and allows many of the smaller insulinomas and gastrinomas to be recognized by a characteristic tumour blush. If arteriography is negative, selective sampling of the pancreatic veins may be helpful. This technique requires percutaneous catheterization of the portal vein with serial sampling of blood from the veins draining the tail, body and head of the pancreas. Immunoassay for the relevant hormone should indicate the position of the tumour within the gland.

Pancreatic Trauma

Trauma to the pancreas is less common than damage to the spleen and liver but carries significant morbidity and mortality. It is often associated with trauma to other abdominal organs and clinical findings are typically non-specific. Diagnosis is difficult whichever technique is used but CT is probably the best option currently available. Pancreatic fracture, laceration, intra- and peripancreatic fluid collections can be shown, but CT interpretation is often tricky. Early scans may be normal or may show only minor changes in the peripancreatic fat and adjacent mesentery. If CT is performed without oral contrast medium, bowel loops can mimic pancreatic enlargement. Streak artefacts may look like pancreatic fractures but the use of intravenous contrast media will help to minimize difficulty in interpretation.

Either ultrasound or CT can be used to monitor the progress or resolution of peripancreatic fluid collections after trauma. Percutaneous aspiration or insertion of drainage catheters with guidance by CT or ultrasound can be helpful.

Further Reading

Cosgrove, D.O. and McCready, V.R. (1982). *Ultrasound Imaging of the Liver, Spleen and Pancreas*. John Wiley, Chichester.

McCort J. (1981). *Abdominal Radiology*. Williams and Wilkins, Baltimore & London.

Megibow, A.J. and Balthazar, E.J. (1986). *Computed Tomography of the Gastrointestinal Tract*. C. V. Mosby, Ontario.

Robinson, P.J. (Ed.). (1986). *Nuclear Gastroenterology*. Churchill Livingstone, Edinburgh.

Simeone, J.F. (Ed.). (1984). Co-ordinated diagnostic imaging. *Clinics in Diagnostic Ultrasound* **14**. Churchill Livingstone, Edinburgh.

6

THE GENITOURINARY SYSTEM

Renal Tract Imaging

The Abdominal Radiograph

The investigation of a patient with renal tract disease, will usually include a KUB (kidneys, ureters and bladder) radiograph as part of the radiological work-up. It is easily performed, inexpensive, and can provide the clinician with a considerable amount of useful information.

Providing that the kidneys are adequately outlined by perirenal fat, the abdominal radiograph will allow assessment of renal size, renal contour, and the respective position of each kidney. Renal tract calcification — both urinary tract calculi and nephrocalcinosis — can be identified, as may the presence of renal tract gas. Indirect evidence of renal tract pathology may also be shown, e.g. a lumbar scoliosis in a patient presenting with renal colic, or effacement of normal soft-tissue structures such as the psoas outline when inflammatory change or a mass is present in the renal fossa.

The KUB provides little information about renal function with the exception of the characteristic appearance of a calcified post-tuberculous 'autonephrectomy'.

A plain abdominal radiograph is a useful first investigation in any patient presenting with loin pain. Renal angle pain may be due to pathology of non-renal origin and an abdominal radiograph may indicate the source of the clinical problem, e.g. gallstones, an appendicolith or bowel obstruction. It is also important to obtain a plain radiograph prior to any contrast study in order that renal tract calcification may be demonstrated, whether it be in kidney, ureter or bladder. The KUB also provides an excellent means of patient follow-up in conditions such as nephrocalcinosis or nephrolithiasis.

Contrast Studies of the Renal Tract

The intravenous urogram (IVU)

Whenever intravenous contrast agents are used, certain precautions must be observed.

Risk of allergic reaction
Approximately 60 per cent of patients undergoing urography experience minor transient symptoms such as nausea or flushing. Allergic reactions, however, occur in less than 2 per cent. The mortality rate using conventional ionic agents is 1 in 50000. The mortality rate using non-ionic agents has not yet been established but it appears to be less. Although the likelihood of a patient who has had a previous reaction to intravenous contrast experiencing a second reaction is no greater than that for the first exposure, it is only prudent that when a patient gives such a history, precautions are observed if a second intravenous urogram is thought clinically necessary. Prophylaxis with antihistamines has not been of proven benefit and most centres will prescribe oral steroids both before and after a planned procedure. Non-ionic rather than ionic contrast should be used in this situation.

The role of dehydration
Before modern contrast agents were available, dehydration was considered necessary in order to obtain an adequate pyelogram. With newer agents, the radiologist is able to use greater volumes of contrast which, if combined with nephrotomography, will demonstrate the renal tract in all but the severely azotaemic. The use of modern agents means that routine dehydration is no longer required. Dehydration is strictly contraindicated in certain patient populations: diabetic patients, children, patients with renal insufficiency, and in multiple myeloma. Problems due to inappropriate dehydration in these patient groups would be removed if the routine practice of prior dehydration was discontinued.

Non-ionic contrast agents
The higher cost of non-ionic media compared with conventional ionic contrast has limited their use for routine intravenous urography. They are indicated for high risk patients, e.g. those with known atopy or a known previous contrast reaction, in diabetic patients, and in patients with renal insufficiency.

A conventional intravenous urogram will include a precontrast abdominal radiograph including the renal and bladder regions, followed by sequential films after contrast which will demonstrate functioning renal tissue, the pelvicalyceal systems, the ureters and the bladder. The IVU will allow some assessment of renal function by observing the density and appearance time of the nephrographic phase, particularly when comparing the two kidneys. Renal cortical thickness can be assessed and any focal or generalized cortical defects identified. Disorders of the renal papillae, renal tract masses, obstruction to the renal tract and developmental abnormalities can all be shown by the IVU.

Although the IVU is a very useful technique in the investigation of renal tract disease it does have certain limitations. The IVU cannot differentiate between a cystic and a solid mass. Any mass shown during urography requires further radiological assessment to determine its nature. The IVU has a very limited role in the diagnosis of renovascular hypertension. Classic urographic descriptions do not apply when a segmental arterial lesion is present and alternative imaging techniques are more appropriate. Although bladder tumours may be demonstrated by intravenous urography, it is important to remember that a urographically normal bladder in a patient presenting with haematuria does not exclude the presence of a bladder neoplasm; cystoscopy is required. Although high dose intravenous urography allows assessment of the renal tract even in patients with moderately severe renal insufficiency, it should be appreciated that alternative forms of imaging, e.g. ultrasound may give the clinician the information he requires while avoiding the necessity for intravenous contrast. This applies particularly when investigating obstructive uropathy.

The main value of urography is to demonstrate the gross appearance of the renal tract, identify renal tract masses, and confirm or exclude the presence of obstruction. It is often the initial approach to a patient presenting with haematuria, a mass in the flank, or recurrent urinary tract infection. The continued use of the IVU as a first investigation is largely traditional, and the greater availability of other forms of renal imaging has begun to erode the position of the IVU as the principle means of investigating the renal tract. This is no bad thing particularly when the small but definite risks associated with intravenous radiographic contrast agents are considered.

Antegrade pyelography

Opacification of the pelvicalyceal system and ureter via an antegrade puncture is performed under fluoroscopic or ultrasound control. When fluoroscopy is used, the collecting systems need to be opacified by intravenous contrast first. This has disadvantages in the non- or poorly functioning kidney. Ultrasound provides a simple means of guidance which does not require the use of intravenous contrast. Using a fine flexible needle (22 G), the collecting system can be punctured and opacified. An examination of the pelvicalyceal system and ureter can then be performed. Should obstruction be demonstrated, the technique can be extended to provide temporary relief by performing a percutaneous nephrostomy (see p. 174).

Antegrade pyelography will demonstrate the level of an obstruction and may suggest the underlying aetiology, e.g. calculus, clot, tumour. The risks of introducing infection by an antegrade approach are significantly less than by retrograde pyelography — especially when obstruction exists.

Antegrade renal puncture is particularly useful in the transplanted kidney. If there is doubt as to whether a transplanted ureter is obstructed or not, an antegrade study will demonstrate clearly whether there is free drainage of contrast into the bladder or not.

Retrograde pyelography

Retrograde pyelography is performed following ureteric catheterization during cystoscopy. It therefore requires both surgical and radiological expertise. Direct inspection of the ureteric orifices during cystoscopy is of particular value when developmental abnormalities of the renal tract are present. Retrograde pyelography will also identify the level of a ureteric obstruction and, like antegrade pyelography, may suggest the underlying aetiology. Retrograde pyelography is also the traditional approach to further investigation of pelvicalyceal or ureteric filling defects shown at intravenous urography.

There is a recognized risk of introducing infection into the renal tract during retrograde pyelography and the examination should be avoided in the presence of a urinary tract infection.

Micturating cystography

Micturating cystography is used primarily to demonstrate the presence and severity of vesicoureteric reflux in patients with recurrent urinary tract infection. It is also used to assess abnormalities of the bladder neck and urethra in developmental, gynaecological and neurological disorders. It may also be used as part of a urodynamic study when bladder pressures and flow rates can be assessed in the investigation of stress/urge incontinence.

Micturating cystography has a clear role to play in the assessment of bladder trauma. Cystography will reveal bladder tears which have not been shown during intravenous urography.

Ascending urethrography

The aim of this technique is to demonstrate the urethral anatomy. This is considerably simpler in the male than in the female. The principle indications are the investigation of urethral stricture and urethral trauma. Urethrography should not be performed within 2 weeks of urethral instrumentation or in the presence of an active urethritis.

The posterior urethra is best demonstrated by descending urethrography. When multiple strictures are present, a combined ascending urethrogram and a micturating cystourethrogram are often required.

Ultrasound

The addition of ultrasound to the imaging options for renal tract investigation has made a significant impact on radiological practice. The avoidance of ionizing radiation and intravenous contrast makes ultrasound the ideal technique both for follow-up studies in patients with chronic renal problems, and for screening relatives of patients with hereditary renal disease. Ultrasound now offers an alternative to the intravenous urogram in many situations, particularly in children, pregnant women and patients with relative contraindications to the use of intravenous contrast.

Ultrasound will establish the presence or absence of renal tissue in the renal fossa. It will give a three-dimensional assessment of renal size, assess the cortical thickness and detect any generalized or focal cortical loss, and will indicate the presence or absence of dilatation of the collecting system. Renal masses will be demonstrated if they are approximately 2 cm or more in diameter and, of particular importance, ultrasound will differentiate between a cystic and a solid mass. Parenchymal abnormalities are recognized by a change in cortical or medullary echo patterns which allows ultrasound to detect calcium in the parenchyma or collecting systems and identify renal tract gas.

Ultrasound does not give any indication of renal function other than that which can be deduced from the gross anatomical appearances. Although less so with real-time ultrasound, the technique is operator dependent and therefore the quality of the information derived from the examination depends on the experience and expertise of the examining radiologist or technician. The clinician should remember that the 'hard copy' record represents only very few sections from the many observed during the actual examination and that the diagnosis is made mainly during the study and not from the 'hard copy'.

Computed Tomography (CT)

Like ultrasound, CT provides cross-sectional images of detailed anatomy, and is particularly suited to the study of the retroperitoneal structures, including both the renal and perirenal tissues. It should be remembered that it is common practice to perform CT scans both before and after intravenous contrast so that the precautions which pertain to an intravenous urogram are also observed when considering a contrast-enhanced scan. All requests should inform the radiologist of any conditions predisposing to contrast reactions.

As well as cross-sectional anatomical display of renal and perirenal disease, the use of dynamic CT scanning allows an 'angiographic' display of the renal vasculature. This is particularly valuable when renal vein assessment is required during the staging of renal cell carcinoma. Some assessment of renal function, both between and within each kidney, is possible following intravenous contrast. This has particular value in renal trauma when vascular pedicle damage can be identified and renal viability assessed.

Like ultrasound, CT provides information concerning the other abdominal structures at the same examination. CT is not compromised by gas and obesity — an advantage when assessing structures in the retroperitoneum. Like ultrasound, CT can differentiate between solid and cystic structures.

Renal Scintigraphy

Radionuclide techniques are useful in the assessment of various aspects of urinary tract pathology including differential renal function, the evaluation of renal space-occupying lesions and suspected vesicoureteric reflux.

Assessment of renal function (renography)

The compound orthoidohippurate (hippuran) is excreted by the renal tubules predominantly and subsequently is not resorbed to any significant extent. By studying its disappearance from plasma and reappearance in the urine, an estimate of effective renal plasma flow can be obtained. When labelled with ^{131}I or ^{123}I, the handling of hippuran by the kidneys can be studied by either a pair of probes or a gamma camera linked to a computer. After bolus injection of the tracer, the activity/time curve for each kidney is derived — this technique being known as renography. By the use of various mathematical models and computer processing techniques, the renogram curves can be used to derive information on the differential function of each kidney. In the case of gamma camera studies, differential function of areas within a single kidney can also be studied and it is possible to derive curves from the cortex of the kidney only. The shape of the renogram curve may be helpful in determining the nature of the renal abnormality.

The renogram curve obtained from labelled hippuran studies has three distinct phases. The initial phase consists of a sudden increase in activity within the kidneys and represents the arrival of the tracer after bolus intravenous injection. The first phase of the renogram curve thus reflects renal perfusion, predominantly. The rapid slope is then followed by a slower rise in activity. This second phase of the curve is dependent on the renal handling of the hippuran. The integrated counts in the kidney 1–2 minutes after bolus injection of the tracer are widely used as a measure of relative differential function in the kidney. At a variable time, usually 2–5 minutes after injection, the activity/time curve shows a fall. The onset of this third phase corresponds to the appearance of activity in the bladder and the principal factor governing this portion of the curve is excretory function. A commonly derived function for the assessment of excretory function is the renal transit time. This is a mathematically-derived value which gives an estimate of the time from entry of the tracer into the kidney to excretion by the kidney. In patients where there is delayed transit through the collecting systems, the distinction between obstruction and non-obstructive dilatation can often be made by assessing the washout response to intravenous diuretic.

In many centres ^{131}I hippuran has been replaced by ^{99}Tc^m-diethyl-enetriaminepentaacetate (^{99}Tc^m-DTPA) for renal function studies. This compound differs from hippuran in being excreted by glomerular filtration without any significant resorption or metabolism by the kidneys. Attention has to be paid to the stability of the labelling and careful quality control by the radiopharmacy is essential at this point. The radiation dose resulting from ^{99}Tc^m-DTPA is considerably lower than that from ^{131}I-hippuran and this allows very much larger activities to be injected. The larger injected dose produces much higher photon fluxes over the kidney enabling quantitative analysis of the renography type. The overall quantitative results obtained with ^{131}I-hippuran and ^{99}Tc^m-DTPA are similar, in spite of handling differences.

Isotope methods are also employed in the evaluation of total renal function. The radiopharmaceutical usually employed is chromium-51-ethylenediaminetetraacetate (^{51}Cr-EDTA) which is excreted purely by glomerular filtration and is not resorbed. A dose is injected intravenously and blood samples obtained at 2, 3 and 4 hours. By calculating the slope of the disappearance curve of blood activity, an accurate estimate of glomerular filtration rate can be calculated. The main pitfalls in the technique are failure to inject the full dose intravenously and the presence of large extravascular fluid collections, such as gross oedema, which prevent equilibration of the tracer occurring during the period of study. Either of these problems invalidate the results. Effective renal plasma flow can also be calculated accurately by injecting radioiodinated orthoiodohippurate and measuring blood disappearance at intervals over a 2-hour period.

Renal imaging

As noted above, when ^{99}Tc^m-DTPA is employed for renography, images of the kidneys can also be obtained. However, when the isotope study is being performed to obtain information on renal cortical morphology, it is more usual to employ the radiopharmaceutical ^{99}Tc^m-dimercaptosuccinic acid (^{99}Tc^m-DMSA). There is no significant glomerular filtration of this tracer. It is fixed by the kidney with high efficiency — up to 70 per cent of the injected activity being present some 3 hours after injection. As with ^{99}Tc^m-DTPA, some problems may be encountered with the stability of the radiopharmaceutical which should be used immediately after preparation. When using ^{99}Tc^m-DMSA, images are usually obtained 3–4 hours after injection. The detection of cortical lesions, which characteristically appear as cold spots, is greatly enhanced by obtaining images in multiple projections. The relative uptake of DMSA by each of the kidneys has been shown to be a good index of differential renal function, reflecting functional renal mass.

In children, DMSA is useful in demonstrating cortical scarring secondary to pyelonephritis. It is particularly suited to follow-up studies as the whole body radiation dose is less than that from an intravenous urogram.

Vesicoureteric reflux

The demonstration and quantitation of vesicoureteric reflux is possible using nuclear medicine techniques.

In the indirect approach, a radiopharmaceutical employed for renography is administered and, following accumulation of the excreted activity in the bladder, the patient strains and voids. Reflux may be detected visually or by an increase in activity in the kidneys. In the direct method, a catheter is inserted and radioactivity, usually in the form of $^{99}Tc^m$-pertechnetate or $^{99}Tc^m$-colloid, is installed into the bladder. With continuous collection of gamma camera data, saline is then infused into the bladder until the bladder is judged to be full. The catheter is removed and the patient voids as imaging continues. The direct method, involving catheterization, may be more accurate than the indirect method, possibly because a significant number of patients will show reflux only during the bladder-filling period. The indirect method has the advantage of avoiding catheterization with its attendant hazards.

Angiography

There are now two approaches to renal tract angiography. Until recently, such studies involved conventional catheterization of the femoral artery followed by the insertion of a 6/7 F French catheter into the aorta or selectively into the renal artery. More recently the development of digital subtraction angiography (DSA) has greatly widened the scope of angiography especially as it can be performed as an outpatient procedure. DSA can use either intravenous or intra-arterial contrast. The original hope for DSA was that a simple intravenous injection would allow adequate demonstration of the arterial system avoiding arterial catheterization. Although the injection of contrast through a central venous catheter is certainly adequate to show the gross appearance of the renal arterial tree, the degree of opacification is inadequate to demonstrate finer detail. DSA does, however, allow smaller (4 F) intra-arterial catheters to be used along with reduced amounts and lower concentrations of contrast agent. This then provides the option of an outpatient procedure and also allows angiography to be performed in situations where the conventional approach would be contraindicated, e.g. in patients on anticoagulants.

There are only a few conditions which require renal angiography. Firstly, assessing the circulation to a renal mass demonstrated by another imaging technique, angiography can determine the vascularity of the mass and give indications as to whether the circulation is of a malignant or a benign type.

Other arterial anomalies, such as a renal artery stenosis, arteriovenous malformations, and fibromuscular hyperplasia, can also be demonstrated. Secondly, angiography is often performed prior to surgery for renal cell carcinoma and, although CT with dynamic scanning will provide most of the information available from conventional angiography, many surgeons still insist on a preoperative demonstration of the arterial tree.

Interventions

Angioplasty

Angiographic techniques can be extended to provide a therapeutic role. Transluminal angioplasty has made a significant contribution to the management of patients with renal artery stenosis. It is associated with a low morbidity and excellent results are obtained, particularly in patients with fibromuscular hyperplasia. Because of the low morbidity, the previously held strict criteria for diagnosis in renal artery stenosis are becoming less critical. Generally speaking, when a renal artery stenosis is demonstrated on angiography and the renal vein renins and other biochemical parameters support the clinical diagnosis, then renal angioplasty is a simpler and safer approach than renal artery surgery. Some clinicians are happy to proceed directly to angioplasty should a stenosis be demonstrated at angiography.

Embolization

Arterial embolization has been used in the management of patients with renal cell carcinoma and renal arteriovenous malformations. Although the enthusiasm for preoperative embolization of renal malignancy has waned, the radiological option remains and may be of value when a patient is considered unsuitable for surgery. The arterial supply to the tumour can be occluded by using a variety of materials following conventional angiography.

Percutaneous nephrostomy

Using either fluoroscopically or ultrasound-guided techniques, the renal collecting system can be punctured and a catheter placed percutaneously to allow drainage of an obstructed system. This simple technique allows both clinical and biochemical improvement in a patient with obstruction prior to any surgical intervention. The technique is also of particular value when pyonephrosis is present allowing control of the renal infection prior to surgery.

Percutaneous stone removal

This technique is generally performed by both a urological surgeon and a radiologist. The relevant calyx is punctured by the radiologist, generally under ultrasound control. A tract is created percutaneously using successively larger dilators until the surgeon can insert a nephroscope and remove the calculus under direct vision. When successful, this avoids the need for a nephrolithotomy.

Renal Masses

Renal masses are generally detected either on clinical examination or during radiological investigation of the renal tract. Simple renal cysts are very common in the general population, particularly in the elderly, a fact which has been reinforced by their frequent demonstration on ultrasound and CT. Most cysts are incidental findings and only rarely relate to the patient's symptoms — exceptions being superimposed infection or haemorrhage into the cyst. Although an intravenous urogram will demonstrate the majority of renal masses greater than 3 cm, differentiation between a solid and a cystic lesion is not possible. Ultrasound and CT are both very accurate in differentiating cystic and solid lesions provided strict criteria for diagnosis are observed (Fig. 6.1). When the findings are equivocal, ultrasound or CT-guided cyst aspiration is indicated. A number of renal lesions may mimic cysts because of their predominantly fluid characteristics. These include some haematomata, necrotic carcinomas, intrarenal arteriovenous malformations, some abscesses, and urine collections.

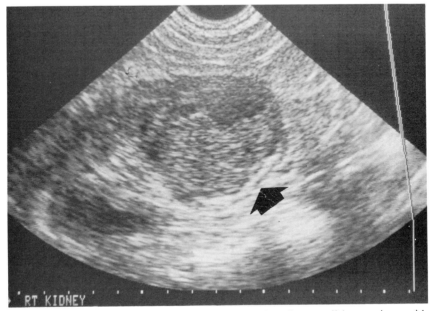

Fig. 6.1 Renal carcinoma. Ultrasound section showing a solid mass (arrow) in lower pole of kidney characteristic of a hypernephroma. (By courtesy of Dr Richard Fowler.)

An anatomical variant which can, on occasion, lead to diagnostic problems is a prominent Column of Bertin which is an extension of normal
cortical parenchyma into the renal hilum. At urography, this may present as
a mass which generally displaces or distorts the adjacent calyces. Ultrasound
may be helpful by showing that the 'mass' and the adjacent cortex are in
continuity and show identical patterns of reflectivity. Occasionally, ultrasound may be inconclusive by demonstrating an area of reduced or even
increased reflectivity. In this situation, scintigraphy will help establish the
diagnosis by demonstrating normal uptake of isotope by the 'mass' which is
comparable to the adjacent cortical activity.

Ultrasound and CT have simplified the diagnostic approach to renal
malignancy. Traditionally, any mass shown by urography required angiography to confirm the diagnosis. Ultrasound and CT, by their ability to
differentiate between cystic and solid masses and to assess surrounding
anatomy are the imaging techniques of choice. Both methods may detect
extrarenal spread to lymph node groups, the liver, the renal vein, and the
inferior vena cava. CT is the more accurate of the two approaches
particularly in demonstrating extrarenal spread to adjacent organs and the
retroperitoneal structures. By using intravenous contrast, dynamic scanning
can demonstrate renal vein patency and any involvement of the inferior vena
cava elegantly (Fig. 6.2). As previously stated, CT can provide most of the
information required prior to surgery, reducing or removing the need for
preoperative angiography. However, many surgeons still demand a preoperative demonstration of the renal vascular tree to aid their surgical
planning.

When renal tumours measure less than 3 cm, they may be missed on
intravenous urography despite the use of nephrotomography. Because of
adjacent gas or patient obesity, ultrasound may also miss these lesions. It has
been shown that CT is particularly good at detecting small renal masses. As
the prognosis of renal malignancy appears related to the size of the renal
tumour at time of diagnosis, it is reasonable to proceed to renal CT in a
patient who presents with haematuria but has negative urography, ultrasound and cystoscopy.

When small or indeterminate masses are demonstrated, ultrasound or CT
guided percutaneous biopsy will enable a histological or cytological diagnosis to be made without resort to surgery. This is of particular value when
doubt exists between an inflammatory or neoplastic aetiology. Some renal
masses which display pathognomonic features, e.g. angiomyolipomata,
because of their high fat content, can be diagnosed with some certainty by
ultrasound and CT.

Calculous Disease

For the purposes of discussion, renal calculous disease can be considered
under the broad headings of calculi and nephrocalcinosis.

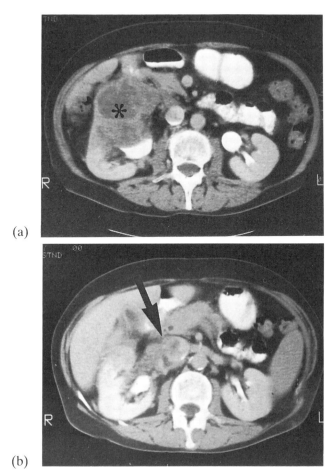

(a)

(b)

Fig. 6.2 Renal carcinoma CT examination following intravenous contrast. (a) Large heterogenous renal carcinoma (asterisk). (b) Tumour involvement of right renal vein with extension into inferior vena cava (arrow).

Urinary Calculi

Some 90 per cent of urinary calculi are radiopaque and most will be demonstrated on a plain abdominal radiograph. Identification of urinary calculi may be difficult in the constipated or gassy patient or when the calculus lies over bone. Generally, two projections in differing phases of respiration are required to localize an intrarenal calculus confidently. If doubt persists, then oblique views may be helpful.

Patients with symptomatic urinary calculi generally present in one of three ways: ureteric colic, haematuria, or a urinary tract infection. Generally, the intravenous urogram is the initial imaging approach to each of these problems. Whether a limited or a full IVU is performed in a patient presenting acutely with renal colic, is contentious. A limited examination generally means a control radiograph and a 15 min full-length film, best taken after micturition. If this series is entirely normal, for management purposes, a clinically significant calculus has been excluded. When obstruction is present, the degree and the level can be shown. Demonstrating the level of obstruction generally requires delayed films, occasionally up to 24 hours. Arguments against this approach include the fact that pathology other than calculus disease can present with renal pain and these may be missed using a limited urographic approach. Moreover, as there is a recognized risk with intravenous radiographic contrast agents, as full a study as possible should be performed in any patient who has been given contrast. Nevertheless, patients with acute renal colic generally have IVUs in the unprepared state and to perform a full examination with nephrotomography etc. is both time-consuming and not always appropriate in a department run by a limited on-call staff. The clinician should therefore remember that when a limited study has been performed, it may not exclude other renal pathology.

Non-opaque calculi, e.g. those composed of uric acid, may show as a filling defect within a calyx or renal pelvis. These must be differentiated from an endothelial tumour and ultrasound or CT may help. Ultrasound may show a strongly reflective focus associated with an acoustic shadow (Fig. 6.3a and 3b). When CT is being used to assess a filling defect in a collecting system, non-enhanced scans must be performed as postcontrast studies may obscure the lesion under investigation. If doubt remains, direct opacification of the collecting system by antegrade or retrograde injection may be required.

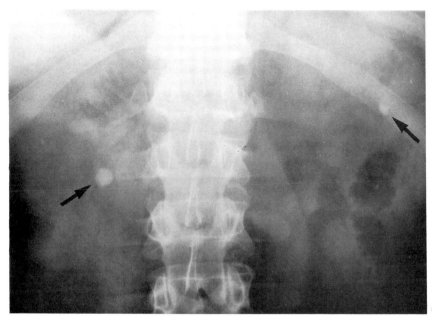

(a)

Nephrocalcinosis

Nephrocalcinosis refers to the deposition of calcium salts within any part of the renal parenchyma. The plain abdominal radiograph will identify the presence of calcification, give some indication as to whether it is located in the cortex or the medulla, and will determine whether it has a focal or generalized distribution. As previously emphasized it is important to obtain an abdominal radiograph prior to any contrast administration. When attempting to determine the aetiology of nephrocalcinosis, identifying the distribution of the calcification may help in reaching a diagnosis. Cortical calcification is most often seen in acute cortical necrosis and occasionally in

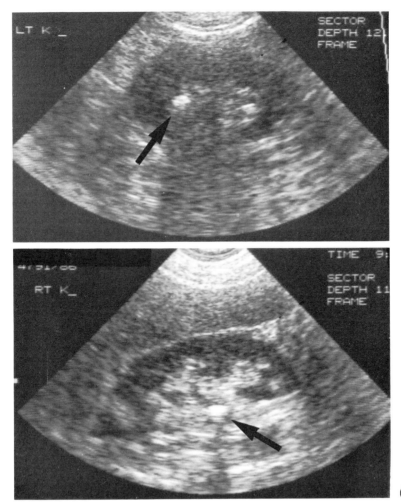

(b)

Fig. 6.3 Urinary calculi. (a) Plain abdominal radiograph of renal area showing bilateral renal calculi (arrows). (b) Ultrasound images of same patient showing bilateral renal calculi (arrows).

nephrotic syndrome. In medullary-sponge kidney, calcification is found in the ectatic tubules located in the medullary pyramids.

When focal renal calcification is shown on the plain radiograph, a number of diagnostic possibilities should be considered. These include infection, e.g. tubercle; post-traumatic damage, e.g. calcified haematoma; neoplastic and vascular calcification. Calcification can occur in both solid tumours and cysts. Calcification which appears very faint or indeterminate on a plain radiograph may be easily identified on CT. Calcified cysts may not show the classic features of a simple cyst on CT and ultrasound, requiring guided puncture with contrast injection to demonstrate a wall irregularity or an associated tumour nodule.

Renal Tract Obstruction

Early detection of urinary tract obstruction is important to avoid the complications of infection and progressive renal failure. The choice of imaging technique depends on the duration of obstruction, i.e. whether acute or chronic, and whether the obstruction is thought to have a physical or functional basis. Although collecting system dilatation is generally due to physical causes, e.g. a calculus, tumour, or aberrant vessels, functional disorders such as renal tract dilatation secondary to vesicoureteric reflux must also be considered.

Acute Obstruction

Ultrasound is excellent for demonstrating dilatation of the renal collecting systems and ureters. However, in acute obstruction, pelvicalyceal dilatation may be less marked in the early stages. Because of this, urography is generally the more useful approach. Urographic diagnosis depends on the demonstration of an increasingly dense nephrogram with delayed calyceal opacification. Delayed films may be required to reveal the level and the cause of obstruction. The time taken for contrast to enter the collecting system depends on the severity of the obstruction. The more severe the grade of obstruction, the greater the delay required to demonstrate a level.

In calculous obstruction, the size of the stone influences subsequent management. Calculi measuring 5 mm or less have a 90 per cent chance of spontaneous passage. This may occur fortuitously during the examination or may occur prior to an IVU being performed. After the spontaneous passage of a calculus, distal ureteric oedema can cause signs of mild obstruction to persist. Occasionally, films taken following the passage of a calculus reveal reduced contrast density of the ipsilateral collecting system and ureter.

Chronic Obstruction

Radiological appearances depend on the severity and the duration of obstruction. As dilatation is a much more prominent feature in chronic renal obstruction, ultrasound is a useful technique for both diagnosis and follow-up, (Fig. 6.4). The degree of hydronephrosis can be determined as can the renal cortical thickness and overall renal size. Ultrasound is particularly helpful when renal function has deteriorated to such an extent that urography is not possible. In this situation, ultrasound will be required to guide any percutaneous nephrostomy attempts.

The urographic appearances of a chronically obstructed system include collecting system dilatation, an increase or decrease in renal size, and a reduction in the extent of functioning renal parenchyma. The classical 'negative pyelogram' will be seen on early films due to dilated calyces containing urine highlighted against the nephrogram of the functioning parenchyma. Unrelieved obstruction leads to progressive renal damage. A previously healthy kidney can tolerate up to one week of complete obstruction without permanent sequelae. Thereafter, irreversible renal damage occurs. As end-stage obstruction approaches, the remaining parenchymal layer will get thinner and thinner as progressive compression of the remaining nephrons occurs by the dilated collecting system. As well as renal collecting system dilatation, ureteric dilatation will develop, associated with progressively less effective ureteric peristalsis.

Obstruction may result from extra-renal pathology. Examples include retroperitoneal fibrosis and pelvic malignancy. In the former, there is, characteristically, bilateral hydronephrosis and hydroureter with medial deviation of the ureters. Periaortic fibrosis can best be assessed on CT when a characteristic appearance will be demonstrated, (Fig. 6.5a and b). When pelvic malignancy compromises one or both ureters, ultrasound may

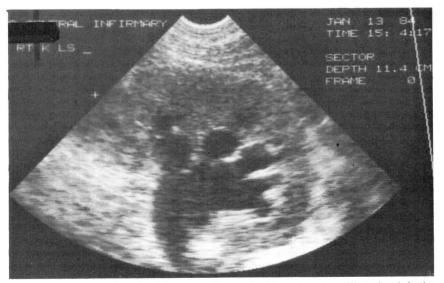

Fig. 6.4 Hydronephrosis. Ultrasound examination showing dilated pelvicaly-ceal system and renal pelvis.

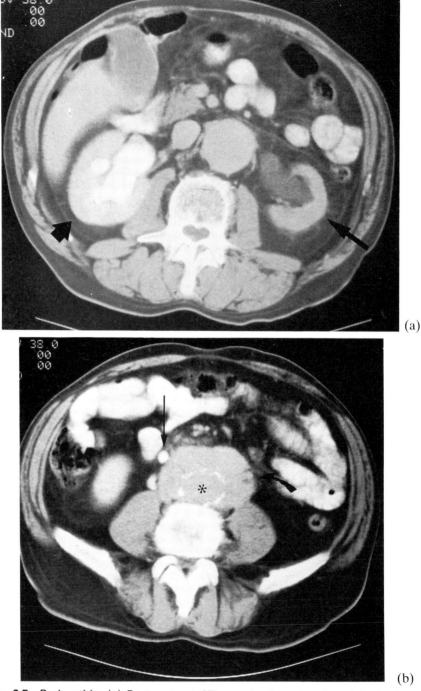

(a)

(b)

Fig. 6.5 Periaortitis. (a) Postcontrast CT examination showing right hydro-nephrosis (short arrow); on the left, the kidney is atrophic and obstructed (long arrow). (b) The calcified aorta (asterisk) is surrounded by a soft-tissue mass characteristic of periaortitis. The right ureter contains contrast (long arrow); the left lies within the fibrosis and is obstructed (curved arrow).

demonstrate the cause, but again CT is more helpful. Intravenous contrast leads to ureteric opacification which allows identification of the level of impairment.

Post-obstructive Atrophy

The gross appearances can be demonstrated by urography, ultrasound or CT. These vary with the severity and duration of obstruction but, generally, the affected kidney is smaller and shows calyceal blunting with varying degrees of parenchymal loss. Assessing the remaining renal function is best accomplished by scintigraphy.

Pelviureteric Junction Obstruction

Obstruction at the pelviureteric junction classically causes intermittent loin pain usually during a period of diuresis. The problem is often intermittent and the IVU may appear normal until conditions which are physiologically similar to those which produce the pain are introduced. Symptoms may be provoked by intravenous diuretics administered during the course of an intravenous urogram. Postdiuretic films will often show a distended ex-trarenal pelvis with a normal calibre ureter on the symptomatic side, while the contralateral system shows contrast washout. When the intravenous urogram is inconclusive, further imaging is required. Dynamic technetium-99m-DTPA scanning combined with an intravenous diuretic allows a quantitative assessment of the rate of wash out of the isotope. The reliability of diuresis renography will be impaired by diminishing renal function because of decreased renal response to diuretics. However, when a reno-gram curve shows no evidence of obstruction, this finding is reliable even in the presence of severe renal compromise.

It should be remembered that hydronephrosis or hydroureter may occur in the absence of mechanical obstruction. Congenital megacalyces and primary megaureter are conditions which may be demonstrated incidentally during an IVU. The former results in dilatation of an increased number of calyces but, in contrast to postobstructive atrophy, the renal cortical thickness is normal as are the renal pelvis and ureter. Primary megaureter or achalasia of the ureter results in ureteric dilatation occasionally associated with pelvicalyceal dilatation. This occurs in the absence of an anatomical obstruction. Various defects in the ureteric wall have been described which impair ureteric peristalsis.

Renal Tract Infection

The use of imaging in the investigation of renal tract infection is to some extent influenced by the patient's age. The priority in infants and children who present with a urinary tract infection is in the exclusion of a congenital abnormality and in the demonstration of vesicoureteric reflux. Ultrasound is gradually displacing the intravenous urogram as the initial investigation of the renal tract in infants and children and is usually combined with a micturating cystogram when the problem is urinary tract infection. Renal duplication, ectopic ureteric insertions and bladder diverticula can all be demonstrated. It should be remembered that urinary tract infection in childhood may be associated with non-specific symptoms, such as failure to thrive, rather than symptoms and signs directly referable to the renal tract and that many children with significant bacteriuria are symptom free.

Vesicoureteric Reflux

Vesicoureteric reflux, by allowing infection from the bladder to reach the kidney, is responsible for significant renal damage in childhood. It has been estimated that approximately one third of children with a history of urinary tract infection will be shown to have reflux. It is the exception for imaging to demonstrate the actual cause of reflux in a particular child. Most are presumed to reflux because of some congenital malfunction at the ureterovesical junction. Imaging is important to grade the severity of the reflux and to identify any existing renal damage. Mild refluxers will have an 80 per cent chance of spontaneous cure while only 20 per cent of children with severe reflux will cease refluxing. These cases require consideration for a surgical correction. The value of scintigraphy in the diagnosis of reflux has already been discussed.

Acute Pyelonephritis

In the majority of cases of acute pyelonephritis, both the intravenous urogram and ultrasound will be normal. If severe infection is present, both techniques may show renal swelling, with impaired concentration and delayed excretion on intravenous urography, and changes in the parenchymal echo pattern on ultrasound. Infection in diabetic patients may produce more significant changes such as gas in the renal tissue, perirenal structures and renal vessels. Occasionally, acute pyelonephritis may cause a local swelling within the renal parenchyma which can be shown by ultrasound or CT.

Chronic Pyelonephritis

Most cases of chronic pyelonephritis are due to the long-term consequences of repeated reflux with infection in childhood and it is the commonest cause of death from renal failure and hypertension in the UK. Imaging will demonstrate shrunken, scarred kidneys with calyceal deformity and loss of cortical substance. Chronic pyelonephritis may affect one or both kidneys.

Renal Abscess

Focal renal abscesses or perirenal abscesses are more common in diabetics or patients who are immunocompromised. Intravenous urography will generally show a mass which cannot be differentiated from other causes of mass lesions in the kidney. On ultrasound, there is a spectrum of appearances varying from a cystic pattern due to a primarily liquid component, to a more solid mass in conditions such as xanthogranulomatous pyelonephritis. If gas is present, highly reflective echoes with acoustic shadowing will be shown.

Renal abscesses on CT have attenuation values greater than a simple cyst but less than that of a solid tumour. Gas or a gas/fluid level can often be identified. Not infrequently, a semi-solid abscess may mimic a tumour and aspiration biopsy will be required to aid differentiation. CT may show perinephric extension, which is a sign in favour of an abscess, rather than focal pyelonephritis. If suspicion of an abscess persists after CT studies and aspiration, indium-111-labelled white cell or gallium-67 imaging will demonstrate focal uptake if an abscess is present. This can be particularly helpful in polycystic renal disease.

Tuberculosis

Renal tuberculosis is a disease of adulthood. Approximately 10 per cent of cases are asymptomatic and the diagnosis may be suggested during imaging for some other pathology. As one third of patients will have tuberculosis elsewhere and over 50 per cent will have evidence of old or current pulmonary tuberculosis, a chest radiograph is mandatory when renal tubercle is suspected. The radiological changes are best shown on intravenous urography and these include calcification of the renal parenchyma, ureters and bladder, and parenchymal changes focused at the papillary level. Early disease may show as minimal erosion to the tip of one calyx, but with progression, the papillary tips become increasingly motheaten and cavities may develop.

When multiple radiological signs occur, e.g. cavitation, multiple ureteric strictures and calcification, the diagnosis is relatively straightforward. It should be remembered, however, that a normal intravenous urogram does not exclude renal tuberculosis and as 90 per cent of early lesions are focused at the corticomedullary junction, the intravenous urogram may be normal in the early stages of the disease. The CT appearances of renal tuberculosis are non-specific but CT is the best technique available to demonstrate an associated psoas abscess or any spinal involvement.

The end-stage of renal tuberculosis is the so-called 'autonephrectomy'. A non-functioning calcified mass will be shown on the abdominal radiograph. The appearances are characteristic but may be mimicked by a large irregular staghorn calculus, particularly when associated with xanthogranulomatous pyelonephritis.

Renal Vascular Disorders

Up to 5 per cent of hypertensive patients will have a potentially correctible renovascular disorder. A modified intravenous urogram formed the traditional approach to the diagnosis of renal artery stenosis. This involved taking a sequence of early films following contrast and identifying the classic features of reduced renal size, delayed pyelogram, and increased contrast concentration on the side of the arterial stenosis. This method of diagnosis has been shown to be unreliable particularly in younger patients who may have segmental arterial lesions. As better techniques are now available the IVU should no longer be requested to diagnose renovascular pathology.

Characteristic scintigraphic features have been described in renal artery stenosis. A dynamic renal study using $^{99}Tc^m$-DTPA in a patient with renal artery stenosis will demonstrate a reduced uptake of radiopharmaceutical, a delay in the time to reach peak activity, and a lengthening of the excretory phase of the curve.

The definitive imaging approach to a possible renal artery stenosis is angiography. Conventional angiography using 6/7 F catheters with mainstream aortic and selective arterial runs will demonstrate a stenosis and the characteristic poststenotic dilatation. Digital subtraction angiography (DSA), with contrast delivered via a central venous line, allows adequate opacification of the aorta and the main renal arteries. For more detailed examination, an intra-arterial study is required.

Traditional clinical practice has dictated that any renal artery stenosis demonstrated on angiography must be accompanied by characteristic biochemical and hormonal changes before surgery is contemplated. Renal vein renin sampling and divided renal function are among the well established investigations in the standard hypertension work up. Recently, the technique of percutaneous angioplasty has been used to treat renal artery stenosis. Surgery is avoided and results are excellent — particularly with fibromuscular hyperplasia although less so with arteriosclerotic stenosis. As the mortality and morbidity associated with angioplasty is very low, the technique can be attempted using less strict criteria than those required for conventional surgery. Ideally, if a suitable stenosis is shown at angiography then dilatation should be performed during the same examination.

Renal Failure

A detailed discussion of the causes of renal failure is beyond the scope of this text. Renal damage occurring at cortical or medullary level or resulting from renal tract obstruction or vascular compromise will give appropriate changes on intravenous urography and other imaging techniques. Some of these are specific for the underlying pathology but often the changes are non-specific and histology is required for a definitive diagnosis.

Acute Renal Failure

For management purposes, acute renal failure is traditionally considered in three broad groupings: prerenal, renal, and postrenal failure. Prerenal failure is due to insufficient renal perfusion, whereas renal is generally due to intrinsic parenchymal disease; postrenal failure is associated with urinary tract obstruction. The most important information which any investigation is required to give the clinician is:

1. The size and position of each kidney
2. Whether either system is obstructed.

Plain abdominal radiographs should be requested to identify any renal tract calcification. As has already been discussed, the type and distribution of this calcification may suggest the underlying cause of the renal failure. Intravenous urography is to be avoided in prerenal failure patients and also in diabetics and young infants. Whenever intravenous urography is used in a patient with renal compromise, dehydration is contraindicated.

Along with a plain abdominal radiograph, ultrasound should now form the initial approach to the problem of acute renal failure. The renal size, cortical thickness with any focal or generalized loss, and the collecting system status will be shown. Intrinsic renal disease may produce parenchymal echo changes. It appears that there is a direct relationship between the severity of interstitial changes (on biopsy) and the intensity of the parenchymal echo pattern of the renal cortex as detected on ultrasound. However, there are reports of failure to detect any increase in echo pattern in up to 40 per cent of patients with advanced renal parenchymal disease. Histology is required for a definitive diagnosis but sequential ultrasound studies can monitor disease progression and treatment response.

When obstruction is demonstrated in a patient presenting with deteriorating renal function, percutaneous drainage of one or both kidneys should be considered to help preserve any remaining renal function while the patient is prepared for definitive management.

Chronic Renal Failure

As previously stated, high doses of modern contrast agents combined with nephrotomography will produce a urogram in most patients with renal failure — but dehydration is a contraindication. Ultrasound is the preferred option for long-term follow-up allowing overall renal size, residual cortical thickness, and the status of the collecting systems to be monitored. Scintigraphy will demonstrate the extent of remaining functional renal tissue.

Secondary hyperparathyroidism produces characteristic changes in the bony skeleton and soft tissues and these can be assessed using conventional radiography. It is now well established that patients on long-term dialysis have a high incidence of acquired renal cystic disease. Of more importance, they are also at risk of developing solid adenomatous and adenocarcinomatous lesions. Both ultrasound and CT are helpful in detecting these changes. CT has a relatively limited role in the assessment of patients with chronic renal failure, but should ultrasound suggest tumour formation, then CT may give further information concerning its extent and spread.

Trauma to the Renal Tract

The principle role of imaging in renal tract trauma is to help the clinician decide which patients can be managed conservatively and which require surgical intervention. As always, plain abdominal radiographs should be obtained. Rib or spinal fractures will be demonstrated and any effacement of psoas or renal soft-tissue outlines may point to a traumatized kidney.

Intravenous urography has for many years remained the standard approach to the patient with renal trauma. The relative function of both kidneys and any extravasation of contrast will be demonstrated. Conventional tomography performed during the intravenous urogram may show segmental defects of the renal nephrogram.

The technique which can provide the most information during a single examination is CT. Not only will the kidneys themselves be imaged, but the perinephric and pararenal spaces, adjacent bony structures and soft tissues. Also by using intravenous contrast enhancement, the vascular supply to each kidney will be shown and any global or segmental ischaemia identified (Fig. 6.6). At the same time, some assessment of renal function can be made and subsequent ureteric and bladder opacification may reveal injuries to these structures.

The commonest type of kidney injury shown by CT is a subcapsular haematoma. Whenever blood is shown by CT, its appearance depends on the length of time from injury. Recent haemorrhage will be dense compared with adjacent soft tissue whereas older blood will be of lower density. Subcapsular haematomas are usually managed conservatively. Larger collections tend to be perirenal in location but are generally confined by Gerota's fascia. CT will also demonstrate renal tears and intrarenal haemorrhage. These are often seen best after contrast and are frequently associated with subcapsular and perirenal collections. CT is very useful in the demonstration of renal fracture, a 'shattered kidney' or a complete renal laceration. Extravasation of contrast will be shown and, in the case of a renal fracture, separation of the two poles will be demonstrated. By displaying the extent of trauma as accurately as possible, unnecessary surgical intervention may be prevented.

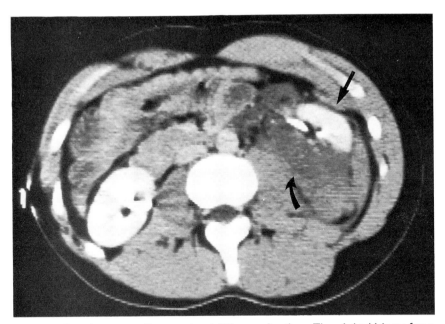

Fig. 6.6 Renal trauma. Postcontrast CT examination. The right kidney functions normally. On the left, the anterior half of the kidney opacifies normally (arrow); the posterior half fails to opacify (curved arrow) indicating ischaemic damage.

Ultrasound will also demonstrate perirenal collections and intrarenal haematoma formation. Renal laceration may be diagnosed but less easily than by CT. Ultrasound cannot provide any functional information about the damaged kidney but is useful in the follow-up of the traumatized patient to assess any improvement or deterioration in a perirenal or intrarenal collection. Scintigraphy may also be of value in the follow-up of a traumatized kidney. It will assess residual function best and identify any post-traumatic obstruction. Scintigraphy is not often requested in the acute phase although it may be useful when assessing damage to the renal vascular pedicle. The only value of angiography in renal trauma is to demonstrate the vascular anatomy prior to any attempts at reconstructive surgery.

The Bladder, Urethra and Prostate

The Bladder

Both intravenous urography and ultrasound are appropriate initial investigations of bladder pathology. Both will assess bladder size and wall thickness, and bladder neoplasms may be detected. Bladder calcification can be demonstrated on a plain abdominal radiograph and can occur in a variety of conditions, including malignancy, tuberculosis and schistosomiasis.

Cystitis

Imaging patients with this common complaint has a fairly poor diagnostic pay-off. Intravenous urography is often requested and this will usually reveal no abnormality. In severe cystitis, the bladder will show a reduction in capacity and thickened folds and mucosal oedema may be demonstrated. Emphysematous cystitis, caused by gas forming organisms — typically in diabetic patients — may be identified on a plain X-ray by gas in the bladder wall or bladder lumen.

Bladder tumours

Typically, the patient presents with haematuria and an IVU is requested. Filling defects in the bladder may be shown and any compromise of the ipsilateral ureter demonstrated. Bony metastases around the pelvis may be identified on the control radiograph and often these are sclerotic or osteoblastic in type. As has already been stated, a normal bladder appearance on intravenous urography does not exclude bladder malignancy, and cystoscopy is always required. Ultrasound will give information similar to the intravenous urogram, but rather than show a filling defect, ultrasound will demonstrate localized thickening of the bladder wall or a polypoid mass projecting into the bladder lumen (Fig. 6.7).

CT is a valuable means of staging bladder neoplasms and therefore enabling the most appropriate method of treatment to be pursued. CT will assess any local extension of tumour through the bladder wall into the adjacent soft tissues of the pelvis and will identify more distant spread to the draining lymph node chains and liver (Fig. 6.8).

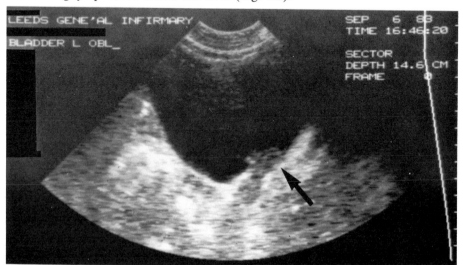

Fig. 6.7 Bladder tumour. Oblique ultrasound section showing soft-tissue mass projecting into bladder lumen and extending through the bladder wall (arrow). (By courtesy of Dr Richard Fowler.)

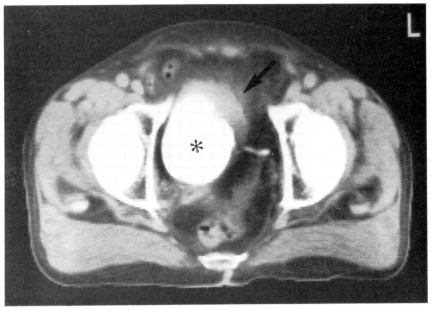

Fig. 6.8 CT bladder neoplasm. Postcontrast examination showing contrast within the bladder lumen (asterisk). Anterolateral bladder tumour (arrow) extending through the bladder wall into the perivesicular tissues.

Bladder trauma

Bladder contusion due to an incomplete or non-perforating tear of the bladder mucosa following blunt trauma, is the commonest form of bladder injury. Generally, both the intravenous urogram and micturating cystogram are normal.

Rupture of the urinary bladder is uncommon and occurs in only 10 per cent of patients with fractures to the bony pelvis. However, approximately 80 per cent of patients with bladder rupture will have a pelvic fracture. Following rupture, extravasation of urine can be extra- or intraperitoneal. Whichever type occurs generally depends on the state of fullness of the bladder at the time of injury. Relatively empty bladders tend to rupture extraperitoneally; full bladders, intraperitoneally.

Intravenous urography is the usual initial approach to bladder trauma. If contrast extravasation is shown, the diagnosis of a bladder tear is confirmed. However, it should be stressed that as many as 85 per cent of patients with a bladder tear will have a normal intravenous urogram. Micturating cystography is the best imaging technique for showing bladder rupture and should always be performed when the IVU is normal against a background of strong clinical suspicion of bladder rupture (Fig. 6.9).

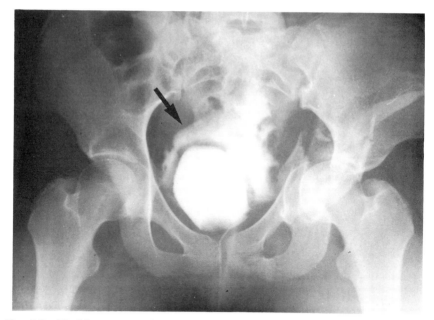

Fig. 6.9 Bladder trauma. Micturating cystogram in young female following severe pelvic trauma. Extraperitoneal extravasation of contrast (arrow) indicting bladder tear.

Bladder problems in the gynaecological patient

Stress incontinence and urge incontinence form a significant proportion of the gynaecological work load. Intravenous urography and micturating cystography have been used, traditionally, to give further information before reconstructive surgery or conservative management. Micturating cystography allows both assessment of the position of the bladder base with regard to the pelvic floor and the angle formed by the urethra at the bladder base. By using a lateral projection, the bladder base and cystourethral angle can be observed fluoroscopically during straining and on micturition.

The Urethra

Urethral strictures

Urethral strictures can be secondary to trauma, inflammation or malignancy. Imaging involves a combination of ascending and descending urethrography to define the extent and character of the stricture.

Urethral valves

This congenital condition occurs in males and requires early diagnosis to prevent progressive damage to the upper renal tract. Urethral valves are generally due to an exaggeration of the folds that run obliquely down to each side of the urethra from the veramontanum. Urethral valves are best shown on micturating cystourography when a distended posterior urethra will be demonstrated during voiding. It should be stressed that valves do not obstruct retrograde flow of contrast and therefore will not be demonstrated by retrograde urethrography.

The Prostate

Prostatism is a common clinical problem and is responsible for a significant percentage of the intravenous urography workload to an X-ray department. Most urologists prefer their patients to have an IVU prior to prostatectomy to assess the degree of bladder outflow obstruction. The extent of pelvicalyceal and ureteric dilatation and the degree of bladder wall thickening and trabeculation will be demonstrated.

As an alternative, ultrasound can provide similar information about the kidneys and ureters and can also measure bladder size before and after micturition, thereby avoiding the use of radiographic contrast media in an elderly population.

Conventional ultrasound cannot usually differentiate between benign and malignant enlargement, and an IVU will give no information as to the underlying cause for the prostatic enlargement. Although rectal examination by the clinician will be helpful, its limitations are well known, particularly when prostatic malignancy is confined within the prostatic capsule. Ultrasound scanning using a rectal transducer can demonstrate some prostatic carcinomas which cannot be detected by digital examination. It can also differentiate between malignancy and benign pathologies which may mimic malignancy on rectal examination. An assessment of tumour spread can be made by differentiating between malignancy which has spread through the capsule and that which remains confined by it. Finally, ultrasound can assess prostatic size before and after treatment which is of value in monitoring treatment response.

At the time of writing, rectal ultrasound is not as widely available as conventional real-time ultrasound. However, in comparative studies, the technique appears superior to transabdominal ultrasound and CT scanning. CT is limited by the fact that it can only demonstrate an abnormality of contour or a change in the overall size of the prostate gland. Any malignancy which does not extend through the capsule and, therefore, does not result in an alteration of contour, cannot be demonstrated by CT as tumour tissue is of similar attenuation to normal prostate. When simple enlargement of the gland occurs, CT cannot differentiate between benign prostatic hypertrophy and malignancy. CT will detect prostatic calcification more readily than the conventional radiograph.

The major role of CT in the management of prostatic carcinoma is in its staging and assessment of treatment response rather than in making the diagnosis. This applies to extra prostatic spread, particularly, where tumour extension into adjacent structures, especially the seminal vesicles, can be shown. Tumour extension into the bladder and rectum can also be identified. Lymphadenopathy, both pelvic and para-aortic, and bony metastases may be demonstrated at the same examination. However, bone scintigraphy is the definitive means of assessing skeletal involvement and the response of bone metastases to treatment. Prostatic metastases on conventional skeletal radiographs tend to be osteoblastic or sclerotic.

Testicular Imaging

Although palpation of the testes forms part of every examination in the adult male, clinical assessment of testicular masses alone is insufficient to differentiate benign and malignant disease. High frequency real-time ultrasound now forms the basis of testicular imaging. Transducers of 7·5 MHz (as compared with 3·5 MHz for general scanning of the abdomen) results in high resolution images which are simple and quick to obtain. A normal testis shows a homogeneous texture allowing focal and diffuse alterations to be identified readily. A variety of pathologies can be diagnosed using real-time ultrasound.

Inflammation

Acute and chronic epididymitis and orchitis (both focal and diffuse) result in swelling and reduced reflectivity of the epididymis or of the testis itself. Focal abscess formation can be demonstrated and serial studies will allow assessment of treatment response.

Hydroceles and varicoceles

The testis is normally only surrounded by a few millilitres of fluid. Hydroceles are therefore readily demonstrated, as are dilated testicular veins (Fig. 6.10).

Testicular trauma/torsion of testis

Trauma ranging in severity from intratesticular haemorrhage to frank testicular fracture can be identified by ultrasound. Testicular torsion remains a difficult diagnosis to confirm using imaging techniques. The ultrasound features may be indistinguishable from orchitis or epididymo-orchitis and on occasion may appear normal. A change in testicular alignment or position may suggest a torsion, but when a clinical diagnosis of torsion is likely, an equivocal or normal ultrasound should not delay surgical intervention. Some centres have found that Doppler ultrasound (to assess blood flow in the testicular vessels) or scintigraphy (to assess testicular perfusion) are helpful in the diagnosis of torsion.

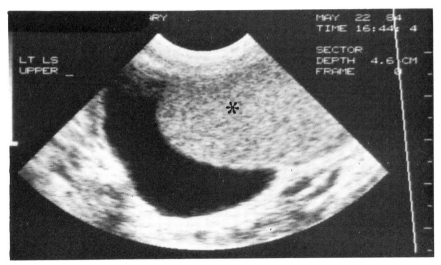

Fig. 6.10 Testicular ultrasound: hydrocele. The normal testis (asterisk) is shown surrounded by a moderate sized hydrocele. (By courtesy of Dr Richard Fowler.)

Testicular tumours

Characteristically, these appear as hypoechoic masses within the testicular substance. Ultrasound can demonstrate tumours which are clinically impalpable. This role of reducing the false-negative clinical diagnosis for tumours is one of the most important aspects of testicular ultrasound. Some authorities attempt to differentiate the various types of testicular tumour — e.g. embryonal cell tumours may contain echogenic foci reflecting calcification or fibrosis. The diagnosis of a solid mass within the testis remains the foremost objective, leaving histological differentiation to the pathologist (Fig. 6.11).

CT has an important role to play in the staging and follow-up of testicular malignancy. As testicular tumours metastasize by both lymphatic and haematogenous routes, CT is excellent at detecting para-aortic lymphadenopathy and lung metastases (Fig. 6.12). The advantages of CT include the evaluation of structures other than lymph node groups at the same examination and allowing metastatic disease to the liver and kidney to be identified. It provides a baseline pretreatment assessment which is readily reproducible to allow follow-up and monitor treatment response. The prognosis and treatment of testicular tumours largely depends on the stage of the disease when first diagnosed.

Prior to the development of CT, lymphangiography was used to assess lymphatic spread and conventional tomography to assess the presence of lung metastases. CT is more accurate in detecting lung metastases and although lymphangiography is accurate in assessing certain lymph node groups, it is poor in demonstrating retrocrural and renal hilar lymph node spread and will not reveal mesenteric or mediastinal lymphadenopathy. The

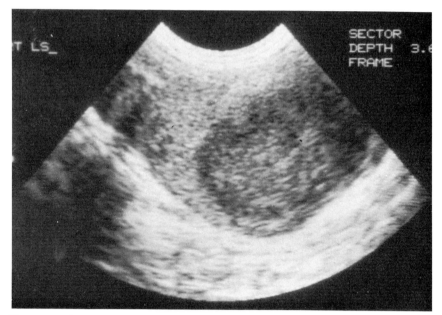

Fig. 6.11 Testicular teratoma. Longitudinal ultrasound section showing a mass of mixed echogenicity within the testicular substance. (By courtesy of Dr Richard Fowler.)

examination itself is technically difficult and rather uncomfortable for the patient.

Undescended testis

The absence of a palpable testis may be due to agenesis, acquired atrophy or incomplete descent. The importance of detecting an undescended testis is to prevent loss of fertility and the development of malignant change. The conventional approach to the problem of a non-palpable testis is firstly complete surgical exploration of the inguinal canal, and secondly along the course of the gonadal veins. Both these procedures are lengthy and may fail to detect a small testis. Several imaging techniques have been employed to detect undescended testis including testicular venography, spermatic artery angiography, ultrasound and CT. CT can demonstrate a soft-tissue mass along the course of testicular descent and relies on the asymmetry of the soft-tissue anatomy for the detection of ectopic testicular tissue. The use of intravenous contrast will help in differentiating a vascular from a non-vascular structure. CT provides excellent cross-sectional anatomical detail but at the cost of the risks of ionizing radiation to young patients. If the undescended testis lies within the inguinal canal, it will often be recognized on ultrasonography. Ultrasound is of little value when the testis is located in the lower abdomen. However, as the majority of undescended testes are found within the inguinal canal, ultrasound should be the initial approach.

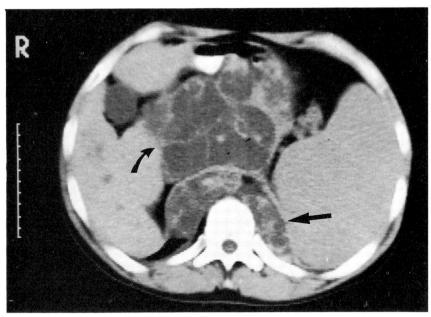

Fig. 6.12   Teratoma lymph node metastases. CT of upper abdomen showing extensive lymphadenopathy involving both the retrocrural (arrow) and mesenteric (curved arrow) groups.

Further Clinical Problems Related to the Renal Tract

The Patient with Haematuria

Haematuria can be the presenting complaint for a variety of renal tract pathologies including infection, calculous disease, malignancy and trauma. History, clinical examination, and basic laboratory investigations may point to the likely aetiology but imaging is generally required at some stage in the diagnostic work up. Whether the haematuria is maximal at the beginning or at the end of micturition may help localize the level of the problem within the renal tract but, whatever the clinical findings, the patient is generally referred to the X-ray department for an intravenous urogram and to the urologists for cystoscopy.

The value of ultrasound has been repeatedly stressed in the preceding sections and is an appropriate first-choice imaging investigation for the patient with haematuria. However, one limitation of ultrasound is its poor detection of small endothelial lesions, e.g. a tumour in the renal pelvis, which may be very difficult to demonstrate. Intravenous urography generally demonstrates urothelial tumours better but should equivocal findings result, a retrograde pyelogram may be required. CT can also be used to assess lesions within the collecting systems and help differentiate between calculus and tumour.

A normal intravenous urogram in a patient presenting with haematuria must be followed by cystoscopy as bladder lesions may not be shown. Should cystoscopy be normal with no evidence of urinary tract infection, then it would be reasonable to consider proceding to renal CT.

When Imaging Demonstrates a Solitary Kidney

It is important to determine whether the patient has a single kidney or some pathology which prevents the demonstration of the contralateral kidney. Intravenous urography generally demonstrates an ectopic kidney whether it is located within the pelvis or adjacent to the contralateral renal bed. Ultrasound may have some difficulty demonstrating a pelvic kidney particularly if the kidney is small and bowel loops obscure access. Scintigraphy is particularly useful for detecting ectopic functioning renal tissue.

The non-visualized kidney on an IVU can have a number of causes. Agenesis is not uncommon and may be suggested by hypertrophy of the contralateral kidney. A calcified mass in the renal bed with no associated function suggests a tuberculous autonephrectomy or, occasionally, xanthogranulomatous pyelonephritis. Both arterial and venous occlusion may result in non-visualization on intravenous urography. Ultrasound may help differentiation by showing a normal or slightly small kidney when acute arterial occlusion is the problem. With venous occlusion, the kidney is globally enlarged, shows a reduced echo pattern, and demonstrates some compression of a non-dilated collecting system. Ultrasound may also identify thrombus in the renal vein or inferior vena cava. CT can provide similar information.

Both ultrasound and CT will demonstrate a cystic dysplastic kidney which appears, typically, as a multiloculated mass in the renal bed. Using ultrasound and CT, there is, generally, little difficulty in differentiating a dysplastic kidney from a chronically obstructed kidney.

Imaging of the Transplant Kidney

Improved surgical techniques, the availability of effective antirejection agents and more sensitive methods of tissue typing have contributed to an optimistic prognosis for those who undergo renal transplantation. Imaging provides the means of detecting postoperative complications which allows appropriate clinical intervention. Some complications are related to the surgery itself: e.g. extravasation of urine causing urinoma formation, obstruction to the transplanted ureter or damage to the renal vascular pedicle. Other sequelae are related to ischaemia or tissue mismatch such as acute tubular necrosis (ATN) and rejection.

A combination of ultrasound and scintigraphy is generally used to assess and monitor the transplant kidney during the postoperative period. Ultrasound will provide a baseline measurement of renal size, will detect any dilatation of the collecting systems and will identify any perirenal collection. Scintigraphy is used to assess renal vascular perfusion and identify any obstruction. Most transplanted kidneys will, initially, show changes on scintigraphy consistent with ATN, i.e. a good nephrographic phase but no evidence of excretion. Follow-up studies will help monitor recovery and response to supportive therapy. Recently, Doppler ultrasound has been used to assess blood flow in both the main renal arteries and the major intrarenal vessels.

Rejection and acute tubular necrosis may give similar appearances on intravenous urography. Some authorities feel that intravenous urography is contraindicated if rejection is considered likely in order to prevent the risk of further tissue damage. Ultrasound should now provide the initial assessment. Signs of rejection include a change in the renal size and alterations of the internal parenchymal echo pattern. In particular, changes in the degree of corticomedullary differentiation occur. The definitive diagnosis of transplant rejection is histological and, again, ultrasound can be used to accurately guide the biopsy needle.

When vascular compromise is suggested by scintigraphy or Doppler ultrasound, angiography may be considered. Renal vein occlusion may be diagnosed and if an acquired renal artery stenosis is demonstrated, angioplasty can be offered. Obstruction to the transplanted ureter can be demonstrated best by ultrasound and/or scintigraphy, and its progression or resolution monitored by sequential scans. It is now well recognized that the use of certain antirejection agents, e.g. cyclosporin, can lead to intrarenal fibrosis. This fibrosis may prevent collecting system dilatation. Should obstruction be considered likely and ultrasound is unhelpful, an antegrade percutaneous puncture of the transplanted kidney with subsequent opacification of the collecting system and ureter is required (Fig. 6.13). The technique is simple because of the superficial position of the transplanted kidney in the pelvis and obstruction can be confirmed or excluded with confidence.

Haematomata, urinomas, and lymphoceles can all, if large enough, compress the transplanted kidney or ureter and cause a subsequent deterioration in renal function. Ultrasound can both make the diagnosis provide a simple means of monitoring progress, and allow accurate puncture of any collection for drainage purposes.

Imaging in Pregnancy

The risks of ionizing radiation to the fetus are well established. Use of conventional X-rays on a pregnant woman is to be avoided whenever possible, especially during the first trimester. Appropriate precautions should be taken to avoid irradiating the uterus during pregnancy or even when menstruation is delayed.

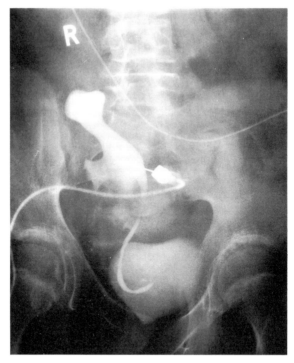

Fig. 6.13 Transplant kidney. Antegrade percutaneous puncture of a transplant kidney showing normal flow of contrast down the ureter and into the bladder.

Conventional radiography in late pregnancy is used, generally, for one of two reasons. The first is to assess fetal maturity by identifying various epiphyses, e.g. around the knee. However, this technique is performed less and less following the availability of ultrasound which allows accurate assessment of fetal maturity without harm to the fetus or mother. Secondly, conventional radiography is used to assess pelvic dimensions. If the obstetrician is concerned that pelvic disproportion may result in a difficult passage for the baby through the birth canal (dystocia), pelvic radiographs in a variety of projections can give an assessment of pelvic dimensions and assist in an informed decision concerning caesarian section.

Ultrasound has completely transformed obstetric practice. It can accurately date pregnancy at an early stage and can assess fetal development and maturity throughout the pregnancy. A variety of measurements are used, the choice of which depends on the trimester under study: fetal femur length, biparietal diameter and abdominal circumference are among the assessments available. Ultrasound can also be used to assess the complications of early pregnancy allowing the diagnosis of fetal death, ectopic pregnancy, and the various causes of vaginal bleeding.

Fetal anatomy, both normal and abnormal, can be identified using ultrasound allowing assessment of the fetus, placenta, and umbilical cord. Congenital abnormalities can be identified *in utero*, e.g. spina bifida, hydrocephalus, hydronephrosis, polycystic kidneys and exomphalos. Fetal development can be monitored by sequential studies and a falling off or an acceleration of fetal growth can be shown. Abnormalities of the placenta, e.g. placenta praevia, can be identified without resort to more invasive techniques.

Amniocentesis is a commonly performed procedure with a low morbidity to both the mother and fetus. This morbidity may be reduced further by using ultrasound to accurately guide the needle to an appropriate position for a diagnostic aspiration. More recently, ultrasound has been used to guide the insertion of intrauterine shunts for fetuses with obstructive uropathy in order to relieve the obstruction until definitive surgery can be performed after birth.

Imaging in Gynaecology

Prior to the advent of ultrasound and CT, investigation of pelvic pathology was largely limited to plain radiographs or hysterosalpingography. Pelvic radiographs were of some value in identifying calcified pelvic masses but little else. By performing an intravenous urogram, further information concerning ureteric compromise was obtained and obstructive uropathy, ureteric deviation or bladder displacement could be shown. Hysterosalping-ography, which involves the introduction of contrast into the uterus and Fallopian tubes, remains part of the investigation of infertility. Before ultrasound, this technique was more widely used in the assessment of a variety of uterine abnormalities.

As with obstetrics, ultrasound has transformed the approach to gynaeco-logical imaging. The patient is examined with a full bladder in order that small bowel loops will be displaced out of the pelvis and allow the uterus, cervix, ovaries and adnexal regions to be examined. During any pelvic ultrasound study, the kidneys will be assessed to detect abnormalities such as obstructive uropathy.

A common use of ultrasound is in the assessment of a palpable pelvic mass. Ultrasound can determine if the mass is cystic or solid, whether it is single or one of several, and the organ of origin may be demonstrated. Uterine enlargement is commonly due to fibroid formation. Although ultrasound will demonstrate solid masses within the uterus and may suggest that fibroids is the most likely diagnosis, it cannot differentiate categorically between simple fibroids and malignant disease of the uterus. This is especially true when cystic degeneration of a fibroid occurs.

The ovaries can usually be identified and measured by ultrasound. Cysts and solid masses will be shown and follow-up on sequential scans may show spontaneous regression of simple cysts. Many solid ovarian masses are malignant and most simple cysts are benign. However, when predominantly cystic structures contain solid components, malignancy may be present and further investigation is required. Dermoid cysts may have a characteristic appearance because of their fat or calcium content. Not infrequently, ovarian carcinoma and uterine carcinoma cannot be differentiated positively.

Recently, ultrasound studies of follicular maturation have lead to interesting developments in the management of female infertility. Follicular size increases as ovulation approaches: by observing ovarian follicle maturation on serial ultrasound scans, the timing of artificial insemination or ovum harvesting can be determined to increase the chances of successful fertilization.

Ultrasound is the definitive technique for demonstrating intrauterine contraceptive devices (IUCDs). Traditionally, a pelvic radiograph was performed but, apart from the unacceptable use of ionizing radiation to a female of a childbearing age, a standard X-ray cannot determine whether the device is *in utero* or *ex utero*. Ultrasound will easily demonstrate the presence or absence of an IUCD and will also identify any coexisting pregnancy without damage to the fetus.

Ultrasound will also demonstrate fluid in the pelvis. Identifying fluid in the Pouch of Douglas is a valuable if non-specific sign. Fluid at this site may be due to ascites but also to other conditions, such as a ruptured ectopic pregnancy, infection or endometriosis. When ascites is responsible, the pelvic fluid can be shown to change in position and shape with changes in the patient's position. An abscess can be suspected if the fluid shows both constancy in position and shape despite changes in patient position. In endometriosis, ultrasound may show fluid collections within the pelvis caused by the haemorrhage. Occasionally, masses of ectopic endometrial tissue can lead to compression of pelvic structures, e.g. the ureter resulting in obstructive uropathy.

CT has an established role in the assessment of gynaecological patients. As with ultrasound, masses can be assessed and their sites of origin identified. The pelvis is a difficult region to assess on CT. Although the cross-sectional anatomy is demonstrated in great detail, loops of bowel may mimic mass lesions and it may be quite impossible to determine whether a mass is of uterine or ovarian origin. The normal ovaries are generally not identified on CT. CT scanning of the pelvis is aided by using adequate small and large bowel contrast, inserting a vaginal tampon to identify the cervix, and examining the patient with a full bladder. CT can assess the extent of gynaecological malignancy and can look for evidence of both local and distant spread. This has particular relevance for radiotherapy planning. Follow-up studies can provide assessment for treatment. Both CT and ultrasound can be used to guide biopsy needles in order to obtain histological specimens.

Further Reading

Friedland, G.W., Filly, R., Goris, M.L., Gross, D., Kempson, R.L., Korobkin, M., Thurber, B.D. and Walter, J. (Eds). (1983). *Uroradiology — an Integrated Approach*. Churchill Livingstone, Edinburgh.

Lang, E.K. (Ed.). (1986). *Interventional uroradiology. Radiology Clinics of North America* Vol. 24, **4**. W.B. Saunders & Co., Eastbourne & Philadelphia.

O'Reilly, P.H., Shields, R.A. and Testa, H.J. (1986). *Nuclear Medicine in Urology and Nephrology* (2nd edn). Butterworth Scientific Ltd, Guildford.

Sherwood, T., Davidson, A.J. and Talner, L.B. (1980). *Uroradiology*. Blackwell Scientific Publications, Oxford.

Taylor, K.J.W. (Ed.). (1985). *Atlas of Obstetric, Gynecologic and Perinatal Ultrasonography*. Churchill Livingstone, Edinburgh.

7

ENDOCRINE DISORDERS

The Pituitary Gland

Pituitary Tumours

Most pituitary disease arises because of space-occupying lesions within either the pituitary, its stalk or in the hypothalamus. Lesions within the pituitary are usually primary adenomas and may secrete one of the trophic hormones. Non-functional adenomas produce their effects by compression of normal pituitary tissue or surrounding structures, notably the optic chiasma which lies above the pituitary fossa. Granulomatous lesions, infective infiltrations, or metastases from extracranial cancers can produce symptoms identical to those of a non-functional adenoma.

In suspected pituitary tumours, the first imaging procedure is a lateral X-ray of the skull to evaluate the size and shape of the pituitary fossa. As pituitary lesions enlarge they cause erosion and thinning of the bony walls of the fossa (sella turcica) (Fig. 7.1). This may be reflected by an increase in the area of the fossa but, in view of the marked variation in normal fossa size, such measurements are often not very helpful on their own. Sequential measurements of fossa size in an individual patient may be more useful in following enlargement of a tumour but care has to be taken to ensure that magnification factors are not varied by changing the distance between the patient and the X-ray machine. This should be excluded by looking for changes in the size of the skull. Erosion of the posterior clinoid processes should also be looked for as a sign of a pituitary tumour. Downward enlargement of the pituitary will produce erosion of the floor of the fossa and this may be manifest as a ballooning downwards of the fossa into the sphenoid bone. An asymmetrical tumour may result in more marked bone erosion of the corresponding side of the fossa. This is demonstrated on the plain radiograph as a double contour of the fossa floor. Circumspection should be employed, however, in assessing this particular sign as it may be

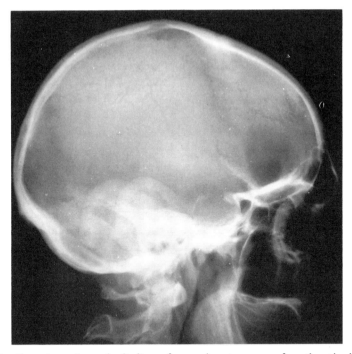

Fig. 7.1 Grossly enlarged pituitary fossa due to a non-functional pituitary tumour. (Courtesy of Dr JA Thomson.)

seen as a normal variant or due to poor radiographic technique resulting in the image not being a true lateral. It should be noted that many pituitary tumours producing profound endocrine disturbance will come to investigation while still of a relatively small size and will therefore not produce any distortion of the sella turcica. A normal pituitary fossa on plain radiology does not exclude a pituitary lesion.

Suprasellar extension of pituitary tumours is extremely important both in terms of the possible compression of local structures and of the type of neurosurgical approach which is required. Suprasellar extension is likely if the posterior clinoids are eroded but it cannot be established with certainty on plain X-ray unless the tumour is calcified. Calcification occurs commonly in craniopharyngioma but is rare in other pituitary tumours. The possibility of calcification in other structures around the sella, such as the carotid syphon or petroclinoid ligament, should be considered but these can usually be identified by position and shape. Other pathologies which may cause parasellar calcification are: meningioma, previous meningitis, especially tuberculous, and an aneurysm; in these cases no intrasellar abnormality should be present.

Tomography of the pituitary fossa has its proponents — especially for assessment of fossa size and demonstration of asymmetry of the fossa — but other authorities doubt whether it adds any information to that obtained from planar X-rays.

Previously, pneumoencephalography was used to study the extent of suprasellar extension but has now been replaced by computed tomography (CT). Pneumoencephalography was not without hazard as it could produce pituitary infarction.

As with most other forms of intracranial disease, the investigation of pituitary disorders has been revolutionized by the advent of computed tomography. CT — usually with contrast enhancement — is especially valuable in demonstrating suprasellar extension, whether the lesion is calcified or not and in determining the downward and lateral extent of tumour growth. The information thus provided by CT is invaluable to the neurosurgeon. Computed tomography has also been employed to image the pituitary itself, especially in patients with a normal fossa on standard X-rays. CT has proven valuable in identifying the location of some pituitary tumours but it is often difficult to differentiate a small tumour from normal variations of pituitary structure.

Computed tomography is of great value in the diagnosis of the 'empty sella syndrome'. In this syndrome, the pituitary fossa is enlarged but the pituitary gland is either normal or reduced in size. The primary empty sella syndrome results from enlargement of an arachnoid diverticulum in the sella causing erosion of the bony walls of the fossa and compression of the pituitary òn the floor of the fossa. An empty sella may also develop after regression or infarction of a pituitary tumour. With the exception of the cases which are due to pituitary tumour, and they are the minority, pituitary function is normal in the empty sella syndrome.

Magnetic resonance imaging (MRI) is still being assessed in pituitary disease and early results suggest it may be helpful in quantifying suprasellar extension and in identifying the presence of small tumours within the pituitary. The relative roles of MRI and CT in pituitary disease are still to be determined.

Angiography has a limited role in pituitary disease, but some neuro-surgeons request preoperative angiography to establish the relationship of the tumour to major cerebral vessels. Ultrasound and scintigraphy have no place in the investigation of pituitary disease.

Acromegaly

When acromegaly is suspected clinically, a lateral X-ray of skull is the first screening investigation as 90 per cent of patients with a growth hormone-secreting adenoma will show enlargement of the sella turcica. In acromegaly, the skull X-ray will often demonstrate characteristic enlargement of the frontal sinuses, enlargement of the jaw and generalized vault thickening (Fig. 7.2). Confirmation of the diagnosis is obtained by detailed anterior pituitary function tests which are aimed at demonstrating excessive and

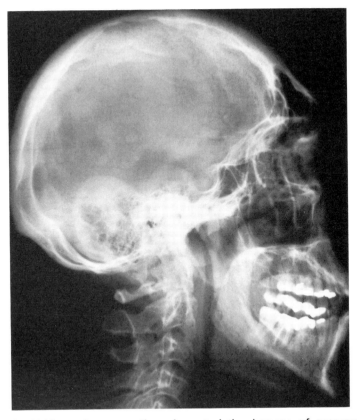

Fig. 7.2 Skull X-ray demonstrating characteristic changes of acromegaly: large frontal sinuses, prognathism and some thickening of the vault. The pituitary fossa is with normal limits. (Courtesy of Dr JA Thomson.)

non-suppressible secretion of growth hormone. As in other pituitary tumours, the effect of the tumour on other elements of pituitary gland function is assessed.

Once the diagnosis of growth hormone excess has been confirmed, a number of imaging procedures are helpful, including detailed assessment of the pituitary and suprasellar area by CT. As previously mentioned, the role of tomograms of the sella is debatable, though they may be helpful in determining whether there has been tumour extension into the sphenoid sinus. A chest X-ray should always be obtained as cardiomegaly is common — partly as the result of hypertension induced by the acromegaly and partly due to a direct effect of growth hormone on the myocardium. Cardiovascular problems are the commonest cause of death in acromegaly and, especially in long-standing acromegalics, detailed non-invasive assessment of cardiac function may be required.

On X-ray, bone proliferation will be seen as cortical thickening and tufting of the distal phalanges. In the vertebral column, periosteal growth may produce loss of the trabecular pattern and increased concavity of the posterior surfaces of the vertebral bodies. Kyphosis may be seen. Osteoporosis is a well recognized feature in patients who progress to hypopituitarism but is not usually marked in acromegaly. Degenerative joint changes, with osteophyte formation, are common in acromegaly. Soft-tissue thickening is usually evident clinically and radiologically in active acromegaly.

Following medical, surgical or radiological treatment for acromegaly, continuing disease activity is monitored by estimations of growth hormone levels. Skeletal changes do not regress after treatment but soft-tissue overgrowth does. This can be assessed radiologically by measurement of the thickness of the soft tissue of the heel pad which usually exceeds 22 mm in acromegalics.

Cushing's Disease

The commonest cause of overproduction of cortisol by the adrenal cortex is an ACTH-secreting pituitary adenoma resulting in Cushing's disease. The investigation of this disorder is considered in the adrenal section of the chapter (see p. 212).

Hyperprolactinaemia

Hyperprolactinaemia has been recognized in recent years as an important, yet reversible, cause of infertility. It may also present as galactorrhoea or disordered menses. Excessive production of prolactin may be due to a prolactin-secreting tumour or to 'functional overactivity' of the gland. In patients with hyperprolactinaemia, a pituitary tumour can be excluded by obtaining a history of drug therapy known to stimulate release of the hormone, or by diagnosing non-pituitary diseases such as hypothyroidism or renal failure which elevate prolactin levels. In many patients, however, even after detailed endocrine testing, the presence of a prolactin-secreting tumour cannot be excluded and imaging may then be helpful. The pituitary fossa should be assessed by plain X-rays and by computed tomography. Prolactin-secreting tumours are often small at the time of presentation. CT can show areas of abnormality in the pituitary of patients with hyperprolactinaemia and bony changes in the sella turcica. These abnormalities sometimes prove to be small prolactin-secreting tumours ('microadenomas') but misleading results are also obtained with no tumour being found at such sites during pituitary surgery. Larger tumours are evident on pituitary fossa X-rays and CT. Suprasellar extension should be excluded in the larger tumours. In patients with a normal pituitary fossa and normal appearances of the gland itself, CT should be employed for careful evaluation of the pituitary stalk and hypothalamus as abnormalities in these areas may induce increased prolactin secretion. Lesions which may produce this effect include primary intracranial tumours such as glioma, craniopharyngioma or pinealoma, granulomatous disorders or pituitary stalk section following trauma or surgery. Magnetic resonance imaging may have a major contribution to make in this area in the future.

Hypopituitarism

Underactivity of the pituitary results from a wide variety of causes. The role of imaging in the hypopituitary patient is primarily to detect the presence of a mass lesion in the hypothalamus or pituitary. Any pituitary tumour may compress normal tissue sufficiently to result in hypofunction. Enlargement of the fossa will usually be evident and exclusion of suprasellar extension is necessary. Craniopharyngiomas usually develop in the suprasellar region, but may also arise or extend into the sella turcica. In children, this tumour is calcified in 80 per cent of cases but in adults the figure falls to around 40 per cent. CT will often demonstrate a cystic component to the lesion.

Patients with pituitary tumours, especially large ones, may develop acute pituitary insufficiency or pituitary apoplexy due to infarction of the tumour and compression of functional tissue in the sella or of the pituitary stalk. A similar picture is occasionally seen in diabetics and in sickle cell disease. An urgent CT scan is required in this condition as it will demonstrate the presence of haemorrhage and oedema within the pituitary or in the suprasellar region.

Posterior Pituitary Disease

Hypofunction of the posterior pituitary produces the clinical syndrome of diabetes insipidus due to inadequate secretion of antidiuretic hormone (ADH). It may result from pressure on the posterior pituitary or stalk by anterior pituitary or suprasellar tumours, or by invasion of malignant or granulomatous lesions. Trauma to the skull or previous neurosurgery are also recognized causes. A further group of patients have idiopathic diabetes insipidus in which no anatomical abnormality can be detected — although on follow-up a proportion develop evidence of a hypothalamic or pituitary lesion. Imaging of the patient with diabetes insipidus consists of skull radiology and CT (as already described for anterior pituitary tumours) with particular attention to the pituitary stalk and hypothalamic regions.

Excessive production of ADH causes marked water retention and hypo-natraemia. The syndrome of inappropriate ADH secretion (SIADH) is almost always due to an ectopic source of ADH. A chest X-ray should always be obtained to look for a bronchial carcinoma (the commonest cause), tuberculosis, pneumonia or thymoma. Imaging may also be required to look for other possible causes such as pancreatic or duodenal tumours or intracranial disorders such as subdural haematoma or cerebral atrophy.

The Adrenal Glands

Imaging Techniques

The adrenals cannot normally be visualized on straight abdominal X-rays. Adrenal calcification — seen as speckled calcification lying at the upper pole of the kidneys — may indicate chronic granulomatous disease (Fig. 7.3). This is usually due to previous tuberculosis but can also result from fungal infection. A similar appearance may be seen from previous adrenal haemorrhage. Calcification and haemorrhage may lead to adrenal hypofunction. Calcification may also occur in an adrenal carcinoma. The plain abdominal X-ray may also be used to demonstrate downward displacement of the kidney by an adrenal tumour.

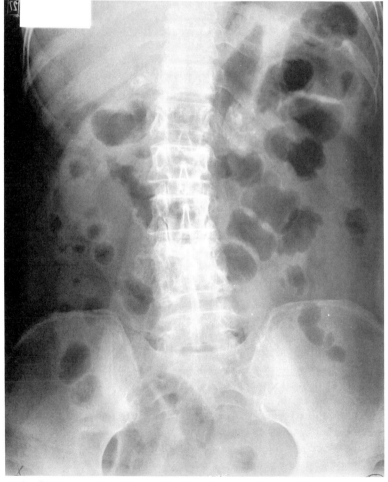

Fig. 7.3 Bilateral adrenal calcification in a patient with hypoadrenalism secondary to previous tuberculosis.

Ultrasound has a major role in the evaluation of possible adrenal tumours. The normal adrenal is difficult to visualize on an ultrasound scan but in the presence of a mass lesion ultrasound may reveal the diagnosis, and bilateral visualization suggests the presence of adrenal hyperplasia.

CT is now the central imaging test for the adrenals. Normal adrenals can almost always be seen on CT, and this modality is extremely accurate in detecting either a unilateral mass lesion or bilateral hyperplasia (Fig. 7.4). In patients with a unilateral adrenal mass CT may show either local invasion or nodal or other metastases, indicating an adrenal carcinoma, rather than the more common adenoma.

Scintigraphy of the adrenal cortex is now rarely required. A cholesterol compound (usually labelled with iodine-131 or selenium-75) is injected intravenously and incorporated into the synthetic steroid pathway. Gamma camera images obtained several days later demonstrate adrenal uptake indicating functional activity.

Recently, more interest has centred on the technique of adrenal medullary imaging with the radiopharmaceutical iodine-131-labelled meta-iodobenzylguanidine (^{131}I-mIBG). This compound will enter into neuro-endocrine tissue and has found a particular role in the localization of phaeochromocytoma.

Adrenal arteriography is carried out in cases of suspected adrenal carcinoma, utilizing a transfemoral approach. The results are useful in demonstrating the presence of a tumour circulation and in planning the best surgical approach. Adrenal venography is also usually performed by a transfemoral approach. In conjunction with sampling at multiple sites for

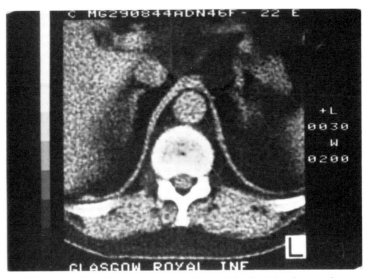

Fig. 7.4 CT scan of abdomen in a patient with Cushing's syndrome. Both adrenals are visible as streak like structures indicating bilateral hyperplasia, rather than unilateral adenoma.

hormone levels, it was previously used widely for the diagnosis and localization of adrenal tumours. It is not without hazard, due both to the general complications associated with angiography and, more specifically, to the risk of causing infarction in the tumour or to the whole adrenal gland. Arteriography and venography have been largely replaced by ultrasound and CT, although venous sampling for catecholamine levels is still required on occasions.

Cushing's Syndrome

Diagnosing the cause of Cushing's syndrome is dependent on hormonal studies, but imaging investigations are of value in demonstrating some of the results of adrenocortical hyperactivity and in determining whether the primary lesion is an ACTH-secreting pituitary adenoma or an autonomous adrenal tumour.

Osteoporosis is common in Cushing's disease and is often both severe and widespread. Skeletal X-rays will demonstrate its presence and frequently also show crush vertebral fractures with increased thoracic spine curvature. Hypertension is also common in Cushing's syndrome and may result in cardiomegaly visible on chest X-ray.

Adrenal scintigraphy may be employed in the rare cases after CT and ultrasound studies where doubt persists as to whether there is bilateral hyperplasia or not. In Cushing's syndrome due to an adrenal adenoma or carcinoma, the function of the contralateral gland is usually suppressed by the high circulating cortisol levels. The radiolabelled cholesterol will be taken up only by the abnormal gland. In bilateral hyperplasia both glands are visualized on scintigraphy. Venous catheterization studies or arteriography may occasionally be needed to resolve the question.

Primary Aldosteronism (Conn's syndrome)

Excessive secretion of aldosterone by the adrenal cortex may be due to an adrenal adenoma, bilateral nodular adrenal hyperplasia or an adrenal carcinoma. Ultrasound and CT examinations will usually reveal a unilateral lesion but may be normal in patients with nodular hyperplasia or with small unilateral lesions. Venous catheterization studies with adrenal vein hormone levels will separate unilateral and bilateral lesions.

Adrenal Insufficiency

The investigation of the patient with suspected primary adrenal failure should always include an abdominal X-ray as adrenal calcification is present in cases secondary to tuberculosis or fungal infection. This finding, together with appropriate clinical and serum electrolyte changes, allows a firm diagnosis to be made without waiting for the results of steroid hormone assays. More commonly, adrenal failure is due to primary (autoimmune) atrophy or atrophy secondary to prolonged steroid therapy. In these cases imaging has little to contribute to the diagnosis. Occasionally, CT and ultrasound demonstrate metastases — most often from a bronchial primary — and so explain the adrenal failure of recent onset.

Phaeochromocytoma

Catecholamine-secreting tumours most commonly arise from the adrenal cortex, but about 10 per cent of phaeochromocytomas are extra-adrenal and may arise at any site where there are sympathetic nerve cells. In addition, about 10 per cent of adrenal lesions are bilateral or multiple.

The presence of a phaeochromocytoma is confirmed by various hormonal investigations, notably estimation of urinary levels of catecholamines. Large tumours may be evident on plain abdominal X-rays or on IVU due to renal displacement or distortion. Computed tomography or ultrasound is usually required, however, to lateralize the lesion. When these techniques fail to demonstrate the location of the tumour a number of courses of action are possible: venous catheterization and multiple venous sampling for plasma catecholamine levels may be helpful in showing the site of adrenal or extra-adrenal tumours, and suspicious sites may then be subjected to venography or arteriography. Catheterization studies in patients with suspected phaeochromocytoma are hazardous because of the risk of provoking catecholamine release and producing a hypertensive crisis. Alpha- and beta-adrenergic blocking drugs should always be administered to minimize this risk.

There has been considerable work recently on ^{131}I-mIBG scintigraphy in the localization of phaeochromocytoma. The tumour will accumulate the radiopharmaceutical resulting in a local hot spot on scintigrams obtained some days after tracer administration (Fig. 7.5). Whole body images are easily obtained allowing detection of extra-adrenal or multiple lesions. In malignant phaeochromocytoma (approximately 10 per cent of all phaeochromocytomas) mIBG images may demonstrate functional metastases in addition to the primary tumour.

It is possible that mIBG scintigraphy will largely replace arteriography and venography for localization of tumours not shown on CT or ultrasound.

The Parathyroid Glands

Imaging Techniques

Ultrasound, CT and scintigraphy are all used in the imaging of the parathyroid gland, usually when searching for an adenoma or hyperplastic gland producing the syndrome of hyperparathyroidism.

Best results with ultrasound are obtained with real-time studies utilizing a

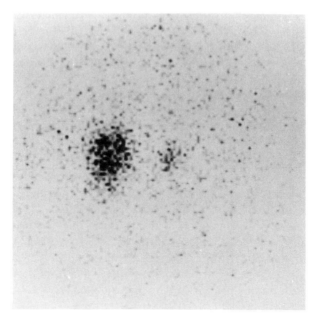

Fig. 7.5   [131]I mIBG scintigram of lumbar area (posterior view) showing tracer uptake in a large left adrenal phaeochromocytoma and a small right adrenal phaeochromocytoma.

10 MHz transducer. It should be remembered that the glands may lie in ectopic sites away from the normal position of the upper and lower poles of the thyroid lobes. Ectopic glands may lie anywhere in the neck and may also be found in the mediastinum. Ultrasound studies cannot usually identify normal-sized parathyroids and attempts at localizing mediastinal glands, even when they are enlarged, are usually unsatisfactory.

Scintigraphy has been introduced recently and utilizes a dual tracer method. After intravenous injection, $^{99}Tc^m$-pertechnetate will be taken up by the thyroid via the iodide trap but not by the parathyroids. The potassium analogue thallium-201 (^{201}Tl) will enter all metabolically active tissue and is widely used for myocardial imaging. In the neck, both thyroid and parathyroid tissue will take up ^{201}Tl via the sodium–potassium pump. If the $^{99}Tc^m$ image of the neck is then subtracted using a computer from the ^{201}Tl, residual activity in the parathyroid will be seen (Fig. 7.6). Normal parathyroids accumulate only a small amount of ^{201}Tl and so are not seen on the subtraction image as a hot spot. It is of course essential that the patient does not move during the period of image acquisition, which usually takes 25–30 minutes.

Computed tomography using second generation machines has been disappointing in parathyroid localization. Better results have been obtained

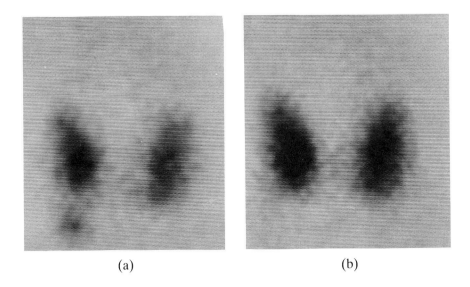

(a)   (b)

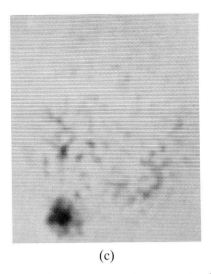

(c)

Fig. 7.6 Parathyroid scan. (a) Anterior neck image after ^{201}Tl injection shows thyroid uptake plus a parathyroid adenoma below the right lower pole of thyroid. (b) ^{99}Tc^m image showing thyroid uptake only. (c) The parathyroid adenoma is clearly seen on the subtraction image.

in some centres using third generation scanners but the numbers of patients studied have been small. CT may be helpful in mediastinal tumours when initial studies in the neck are negative.

Hyperparathyroidism

Many patients with hyperparathyroidism now are diagnosed on a finding of asymptomatic hypercalcaemia; others, now a minority in most series, present with bony or renal tract problems. Primary hyperparathyroidism is almost always due to an autonomous parathyroid tumour, usually an adenoma. A proportion of patients have multiple adenomata. Secondary hyperparathyroidism is due to parathyroid hyperplasia, most often due to renal failure. Tertiary hyperparathyroidism occurs in patients with secondary hyperparathyroidism when one or more of the glands becomes autonomous.

Imaging of the skeleton is helpful in hyperparathyroidism. The characteristic finding is of subperiosteal erosion of bone and this is best demonstrated in the phalanges on an X-ray of the hands. Diffuse osteoporosis is also frequently seen, but in advanced cases areas of osteosclerosis may also be apparent on bone X-rays. This latter finding is rare in primary hyperparathyroidism. In severe cases, the skull develops multiple punched out areas of bone resorption — the so-called 'salt-and-pepper' appearance on skull X-ray. Loss of the lamina dura of the teeth occurs but is less specific. Large cystic lesions in the bone develop late in the course of the disease due to the presence of 'brown tumours'. These lesions are now rarely seen but they may be the site of pathological fracture.

Soft-tissue calcification may occur in hyperparathyroidism and can be visualized radiologically. Extensive soft-tissue calcification is rare in primary hyperparathyroidism. Nephrocalcinosis and calcium-containing renal tract stones are recognized features, though they rarely both develop in the same patient. A straight X-ray of the abdomen and pelvis should be obtained to exclude their presence. Pancreatic calcification may also be seen.

The bone scan in mild hyperparathyroidism is normal but more severe cases show the 'superscan' appearance with diffusely increased tracer uptake throughout the skeleton and absent renal images. There are no features on the bone scan which allow separation of hyperparathyroidism from other metabolic bone diseases. If brown tumours are present they produce a local 'hot spot'.

The diagnosis of hyperparathyroidism is essentially a biochemical one. The role of parathyroid imaging is in locating the causative overactive gland so that it can be removed surgically. In the past, barium swallow, arteriography and venography were employed but these are no longer used. Cervical venous catheterization with sampling at multiple points in the neck and mediastinum for parathyroid hormone levels is still widely utilized. It locates around 80 per cent of adenomas accurately. It is, however, invasive and may be misleading because of anomalous venous anatomy.

Both ultrasound and scintigraphy have been reported to be around 70 per cent accurate in detecting adenomas. Each has some disadvantages. Ultrasound cannot be used in the mediastinum and may be difficult to interpret in patients with previous neck surgery. Scintigraphy can detect mediastinal adenomas. It is unreliable in the presence of thyroid disease, because iodide trap and sodium–potassium pump function do not necessarily change in parallel and a thyroid adenoma may produce a hot spot on the subtraction image indistinguishable from that of parathyroid adenoma. Scintigraphy is poor in demonstrating hyperplasia for reasons that are not yet understood and ultrasound is more likely to be effective in this setting.

CT has a lower localization rate than ultrasound and scintigraphy: around 40–50 per cent in most reported series.

It is still a matter of debate as to when parathyroid localization methods should be used in hyperparathyroidism. An experienced parathyroid surgeon can find an adenoma in 95 per cent of cases without using any preoperative localization techniques. It has thus been suggested that they are unnecessary unless the patient has had a previous failed neck exploration for a parathyroid adenoma. In view of its invasive nature and cost, cervical vein catheterization should be reserved for patients being re-explored after a previous unsuccessful operation and not used routinely prior to initial neck exploration. The non-invasive nature of ultrasound and scintigraphy make them more suitable for all patients undergoing neck exploration. They will identify the site of the adenoma in some of the cases which would otherwise have been missed. Additionally, by indicating the site in other cases they will reduce the operative time — an important consideration in patients who are not infrequently elderly. Which of the two techniques is used in the individual patient will depend on local expertise and on the factors discussed above.

Hypoparathyroidism

Hypoparathyroidism may rest from parathyroid removal at thyroid surgery, from an autoimmune disorder or from congenital hypoplasia or aplasia of the gland.

The skeletal X-rays are frequently normal in hypoparathyroidism although occasionally the bones show generally increased density. In affected children, increased thickness of the calvarium may be a feature. Basal ganglia calcification is common. Subcutaneous calcification may occur but is rare.

Some patients have tissue resistance to parathyroid hormone — pseudohypoparathyroidism. They may show any of the radiological features described above, with subcutaneous calcification being very common. An X-ray of the hands and feet in pseudohypoparathyroidism frequently, though not always, shows shortening of the meracarpals and metatarsals, most often the fourth and fifth.

Parathyroid Carcinoma

This rare tumour has no specific features on imaging. Evidence of local or distant metastases should be sought in the usual fashion. The tumours are often functional and cause hyperparathyroidism resulting in any of the imaging abnormalities discussed above.

The Thyroid Gland

Imaging Techniques

Plain radiology

Straight X-rays have a relatively small part to play in the assessment of the patient with thyroid disease. On an X-ray of the chest or neck a soft-tissue mass due to a goitre may be noted as an incidental finding. Such goitres will usually be apparent clinically. The plain X-ray may give some information on substernal extension of a goitre, while a mediastinal goitre is usually first detected as a superior mediastinal mass on chest X-ray (Fig. 7.7). Plain X-rays should be inspected for evidence of calcification within a goitre. Calcification is frequently due to benign pathology but can be associated with carcinoma and should always be viewed with suspicion.

Scintigraphy

Scintigraphy is of considerable value in the investigation of thyroid disorders. When iodine is given orally or intravenously a proportion is trapped by the thyroid follicular cell and incorporated into the thyroid hormone synthetic pathway. Thyroid imaging was first performed using iodine-131. This has a relatively long half-life ($T\frac{1}{2}$=8·05 days) and emits gamma rays above the optimal range for the gamma camera. Iodine-123 has a shorter half-life (13·3 h) and emits photons that are much more suitable for gamma camera imaging. Iodine-123, however, is cyclotron produced and thus is expensive. For this reason, $^{99}Tc^m$ pertechnetate is used in many centres for routine thyroid imaging. The mechanism of thyroidal trapping of $^{99}Tc^m$ appears to be the same as that of iodine. Unlike iodine, however, $^{99}Tc^m$ is not organified — i.e. it does not enter into the thyroid hormone synthetic pathway. Occasionally, this may lead to discrepancies between $^{99}Tc^m$ and iodine scintigraphic images, though usually the appearances are similar.

Because $^{99}Tc^m$ is rapidly trapped by the thyroid, images are obtained 20 minutes after IV injection. Good quality thyroid scans can be obtained in

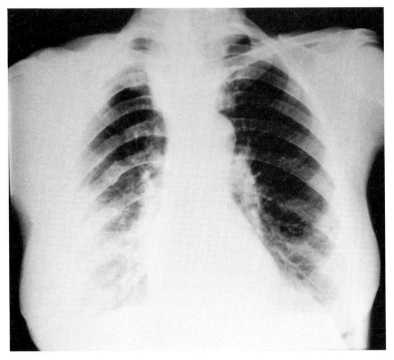

Fig. 7.7 PA X-ray of chest demonstrating a superior mediastinal mass due to a retrosternal goitre.

patients receiving carbimazole or propylthiouracil as these drugs block thyroid hormone synthesis only and do not influence the trapping mechanism. By comparison, potassium perchlorate blocks the iodine trap and good quality scans cannot, usually, be obtained in patients on this drug. A similar effect is seen if the thyroid has been blocked with iodide. This may have been taken orally, as potassium iodide tablets or as a component of many cough mixtures, or may have been given parenterally in the form of a radiographic contrast medium. Prolonged thyroid blockade can be seen after an intravenous urogram or a myelogram. It is usual to obtain anterior views of the thyroid on $^{99}Tc^m$ scintigraphy supplemented by oblique views if nodules are suspected. Iodine-123 scintigraphy, if not used as a routine, may be employed after the $^{99}Tc^m$ scan if the latter is equivocal regarding nodule formation or if the degree of function of a nodule is difficult to establish. Iodine-123 imaging may also be useful in superior mediastinal masses. On $^{99}Tc^m$ images it may be difficult to decide whether activity seen in this area represents blood pool activity in the great vessels or active uptake in the mass, implying that there is a substernal goitre. As no significant blood pool activity is present at the time of imaging with ^{123}I (usually 2–6 hours after administration) this problem does not arise. It is essential to use ^{131}I for scintigraphy when screening for metastases from well differentiated thyroid carcinoma.

Ultrasound

Ultrasound examination of the thyroid has increased in importance in recent years and in some cases has replaced thyroid scintigraphy. The examination yields best results if a real-time instrument with a high frequency transducer (7·5–10 MHz) is employed. Ultrasound is used to determine the size of the thyroid, to detect the presence of thyroid nodules and to determine whether the nodule is cystic or solid. In addition in thyroid cancer patients, ultrasound may be useful in detecting involvement of cervical nodes, both at the time of presentation and during follow-up.

Computed tomography

CT is not used to any extent in the evaluation of the thyroid itself except in assessing the extent of mediastinal thyroid tissue preoperatively. It can be useful in detecting suspected lung metastasis from thyroid carcinoma.

Diffuse Non-toxic Goitre

Thyroidal radioiodine turnover and uptake measurements (usually of ^{131}I) were widely used in the past for the study of thyroid function. They have been replaced in clinical practice by direct estimation of serum thyroid hormone and thyroid stimulating hormone levels, usually by radioimmunoassay. Occasionally, thyroid uptake measurements are used when thyroid function tests are equivocal or out of keeping with the clinical state.

Imaging techniques have only a small place now in the assessment of patients with a diffuse non-toxic goitre. Either scintigraphy or, perhaps more accurately, ultrasound can be used to measure the size of a goitre, though it is doubtful in most cases whether this gives any information additional to that obtained from careful clinical examination.

Scintigraphy is of help when substernal extension of the thyroid is suspected, either clinically or due to the presence of a soft-tissue mass on plain radiology. Extension of the ^{99}Tc^m or ^{123}I uptake area to beyond the suprasternal notch confirms the diagnosis. It should be noted that minor degrees of substernal extension may be missed if the thyroid images are obtained in the standard fashion with the neck hyperextended as this may cause the thyroid to rise up from the substernal space. Failure of tracer uptake in a substernal or retrosternal mass does not exclude non-functional thyroid tissue in this site; this is particularly a problem in patients with multiple non-functional nodules. In these circumstances, ultrasound or CT can sometimes be used to demonstrate continuity between the thyroid and a substernal mass.

Patients with a large goitre or with substernal extension may develop compression of the oesophagus causing dysphagia, or of the trachea producing stridor or dyspnoea. Barium swallow and thoracic inlet films are valuable in such patients.

Occasionally, imaging is employed in diffuse goitre to help elucidate the likely cause. Both scintigraphy and ultrasound can show the presence of multinodular change. Some patients with autoimmune (Hashimoto's) thyroiditis have diffusely increased tracer uptake on scintigraphy, often with a prominent pyramidal lobe.

Ectopic Thyroid Tissue

The thyroid arises from the fourth and fifth branchial pouches and migrates to the neck in fetal life. The gland may come to rest anywhere between the root of the tongue ('lingual thyroid') and the mediastinum. Additional sites of functional tissue may develop along the line of descent, classically forming a thyroglossal cyst.

When ectopic thyroid tissue is suspected, thyroid scintigraphy should be used: if the mass concentrates $^{99}Tc^m$ or iodine, functional thyroid tissue is present. When excision of the ectopic thyroid is planned, the scintigram is also useful in determining whether there is functional thyroid tissue elsewhere. Should the ectopic tissue be the only functional thyroid present, hypothyroidism is inevitable postoperatively and thyroid hormone replacement therapy should be started.

Thyrotoxicosis

Thyroid imaging is not necessary in many patients with thyrotoxicosis and should be reserved for certain restricted indications.

Thyroid scintigraphy can be helpful in differentiating thyrotoxicosis due to a single toxic adenoma with suppression of the rest of the gland ('hot nodule') (Fig. 7.8) from cases caused by diffuse hyperplasia (Graves' disease) (Fig. 7.9) or by toxic multinodular goitre. In toxic adenoma the scan will show intense tracer uptake confined to the lesion with little or no uptake in the rest of the thyroid. In Graves' disease there is uniformly increased uptake throughout the thyroid. The distinction is important. While a significant proportion of Graves' disease patients have permanent cure of thyrotoxicosis after a prolonged course of antithyroid drugs, those with a toxic adenoma almost inevitably have recurrence of thyrotoxicosis when the drugs are stopped. The distinction between a toxic adenoma and Graves' disease also has some relevance when planning radioiodine therapy as toxic

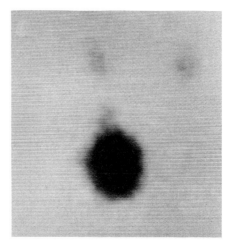

Fig. 7.8 $^{99}Tc^m$ thyroid scan from a thyrotoxic patient showing a toxic nodule in the right lobe and suppression of the rest of the gland.

adenomas tend to be more resistant to ^{131}I. Following ^{131}I therapy, hypothyroidism is less common with a toxic adenoma, presumably due to recovery of normal function in the suppressed remainder of the gland which will not receive much of a radiation dose. Often a confident diagnosis of diffuse hyperplasia or of toxic multinodular goitre can be made clinically, but scintigraphy is indicated if a thyrotoxic patient has either a single thyroid nodule or no palpable goitre.

Prior to giving ^{131}I therapy, a thyroid radioisotope uptake measurement on a thyroid scan should be obtained to establish whether tracer uptake within the gland is blocked, in which case the ^{131}I administration is pointless as it is unlikely to be effective. In some centres, the thyroid size — estimated from scintigraphy or ultrasound — and radioisotope turnover measurements are employed to calculate a precise dose of ^{131}I therapy. Such an approach is not carried out in most institutions, however, because of the lack of precision of the estimates of thyroid size and because of individual variation in response to a given radiation dose delivered to the thyroid.

Uptake measurements or scintigraphy are very valuable in the diagnosis of acute or subacute thyroiditis. This condition is thought to be of viral aetiology usually and should be suspected when the patient has a diffusely tender thyroid, often with fever and leucocytosis. Release of stored hormone from the damaged gland can result in the patient being clinically and biochemically thyrotoxic at the time of presentation, although both euthyroidism and, later in the disease, hypothyroidism can be seen. The thyroid scan shows little or no uptake of tracer in the gland in acute thyroiditis compared with the increased uptake seen in Graves' disease or toxic adenoma. Acute thyroiditis usually recovers spontaneously over a

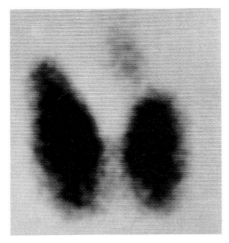

Fig. 7.9 $^{99}Tc^m$ thyroid scan showing diffuse hyperplasia (with pyramidal lobe uptake) in a patient with Graves' disease.

period of weeks or months, and this can be monitored by the improvement in tracer uptake. Most patients are left with normal thyroid function.

Ophthalmopathy is a well recognized complication of Graves' disease, and both CT and ultrasound have an important place in its investigation. In thyroid ophthalmopathy the retro-orbital tissues, including the extra occular muscles, are subjected to an inflammatory infiltrate, predominantly of lymphocytes, plasma cells and mast cells. The increase in tissue produced by the inflammation results in protrusion of the globe. Clinical evidence of this infiltrative ophthalmopathy is present in only a small proportion of Graves' disease patients, but orbital CT and ultrasound studies have shown that almost all have some degree of ophthalmopathy.

Graves' ophthalmopathy is usually bilateral but may be unilateral. When this occurs, orbital CT or ultrasound should be obtained to demonstrate whether an unrelated retro-orbital mass, such as a tumour, is the cause of the unilateral proptosis or whether there is typical thickening of the retro-orbital muscles due to Graves' disease, often with subclinical changes in the other orbit. In severe Graves' ophthalmopathy, CT and ultrasound can be useful in measuring the mass of retro-orbital tissue and can be used as one of the parameters of response to the steroid therapy employed in this condition.

Thyroid Nodule

Scintigraphy and ultrasound are widely employed in the investigation of the patient with a solitary thyroid nodule. Such nodules are usually due to an adenoma, but others are malignant.

Thyroid scintigraphy is employed to determine whether the nodule is functional (or 'hot', i.e. takes up tracer) or is non-functional ('cold'). Functional nodules are almost certainly benign, whereas cold nodules may be malignant and require further investigation. Some cases have been reported of thyroid carcinomas which showed some, though reduced, uptake of $^{99}Tc^{m}$ on the scan but were completely cold on ^{123}I images. Therefore, if $^{99}Tc^{m}$ is used as the primary technique, some authorities suggest that all nodules which show reduced function should be re-imaged with ^{123}I. The scan should also be assessed for the presence of other nodules as carcinoma is more likely in a solitary nodule than in a multinodular gland.

If there is a single non-functional nodule on scintigraphy, thyroid ultrasound is often employed to decide if the lesion is cystic or solid. Purely cystic lesions are likely to be benign whereas solid or mixed cystic and solid lesions may be malignant.

Biopsy is required if there is still doubt after scintigraphy and ultrasound as to whether the lesion is malignant. In recent years, fine-needle aspiration cytology of the thyroid nodule has been used in some centres in place of open thyroid biopsy. Interpretation of thyroidal fine-needle aspiration is crucially dependent on a skilled cytologist. If such an individual is available diagnostic answers can be obtained in a large proportion of patients and the thyroidectomy rate considerably reduced. Where reliable fine-needle aspiration cytology is available, it is often used as the first investigation of a solitary nodule with thyroid imaging by scintigraphy and ultrasound being omitted.

Thyroid Cancer

Prior to thyroidectomy for thyroid cancer, a chest X-ray should be obtained to exclude lung metastases. In some centres, ultrasound of the neck is performed to look for enlarged lymph nodes which cannot be palpated. In the absence of specific symptoms or biochemical features suggesting distant metastases, no other preoperative staging by imaging techniques is required.

Postoperatively, in patients with well differentiated (papillary or follicular) cancer, there is considerable debate as to the role of imaging in follow-up. Some physicians believe that because of the good prognosis in this disease and the fact that the tumour is TSH dependent, patients who have disease strictly confined to the thyroid (i.e. no distant metastases, no nodal involvement and no vessel invasion on histology) should be placed on suppressive thyroid hormone therapy and followed by clinical examination only. Other physicians advocate that all patients with well differentiated thyroid carcinoma should be screened intermittently for disease recurrence using either serum thyroglobulin levels or whole body ^{131}I scans, or both. Both tests are based on the fact that well differentiated thyroid tumours, especially those with follicular elements, retain many of the functions of normal thyroid tissue. They are thus able to trap ^{131}I and synthesize thyroglobulin. If all normal thyroid tissue has been ablated, a rise in serum thyroglobulin signals a recurrence of the tumour. Functional thyroid cancer will show up as an area of localization of ^{131}I on a whole body scan obtained 72 hours after administration of the tracer. Two points of caution should be noted. First, even when functional, thyroid cancer tissue is usually much less able to concentrate ^{131}I than normal thyroid tissue and functioning metastases will show on the whole body scan only if all normal tissue has been ablated. Second, tumours are well visualized only when TSH secretion by the pituitary is not suppressed. The patient should therefore come off suppressive thyroid hormone therapy for long enough to become hypothyroid. This takes approximately 14 days if the patient is receiving triiodothyronine (T_3) and 6 weeks if thyroxine (T_4) is being administered.

Metastases which concentrate ^{131}I can be treated by administering a large therapeutic dose (4000–8000 MBq) of the tracer. This often results in good disease control for a period, though not in a cure, usually.

It is probable, though not entirely established, that suppressive therapy also needs to be stopped before checking serum thyroglobulin levels.

Other imaging investigations are not indicated routinely in the follow-up of well-differentiated thyroid carcinoma but should be employed when clinically required. The role of ultrasound in detecting tumour in cervical lymph nodes, a common site of recurrence has already been mentioned and is a useful technique when questionable masses are found clinically.

A small proportion of thyroid tumours are lymphomas. Such patients require full staging for the extent of their disease as discussed in Chapter 10.

Anaplastic thyroid cancers cannot be followed using thyroglobulin or whole body ^{131}I images, but a specific tumour marker (serum calcitonin) is available for medullary cancer of the thyroid. Some recent work has suggested that the radiopharmaceuticals iodine-131-labelled meta-iodobenzylguanidine (^{131}I-mIBG); and technetium-99m-labelled penta-valent dimercapto succinic acid (^{99}Tc^m-DMSA(V)) are taken up by primary and metastatic medullary cancer. Their role in initial diagnosis and in management is still unclear. Medullary cancer of the thyroid has an association with phaeochromocytoma. All patients with medullary cancer should have urine collections for catecholamines and CT studies of the adrenal glands to exclude phaeochromocytoma prior to thyroidectomy.

Further Reading

Johnson, C.M., Sheedy, P.F., Welch, T.J. and Hattery, R.R. (1985). CT of the adrenal cortex. *Seminars in Ultrasound, Computed Tomography and Magnetic Resonance* Vol. 6. **3**.

Johnson, C.M., Welch, T.J., Hattery, R.R. and Sheedy, P.F. (1985). CT of the adrenal medulla. *Seminars in Ultrasound, Computed Tomography and Magnetic Resonance* Vol. 6. **3**.

Leopold G.R. (1984). Ultrasound in breast and endocrine diseases. *Clinics in Diagnostic Ultrasound* Vol. 12. Churchill Livingstone, Edinburgh.

Sandler, M.P. and Patton, J.A. (1987). Multimodality imaging of the thyroid and parathyroid glands. *Journal of Nuclear Medicine* **28**, 122.

8

SKELETAL DISORDERS

Imaging Techniques

Plain Radiographs

Plain radiographs remain the most commonly used imaging technique in disorders of the bones and joints and in many situations are sufficient to allow a firm diagnosis to be reached when clinical and biochemical findings are taken into account. In interpreting bone X-rays a number of features are assessed.

First, the density of bone should be evaluated. Loss of bone density occurs because of a decrease in the amount of calcified bone matrix present. Decreased bone density may be generalized throughout the skeleton. Rarely, decreased bone density may occur because of an inborn error of metabolism, such as hypophosphatasia. Thin bones are also a feature of osteogenesis imperfecta and are present throughout the body from birth, although the severity of the condition varies greatly in individual patients. More commonly, a general decrease in bone density is an acquired feature, developing later in life. Acquired generalized loss of bone density is usually classified as being due to osteoporosis or osteomalacia. In osteoporosis, there is decreased osteoid, but that which is present is adequately calcified. In osteomalacia, there is a normal amount of osteoid present but its calcification is impaired. In developed countries osteoporosis is common, while osteomalacia is relatively rare. If only the change in bone density is considered, it is not possible to differentiate between osteoporosis and osteomalacia with any certainty.

A localized decrease in bone density may be due to congenital defects in bone production, but is more often acquired. It appears that approximately 50 per cent of bone mineral must be lost before the abnormality is visible on plain radiographs. Local bone destruction may result from a wide variety of processes, both benign and malignant. The nature of the abnormality may allow a conclusion to be drawn about its likely cause. In general, the presence of a well-defined rim of increased bone density (sclerosis) around the area of bone loss implies a benign pathology while a 'moth eaten' appearance with ill-defined margins is more likely to be due to an aggressive malignant process. The finding of multiple areas of localized bone destruction is suggestive of a malignant rather than benign process, although it may occur in widespread bone infection, in multicentric benign bone tumours, or in some forms of metabolic bone disease.

A generalized increase in bone density is seen in congenital conditions such as osteopetrosis (Albers-Schonberg disease), various other bony dysplasias or hyperphosphatasia. A detailed discussion of these congenital problems is beyond the scope of this book.

Acquired generalized increase of bone density is a feature of some endocrine and metabolic bone disorders such as fluorosis, acromegaly and renal osteodystrophy.

Localized increases in bone density occur in a wide variety of pathologies. Bone islands are localized areas of increased bone density due to abnormal bone remodelling and are of no pathological significance. Local bone sclerosis may also occur due to invasion of the skeleton by metastatic disease, though this pattern (osteoblastic metastases) is less common than bone destruction (osteolytic metastases). Locally increased bone density is a feature of Paget's disease and may also be seen late in the course of bone necrosis due to infection or avascular necrosis.

Alteration in the bone trabeculation pattern can occur early in the course of various skeletal diseases, either locally in infiltrative disorders such as metastases, Paget's disease, or more generally in osteoporosis.

Discontinuity of the bony cortex is the major sign of a fracture. Such discontinuity is easily visualized in major fractures, especially when accompanied by malalignment of the bone. Fractures of smaller bones may only show up as a hairline crack of bone and may not become visible until some time after the acute event when remodelling and repair processes around the fracture make the fracture line more obvious. It is important to obtain X-rays in a variety of projections to demonstrate the presence, extent and effects of a fracture fully.

Periosteal changes may be the earliest radiographic manifestation of a wide variety of skeletal disorders including trauma, infection and malignant disease. Periosteal reaction is initially seen as a linear opacity adjacent to the bone and sometimes separated from it by a clear zone. Later, the clear zone is filled in by calcification. If the periosteal reaction is continuous and uniform, the underlying process is often benign, whereas a broken or an uneven pattern of calcification, such as lamellation or spiculation, makes a malignant condition more likely. Periosteal calcification is more common in primary than metastatic bone tumours, but may occur in the latter especially when they are osteoblastic or aggressive. New periosteal bone formation in the limbs, especially in the forearms and shins, is a characteristic feature of hypertrophic pulmonary osteoarthropathy. This condition is considered further on p. 258.

In arthritis, the plain X-ray is used to evaluate the articular surfaces and the subarticular area of the bone, the joint space and the surrounding soft tissues. Soft-tissue changes on bone X-rays may provide an early clue to skeletal disease. This is seen with the soft-tissue swelling around joints early in rheumatoid arthritis or local soft-tissue swelling in the initial stages of osteomyelitis. In general, however, clinical examination is superior to radiology for demonstrating such changes.

Soft-tissue calcification may be seen in a wide variety of conditions: hypercalcaemia — especially if prolonged; tissue damage due to trauma or infarction; in connective tissue disorders following muscle or soft-tissue inflammation — especially dermatomyositis, or in scleroderma where calcification may be part of the CRST syndrome — Raynaud's phenomenon, sclerodactyly and telangiectasia in addition to calcinosis (see also p. 84). Such soft-tissue calcification may occur at any site in the body but is often most marked in the vascular tree, the kidney, and around the joints.

Tomography

Tomography has a useful role in the evaluation of musculoskeletal disorders, though it has been partly replaced by computerized tomography where this is available. It can be of value in clarifying a dubious abnormality on a plain film or, more precisely, in documenting the extent of bone destruction or the nature of a periosteal reaction. Sometimes, it can also help in assessing joint changes, especially in the spine.

Scintigraphy

Since the early 1970s bone scanning has been carried out using a number of phosphate compounds labelled with technetium-99m. When these labelled phosphates are injected intravenously a proportion of the radiopharmaceutical becomes bound to the skeleton while unbound activity is cleared by the kidney. Binding to the skeleton appears to be due to incorporation of the whole technetium-99m-phosphate complex into the crystal lattice of the bone. The precise factors governing distribution of the bone scanning agents within the skeleton are not yet fully understood but both bone blood flow and metabolic activity in the bone are thought to be important.

Increased bone blood flow alone may result in increased uptake of bone scanning agents. Decreased or absent bone blood flow may produce a local cold spot due to decreased tracer uptake in aseptic necrosis of bone. Changes in bone blood flow, however, are probably less important than metabolic changes in the bone in determining bone scanning agent distribution. In particular, an increase in osteoblastic activity will produce a local increase in bone scanning agent deposition and a hot spot on the scan.

Bone scanning is now normally carried out using a diphosphonate compound, of which ^{99}Tc^m-methylene diphosphonate is the most often employed. For an adult, the usual injected activity is 600 MBq. Images are obtained 2–4 hours after injection as, by this time, there is good uptake of the tracer in the bone and unbound soft-tissue activity has been largely cleared by the kidneys. Between injection and imaging, the patient is encouraged to drink a large volume of fluid (at least 1 litre) unless there is a clear contraindication to this. The high fluid intake maintains a high urine flow which may augment renal clearance of the unbound activity. More importantly the high urine flow reduces the radiation dose to the bladder both by dilution and diuresis.

The bone scan protocol is modified when information is sought on the vascularity of the lesion. This is most often required when a differential diagnosis exists of soft-tissue inflammation or osteomyelitis. Under these circumstances a 'three-phase' study is performed. The bone tracer is injected as a bolus with the affected area under the gamma camera and a first-pass flow study obtained. This is followed by a static blood pool image several minutes after injection and standard delayed images at 2–4 hours (Fig. 8.1). The use of three-phase bone scintigraphy is discussed further on p. 250.

The normal bone scan is characterized by uniformity of activity within a structure and symmetry about the midline. Apparent areas of increased uptake are normal at the sternoclavicular joints, the inferior angles of the scapulae, the sacroiliac joints and the ischia. The kidneys are often visualized and, occasionally, renal pathology, such as hydronephrosis may be detected (Fig. 8.2). In patients with actively growing bones, increased activity is seen in the epiphyses.

Most skeletal pathologies appear to be associated with some local increase in osteoblastic activity and thus the characteristic bone scan abnormality is a hot spot. The changes in osteoblastic activity usually occur before any structural change is visible on bone X-ray, thus accounting for the greater sensitivity of the bone scan in various pathologies. The osteoblastic response will occur in many different pathologies so that a focal hot spot on a bone scan is not specific for a particular disease process. Sometimes the pattern of scan abnormality will suggest a particular diagnosis but often the bone scan findings have to be interpreted in the light of other diagnostic information, including bone X-rays.

Advantages and disadvantages

The major advantage of the bone scan over the bone X-ray is its increased sensitivity in many skeletal pathologies. It is also possible to evaluate areas of the skeleton which are difficult to visualize well on standard X-rays, such as the ribs, the sternum, the scapulae, the base of the skull and the sacrum. A further major advantage of the bone scan is that a whole body image can be obtained at the same radiation dose as the examination of a single anatomical area and the dosage is considerably lower than that resulting from a full body X-ray skeletal survey.

The main disadvantage of the bone scan is the lack of specificity: the same scan appearance can result from a variety of disease processes. Thus, it is often necessary to follow a whole body scan with localized X-rays of the area of scintigraphic abnormality. If the diagnosis is still not clear, CT or bone biopsy may be required. Because of the excretion of unbound activity by the kidney, problems may be encountered sometimes with the pelvic images on the bone scan. For example, in a patient who is unable to empty a distended bladder, the radioactive urine present may obscure a large part of the pelvis. To minimize this problem patients are encouraged to empty the bladder as much as possible immediately before the pelvic views are obtained. If this is not possible, and good images of the pelvis are critical, the insertion of a urinary catheter and the acquisition of special views or images obtained 18–24 hours after injection may help. In patients who are incontinent, the leakage of urine may produce apparent hot spots.

A final problem with skeletal scintigraphy is the purely lytic lesion. Most lesions which are osteolytic on X-ray have some associated osteoblastic activity and produce a hot spot on scintigraphy. A purely lytic lesion will not. If large enough, a lytic lesion may show up as an area of decreased tracer uptake on the bone scan, but often is too small to be seen at all. The most common example of this is in multiple myeloma where the bone scan is often relatively normal despite extensive skeletal involvement seen on X-rays.

Gallium-67 and indium-111-labelled leucocyte scintigraphy is of value in some musculoskeletal disorders and will be discussed under the individual conditions.

Photon absorptiometry

The assessment of bone mineral density from standard radiographs is subjective. This makes the identification of early osteoporosis difficult and means that quantitative sequential measurements are impossible. In recent years there has been considerable interest in the use of photon absorptiometry in obtaining this information. In this technique the absorption by the bones of radioactive photons emitted from a source is measured. The measurements are usually made in the radius or in the lumbar spine. The radial measurements are possible using a single source ('single photon') but for the vertebral measurements two sources of different energies ('dual photon') are required. The measurements of bone mineral density obtained at these sites and by these techniques correlate well with direct measurements on biopsy specimens. Measurements at other sites, such as the femoral neck, have been attempted with varying success. A number of commercially produced single and dual photon absorptiometry devices are now available. The clinical value of photon absorptiometry is discussed on p. 247.

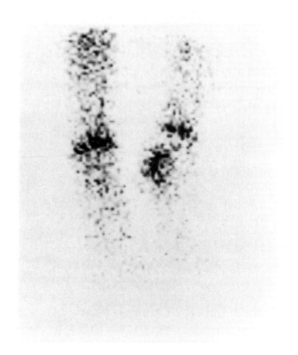

(a)

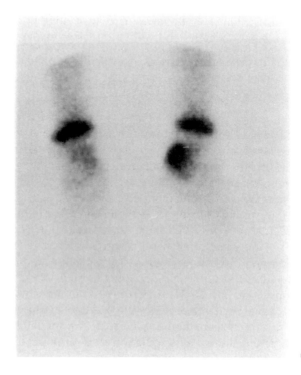
(b)

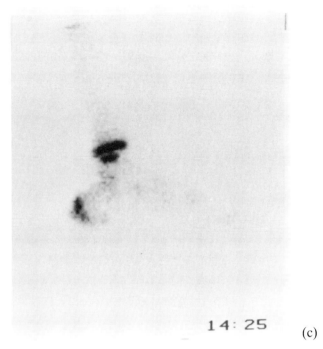

14:25 (c)

Fig. 8.1 Three-phase bone scan from a 15 year old boy with osteomyelitis of the calcaneum. (a) The flow image shows increased linear uptake bilaterally in the epiphysis (a normal finding) and locally in the left calcaneum. (b) The blood pool image obtained 5 min later shows generalized soft-tissue activity and more localized uptake in the epiphysis and left calcaneum. (c) On the 4 h delayed image of the left ankle soft-tissue activity has cleared, normal bone uptake is faintly visualized and there is intense uptake in the epiphysis (normal) and in the calcaneum.

Radionuclide whole body retention techniques using $^{99}Tc^m$-diphosphonates and radioisotopes of calcium have been applied to the study of skeletal disorders. As they do not involve imaging they are not within the remit of this book and, in any case, are of research interest only at present with no role in routine clinical practice.

Bone marrow scintigraphy using ^{111}In-chloride or $^{99}Tc^m$-colloids has been used to demonstrate marrow involvement by malignant disease. The area of affected marrow will have lost its normal haemopoietic and reticuloendothelial tissue and will show up as a cold spot. This can be of value when seeking a biopsy site in disorders such as lymphoma or in following the course of marrow regeneration after high dose cytotoxic therapy or radiotherapy. The use of marrow scanning agents in detecting metastatic disease confined to the marrow and thus not shown on bone scintigraphy is currently being examined. Initial results in some tumours, such as breast and lung cancer, suggest that this may be a useful approach.

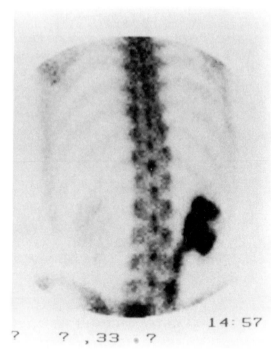

Fig. 8.2 Bone scan of lower thoracic and lumbar spine (posterior view). The bony structures are normal but a right hydronephrosis and right hydroureter are apparent. This was clinically unsuspected and was the cause of back pain in a patient with a past history of pelvic malignancy.

Computed Tomography

CT has played an increasing role in recent years in the study of bone disorders. Because of the high density discrimination possible with CT, bone destruction can be visualized earlier than on conventional radiographs. The prolonged imaging time and the radiation dose make whole body CT impractical in screening for bone disease, but CT examination of limited areas is helpful in following-up and clarifying conventional X-ray or scintigraphic abnormalities which are equivocal or of uncertain aetiology. This is particularly the case in areas such as the spine or pelvis which are difficult to assess by conventional techniques. Percutaneous bone biopsy, guided by CT, can establish the nature of skeletal lesions of uncertain aetiology.

As well as demonstrating cortical lesions in bone, CT can, on occasions, show intramedullary abnormalities due to local spread of tumour. This can be helpful when planning surgery for primary bone tumours, but as similar appearances can be produced by blood, pus or marrow fibrosis it is a guide only and does not replace histological examination of biopsy material.

Contrast-enhanced CT is valuable in assessing the vascularity of skeletal lesions. This can be contributory in making a diagnosis of primary bone tumour and preoperatively when excision or biopsy is planned. Computed tomography in conjuction with myelography has a major role in investigating spinal column disease including degenerative disc disease and spinal canal tumours (see Chapter 9).

In addition to demonstrating bony changes, CT can also show associated soft-tissue masses and indicate their vascularity. This may be helpful in vertebral osteomyelitis with paraspinal abscess formation or in malignant disease, particularly primary bone tumours. CT can be employed to detect local bone invasion by some soft-tissue tumours such as pharyngeal and intraoral cancers and carcinoma of the bronchus. Such information aids in deciding on the operability of the lesions and in planning the surgical approach or radiotherapy fields.

Attempts have been made to measure bone density by CT techniques. Specialized machines or sophisticated software adaptations are used in some centres, but others have attempted to produce information on vertebral body density using a standard CT whole body scanner and minor software changes only. The reliability of these methods is still to be established, but in view of the restricted availability of scanning time it is possible that, even if they are successful, they will be used primarily for research purposes.

The use of CT in degenerative disease of the spine has already been mentioned. Recent reports suggest that it may also be of value in patients with possible sacroiliitis in whom standard radiology is unhelpful.

Ultrasound

Ultrasound has only a small role to play in skeletal diseases. In skeletal tumours it may be used to demonstrate soft-tissue masses but, when available, the use of CT is preferable. Attempts have been made to measure the width of the spinal canal by ultrasound in spinal stenosis. Spinal stenosis is covered fully in Chapter 9.

There are a limited number of applications for ultrasound in joint disease. The width of hyaline cartilage in large joints can be assessed by ultrasound but the clinical utility of this is not established. Ultrasound examination can, occasionally, be useful in confirming the presence of a joint effusion when the clinical signs are equivocal. The strongest indication for ultrasound in rheumatology is the confirmation of a ruptured Baker's cyst as the cause of calf pain mimicking a deep vein thrombosis (see Chapter 3). Recent experience suggests that ultrasound may be a useful aid in diagnosing congenital dislocation of the hip.

Arthrography

Arthrography may be used in both small and large joints. A strict aseptic technique is essential and local sepsis is an absolute contraindication because of the risk of producing a septic arthritis. At the time of arthrography, fluid should be aspirated from the joint, sent for culture and examined microscopically for cell content and crystals. For small joints, a single-contrast method is used, but in larger joints, air as well as radiographic contrast medium is injected to allow a double-contrast technique. As with all contrast examinations there is a risk of allergic reaction, but the incidence with arthrography is very low, probably because of the poor absorption of the material from the joint space. Multiple projections of the joint under study should be aspirated from the joint, sent for culture and examined microscopically for cell content and crystals. For small joints, a single-contrast

In the knee, arthrography can be used to confirm rupture of a Baker's cyst but ultrasound, which is non-invasive, should be used first when this diagnosis is suspected. Arthrography can be used to confirm the presence of meniscal tears, and has an accuracy approximately equal to that of arthroscopy. Arthrography is better than arthroscopy in diagnosing tears of the posterior horn of the medial meniscus while arthroscopy is better for demonstrating lesions of the articular cartilage and cruciate ligaments. In the hip, arthrography can be used to monitor progress in Legg-Perthe's disease and can be helpful in doubtful cases of congenital dislocation of the hip (CDH) or when attempts to reduce CDH have failed. Hip arthrography can also be used to demonstrate loosening of a prosthetic joint. The other main indications for arthrography are to assess rotator cuff damage in the shoulder, to verify articular surface damage or the presence of a loose body in the elbow, and in patients with facial pain, examination of the temporomandibular joint.

Angiography

Angiography was previously used in the diagnosis of suspected bone tumours and in planning surgical management. While still used occasionally (e.g. in pelvic trauma) it has largely been replaced by contrast CT studies.

Magnetic Resonance Imaging

The role of MRI in assessing bone conditions is developing rapidly. Early results suggest that it allows elegant anatomical demonstration of spinal disorders and internal derangements of the knee, and that it may be of major clinical value in patients with such lesions.

Bone Tumours

Differential Diagnosis

The patient who presents with a suspected bone tumour may have either a primary or secondary lesion. Clinical features, including the age of the patient, the presence of premalignant conditions such as Paget's disease of the bone or a past history of malignancy known to metastasize to bone may help in differentiating primary bone tumours from metastases. Imaging investigations are often useful in this decision. Some of the X-ray characteristics which help to indicate a benign or malignant lesion have already been discussed.

The presence of multiple lesions is highly suggestive of metastatic tumour, though primary bone tumours may occasionally arise in multicentric sites, especially multiple exostoses or multiple enchondromata. Rarely, an osteogenic sarcoma may present with distant bony metastases in addition to the primary lesion. Bony metastases are seen at presentation more commonly in Ewing's sarcoma. The presence of multiple lesions can be detected by bone radiographs in some cases but generally this information can be obtained more easily from bone scintigraphy.

The site of the lesion may also aid differentiation of primary bone tumours from metastases. While primary tumours may arise anywhere in the skeleton, they are more common in the periarticular regions, especially around the hips and knees. By contrast, lesions of the axial skeleton and areas of haemopoietic activity are more likely to be due to metastases from extraskeletal tumours — which characteristically spread from the bone marrow — or to haematological malignancies.

The nature of the radiological change may help in the differential diagnosis. Multiple osteolytic lesions are likely to be due to metastatic disease as are mixed osteolytic and osteoblastic lesions.

Metastases from prostatic cancer are characteristically osteoblastic on X-ray. Most other common tumours produce predominantly osteolytic lesions or mixed lesions, although about 10 per cent of breast cancer metastases are osteoblastic. Metastases which appear osteolytic radiologically almost always have some associated osteoblastic response and thus show as hot spots on bone scintigraphy.

Marked bone expansion and an associated soft-tissue mass are highly suggestive of a primary lesion rather than a metastasis and, as already mentioned, the presence of subperiosteal calcification is more often due to primary tumours. Certain primary bone tumours show characteristic changes on X-ray. A detailed discussion of these features is beyond the scope of this book but it is worth mentioning the bone destruction and expansion associated with spicular calcification seen in some cases of osteogenic sarcoma (Fig. 8.3), the onion skin pattern of calcification which may occur in Ewing's sarcoma, and the radiolucent appearance with small contained areas of calcification seen in tumours of cartilaginous origin.

Even after detailed radiological evaluation, including additional techniques such as CT, doubt may remain as to the nature of a lesion and, in particular, whether it is benign or malignant. Under these circumstances biopsy is mandatory.

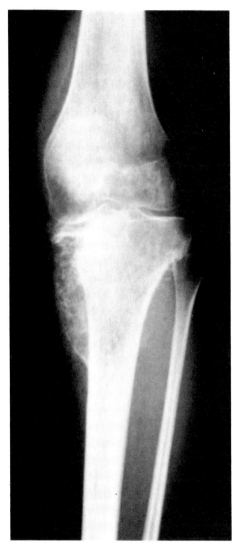

Fig. 8.3   Osteogenic sarcoma of upper tibia showing bone destruction and a calcified expanding mass.

Primary Bone Tumours

The role of bone imaging in the differential diagnosis of primary bone tumours has been considered in the preceding section. In patients with a radiological diagnosis of a primary malignant bone tumour, a biopsy of the lesion is always required for histological confirmation.

Imaging investigations are useful in planning therapy for primary bone tumours. The presence of a soft-tissue mass can be shown on standard X-ray but precise delineation of its extent can be better achieved by CT. Local bony extent can also be assessed by standard radiographs, obtained in multiple projections. Bone scintigraphy probably offers no advantage here and may overestimate the extent of the primary lesion. Gallium-67 scintigraphy has been used in some centres for assessing local extent of primary bone tumours but has not gained widespread acceptance. Some primary bone tumours, notably osteogenic sarcoma, may produce local 'skip' metastases within the medulla of the affected bones. When present they may radically affect the treatment by, for example, altering the amputation level or necessitating a more extensive radiotherapy field. CT can detect skip metastases but false-positive results can be produced by normal variations or benign lesions within the medullary cavity and biopsy confirmation of the CT abnormality is always necessary.

In some primary bone tumours, metastatic spread is common at presentation. The lung is a common site and a chest X-ray is part of the initial investigation. In patients thought to have potentially resectable primary bone tumours, preoperative CT of the lungs should always be obtained, even when the chest X-ray is normal and the patient has no lung symptoms. Distant bone metastases occur in about 2 per cent of patients presenting with osteogenic sarcoma. A bone scan at presentation is worthwhile in order to identify this group and avoid unnecessary amputations. Confirmation of metastases by bone biopsy should always be obtained as false-positive results do occur. Soft-tissue metastases from osteosarcoma may show up as hot spots on the bone scan, but this is unpredictable and the bone scan cannot be used as a reliable method of demonstrating these lesions. The incidence of distant bone metastases at presentation in Ewing's sarcoma is higher, ranging from 5 per cent to 25 per cent in reported series. A bone scan at presentation is essential in this tumour.

Following primary therapy for the tumour, cytotoxic treatment is now often employed and appears to have produced an improved survival rate. Regular CT of the lung during follow-up is indicated to allow early detection of lung metastases which may respond to a change in chemotherapy or, when they are localized, to surgical resection. Local recurrence at the primary tumour site can be detected clinically or by straight X-rays, in some cases, but both bone scintigraphy and gallium-67 imaging have also been shown to be of value. In patients receiving chemotherapy for osteosarcoma, bone metastases are a not infrequent site of recurrence and are best demonstrated by bone scintigraphy.

Bone Metastases

Bone scintigraphy is now generally accepted as the best method of demonstrating bone metastases from most tumours (Fig. 8.4). It must be emphasized again, however, that the bone scan findings are often non-specific and frequently require correlation with results of other imaging investigations and occasionally with bone biopsy.

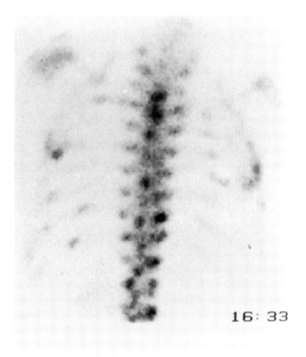

Fig. 8.4 Bone scan of thoracic spine and ribs (posterior view) showing multiple hot spots due to bone metastases from breast cancer.

Indications for bone imaging in patients with extraskeletal malignancies can be summarized as:

1. Staging at the time of presentation
2. Routine follow-up
3. Clinical suspicion of bone metastases
4. Assessment of therapy for bone metastases.

The role of bone imaging at the time of presentation is controversial. A full X-ray skeletal survey is indicated at presentation in patients with myeloma; in most other tumours the primary screening test for bone metastases is bone scintigraphy. It is generally accepted that a bone scan is indicated in all patients presenting with prostatic cancer because of the high incidence of bone metastases which are clinically occult and the existence of effective therapy for these secondaries. A preoperative bone scan is also justified in patients with potentially operable lung cancer to avoid the mortality and morbidity of unnecessary thoracotomy in those shown to have metastases.

The place of bone scanning in carcinoma of the breast is more debatable. The incidence of bone metastases in clinically advanced (Stage III or IV) breast cancer is high (25 per cent) and all such patients merit bone scintigraphy at presentation. The yield of positive bone scans in early (Stage

I or II) breast cancer is lower (around 2–4 per cent) and must be balanced against false-positive results. The low yield, the trend to less radical surgery and the poor efficacy of therapies for asymptomatic bone metastases have resulted in many centres dispensing with a bone scan at presentation in clinically early breast cancer, especially if other factors such as axillary node sampling and oestrogen receptor status indicate a relatively good prognosis.

In children with neuroblastoma there is a high incidence of bone metastases at presentation. Bone scintigraphy and standard radiology may each detect metastases missed by the other technique. They are thus complementary in this group of patients and a bone scan and full X-ray skeletal survey should be performed at presentation. Recent work has indicated that whole body scintigraphy with ^{131}I-mIBG is of great value in showing metastases from neuroblastoma. A case can also be made for bone scintigraphy in all children presenting with Wilms' tumour. In adults with renal carcinoma, bone scintigraphy should be reserved for those with bone pain or biochemical evidence of metastases as the return on the investigation is otherwise very small. In other extraskeletal malignancies, routine bone scanning at presentation is not justifiable and should be reserved for patients with clinical or biochemical suspicion of metastases and for patients with locally advanced disease.

With the exception of carcinoma of the prostate, there is no general agreement on the need for routine bone imaging in the follow-up of patients with extraskeletal malignancies, and in most centres patients are now studied only if they have clinical features suggesting recurrence. If a patient develops bone pain or biochemical evidence of bone metastases, a bone scan is now usually obtained to confirm the diagnosis. As always, the results must be correlated with bone X-rays. Particularly in the case of solitary abnormalities, sites which are abnormal on bone scan but normal on X-ray should be studied by special coned views, tomography, CT, and, if necessary, bone biopsy. Local X-rays should always be obtained of sites of bone pain as this may demonstrate osteolytic lesions missed by scintigraphy. Additionally, X-rays of the long bones will be valuable in showing potential sites of pathological fracture which may benefit from prophylactic pinning. Even when bone X-rays show an obvious metastasis, the bone scan is valuable in detecting additional lesions, either locally or at distant sites in the skeleton.

The assessment of response of bone metastases to therapy can be partially achieved by clinical and biochemical features but imaging can add useful information. Both standard radiology and bone scintigraphy are employed though each has some limitations.

The International Union Against Cancer (UIAC) has defined radiological criteria for the response of breast cancer to systemic therapy. A complete objective regression consists of calcification of all osteolytic lesions without the appearance of new lesions. A partial objective regression has occurred when there are similar changes in some lesions, with no new lesions appearing. Similar criteria can be used for metastases from other tumours. The main problems with these criteria are that the changes are open to considerable interobserver variation, that they take several months to become apparent with restoration of a normal trabecular pattern requiring several years, and finally that they do not lend themselves to evaluation of lesions which are sclerotic prior to therapy.

In bone scintigraphy a reduction in the number of lesions is an indication of response to therapy; an increased number of lesions generally implies disease progression. However, osteoblastic activity produced by healing metastases in the months after institution of chemotherapy may initially result in an apparent worsening bone scan ('the flare response') before the improvement is seen. Changes in the intensity of bone scan agent uptake in metastases are an unreliable guide to response to therapy.

Bone Trauma

The majority of injuries to the musculoskeletal system can be satisfactorily imaged using conventional radiography alone. At least two projections at right angles to each other are obtained as fractures can be easily missed if only one view is taken. Fractures, dislocations, epiphyseal and metaphyseal trauma, and joint effusions can generally be identified using standard projections. Occasionally, oblique views or special views for a particular site, e.g. the radial head, will afford further information. It is often helpful to radiograph the contralateral anatomical site when assessing the immature skeleton, particularly when epiphyseal injury is suspected. This can be particularly useful around the elbow joint if doubt remains after conventional views. In this situation, it is imperative not to miss a displaced epiphysis within the joint.

There are certain situations where a fracture may not be apparent at the initial presentation. A well recognized example of this is the scaphoid fracture. If not seen on the standard four-view examination, then a repeat X-ray 10 days later is appropriate when there is good clinical evidence of injury. Stress fractures, also, may not be obvious on initial presentation. Subsequent radiographs over a period of time may show the development of a periosteal reaction which may be the only evidence of fracture. Both scaphoid fractures and stress fractures, at any site, can be demonstrated elegantly by scintigraphy. Scintigraphic changes are typically present before radiographic change, e.g. within 2 days for a scaphoid fracture. It should be noted that a bone scan image can be obtained through a plaster of Paris cast. Whether scintigraphy or follow-up radiography is performed depends on availability.

Conventional tomography is another technique which can be helpful in the assessment of skeletal trauma. It is particularly useful in spinal fractures; these are discussed in more detail in Chapter 9. Tomography can also be useful in the demonstration of bone healing and in the assessment of the success or failure of bone graft procedures.

Plain radiographs should also be used to assess the soft tissue. Most types of glass are opaque to X-rays and therefore when there is clinical suspicion of a foreign body, relevant views should be obtained. It is helpful to the radiologist to ensure that the suspected site of the foreign body is clearly indicated by a simple diagram on the request card.

Aseptic bone necrosis may develop following bone trauma producing an appearance of local sclerosis and structural collapse on X-ray. Bone scintigraphy can be used to demonstrate aseptic necrosis before X-ray changes are seen. The initial bone scan abnormality is an area of decreased tracer uptake which is gradually replaced by increased activity as healing occurs. The femoral head is the commonest site for aseptic necrosis. Other causes of aseptic necrosis include high dose steroid therapy (e.g. in kidney transplant recipients), diabetes mellitus, fractured neck of femur, Legg-Perthe's disease, sickle cell disease and decompression sickness.

The clinician should be alerted to the possibility of non-accidental injury when a child demonstrates fractures of varying age and healing, inappropriate trauma for the history given, and more subtle evidence such as periosteal reaction along long bones, metacarpals and phalanges etc. In the very young, scintigraphy is a useful alternative to skeletal surveys.

As has already been stated, conventional radiographic techniques usually suffice when imaging skeletal trauma. However, ultrasound and CT can both have useful roles. Ultrasound is particularly helpful in detecting joint effusions, it can identify dislocated hips in infants, and it can assess the adjacent soft-tissue organs which may have been damaged at the time of trauma. CT affords not only unique cross-section information of both bone and soft-tissue structures, but it can also assess the adjacent vascular tree using dynamic postcontrast scanning. CT is particularly useful in the assessment of spinal trauma and fractures involving the acetabulum when bony fragments can be displayed within the spinal canal and joint space, respectively. At the time of writing, the place of MRI has yet to be established.

One of the more contentious issues in any discussion of the role of imaging in skeletal trauma is the medicolegal aspect. When to X-ray and when not to X-ray is a problem faced by all junior hospital doctors. On the one hand, Health Authorities are attempting to save money by reducing unnecessary radiographs while on the other hand, medical defence societies emphasize: 'if in doubt, X-ray'. Concern over medicolegal consequences results in a large proportion of the X-rays taken in an Accident and Emergency Department being of doubtful clinical consequence. As far as fractures to the skull are concerned, the Royal College of Radiologists have issued guidelines as to when radiographs are indicated. The reader is referred to Chapter 9 for further discussion on this subject.

Metabolic Bone Diseases

Paget's Disease

Paget's disease (osteitis deformans) is a common disorder in older people in Western countries, occurring in about 3 per cent of the UK population over the age of 40 years. It is usually asymptomatic but may produce bone pain, compression neuropathy, pathological fracture or, rarely, osteosarcoma. The aetiology of the disease is unknown but it may be due to a viral infection. The pathological changes in the bone are of excessive osteoclastic resorption with a secondary increase in osteoblastic activity, which, however, is disorganized. Paget's disease is not strictly speaking a metabolic bone disease but clinically resembles one and is therefore considered with them.

Paget's disease produces both osteolytic and osteosclerotic changes on bone X-rays (Fig. 8.5). The osteolytic changes consist of advancing V-shaped zones of resorption in the cortex of long bones, as areas of bone resorption in the skull (osteoporosis circumscripta) or as small areas of cortical bone resorption at any site in the skeleton. The osteolytic areas often have osteoblastic changes associated with them. The osteoblastic changes produce bone deformity and expansion with widening of the cortex and disruption of the normal trabecular pattern. Small fissure fractures of

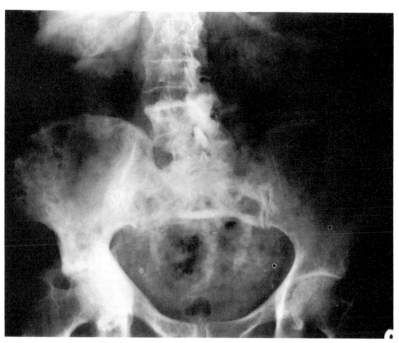

Fig. 8.5 Paget's disease of the right hemipelvis and lower lumbar spine showing the characteristic mixture of osteosclerosis and osteolytic changes.

the cortex are frequently seen. Secondary joint changes of osteoarthritis are commonly seen in Paget's disease.

The bone scan changes of Paget's disease consist of intensely increased uptake in a uniform fashion in the affected bone (Fig. 8.6). In the long bones the scan shows increased uptake progressing from the end of the bone. The scan changes are often characteristic but on occasions it may be difficult to differentiate Paget's disease from sclerotic bone metastases, especially from prostatic cancer.

Comparative studies have shown that both radiology and scintigraphy will detect Paget's lesions missed by the other modality, but that scintigraphy is by far the more sensitive technique. Areas of Paget's disease picked up by radiology and missed by scintigraphy often turn out to be inactive or burnt out.

Paget's disease involves a single skeletal site (monostotic Paget's disease) in around 25 per cent of cases and is multicentric (polyostotic) in the remainder. Any area in the skeleton may be involved, but common sites include the pelvis (often in a hemipelvic distribution), the femora, the lumbar spine and the skull.

Asymptomatic Paget's disease may be diagnosed as an incidental finding on X-ray. It may also be suspected from a finding of elevated serum alkaline phosphatase. In the latter situation, a bone scan followed, if doubt persists, by local X-ray, will confirm the diagnosis.

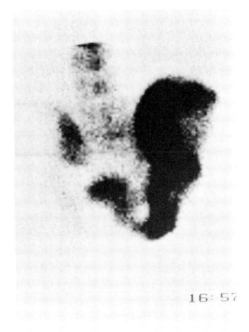

Fig. 8.6 Bone scan of pelvis (posterior view) showing diffusely increased uptake in the right hemipelvis due to Paget's disease.

A patient with bone pain from Paget's disease may require a bone scan to determine whether other areas of skeletal involvement are present but local X-ray is essential. X-ray is required to detect the presence of fissure or pathological fractures (Fig. 8.7) as these may be obscured by intense local activity on bone scanning. Bone pain may also result from the, fortunately rare, complication of sarcomatous degeneration. This can be detected on bone X-ray: initially as a subcortical area of bone destruction, later increasing to produce an osteolytic lesion with an ill-defined boundary and little or no associated osteoblastic reaction. CT may be useful in defining the

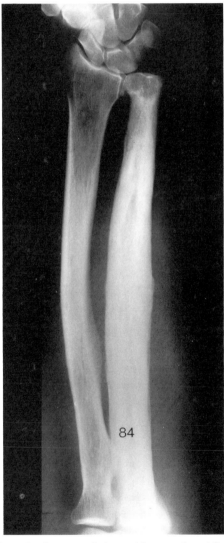

Fig. 8.7 Severe Paget's disease of ulna with associated fissure fracture in the mid shaft.

extent of the lesion and any associated soft-tissue mass. Biopsy is required for confirmation of the diagnosis. Scintigraphy is unreliable in the diagnosis of Paget's osteosarcoma, though it may be suspected if the scan shows an area of relatively decreased tracer uptake within a Pagetic lesion, particularly if that area also shows increased gallium-67 uptake.

The bone scan may show decrease in tracer uptake with successful treatment of Paget's disease by agents such as calcitonin or diphosphonate, while X-rays may show a reduction in the osteolytic areas and normalization of the trabecular pattern. Imaging, however, has only a limited role in evaluating the response of Paget's disease to treatment as this is usually achieved by a combination of clinical assessment and changes in biochemical parameters.

Osteoporosis

Osteoporosis is a common condition — predominantly affecting postmenopausal women — in which there is loss of bone mass. It is important because it increases the risk of fracture, particularly in the femoral neck, the vertebrae and the wrist. The presence of osteoporosis is often suspected in patients who present with a clinical fracture.

Radiologically, osteoporosis is shown as a loss of bone density with decreased contrast between the bone and soft tissue. In the spine, osteoporosis often produces prominence of vertical bone trabeculae due to loss of the horizontal trabeculae. Less commonly, there may be a change in the vertebral shape on the lateral view of the dorsolumbar spine with the bodies assuming a biconcave shape. Vertebral collapse is common in advanced osteoporosis (Fig. 8.8) and may consist of either complete collapse or anterior wedging of the vertebral body, with preservation of the pedicles — a useful differentiating feature compared with metastatic disease.

The assessment of bone density from standard X-rays is very subjective, leading to interobserver variations and to difficulty in following the progress of the disorder and the response to treatment. Quantitation of bone mass at specific sites in the skeleton by single or dual photon absorptiometry is increasingly employed and is now used for the diagnosis of osteoporosis, the evaluation of treatment and in population studies of the frequency of the disease. The precise role of these techniques is not yet established but it is hoped that in individuals at high risk of osteoporosis sequential measurements of bone mass will identify at an early stage when therapies such as oestrogen replacement might be efficacious.

The bone scan does not have a role in the diagnosis of osteoporosis. In severe cases, the bone uptake of the tracer may be poor with a 'washed out' appearance in the appendicular and axial skeleton. In most patients with osteoporosis, however, the intensity of uptake of the bone scan agent appears normal. The bone scan can be helpful, on occasions, in detecting fractures not visible on X-ray. In addition, in an osteoporotic patient with back pain and vertebral collapse the finding of increased activity on

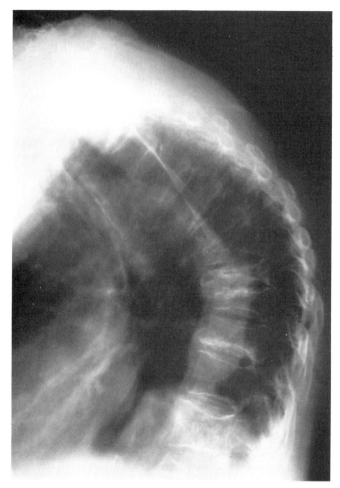

Fig. 8.8 Advanced osteoporosis of thoracic spine. There is biconcavity of the vertebral bodies and collapse with anterior wedging of several vertebrae.

bone scan in the collapsed vertebra indicates that the fracture is recent and is likely to account for the patient's symptoms.

Disuse of a limb, either temporarily because of immobilization following fracture or, more permanently, for example, after a stroke, often produces local osteoporosis. In severe cases, there is marked loss of bone density and cortical thinning accompanied clinically by soft-tissue swelling, smooth, shiny skin and considerable pain. This condition is known as Sudeck's atrophy. Clinical improvement and restoration of a normal radiological appearance occur if the limb is remobilized, but this may take months.

Rickets and Osteomalacia

Vitamin D deficiency produces the syndrome of rickets in children and

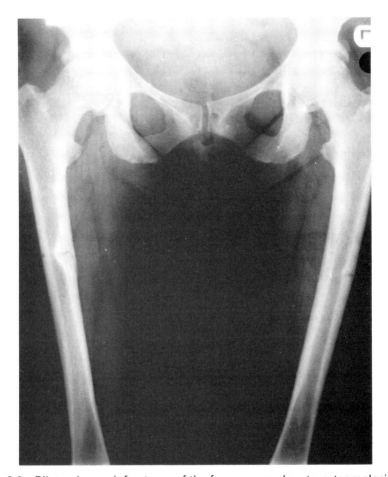

Fig. 8.9 Bilateral pseudofractures of the femur secondary to osteomalacia.

osteomalacia in adults. The deficiency may occur due to inadequate dietary uptake, malabsorption or renal disease. Anticonvulsant treatment may also cause osteomalacia.

Osteomalacia is due to the inadequate mineralization of the osteoid matrix of the bone. Radiologically, it is characterized by some decrease in bone density, with loss of trabeculae and some cortical thinning. The X-ray appearances are often indistinguishable from osteoporosis. A more specific feature suggesting osteomalacia is the presence of pseudofractures (Looser's zones) consisting of radiolucent bands up to several centimetres long. The common sites for these lesions are the femur (Fig. 8.9), especially the inner aspect near the femoral neck, the pelvis, the outer edge of the scapula, the upper fibula and the metatarsals. Bone scans in osteomalacia generally reveal increased tracer uptake in the skeleton, with poor or absent renal images and increased uptake in the periarticular areas (the 'superscan'). Prominence of the costochondral junctions may also occur. The superscan is not specific for osteomalacia and may occur in any metabolic bone disease. Occasionally, diffuse skeletal metastases, especially from prostatic cancer, may produce a superscan appearance. Pseudofractures produce local hot spots on the bone scan. The bone scan may be useful in confirming the presence of a pseudofracture at a site of bone pain when X-ray is normal or equivocal. This is particularly true in the ribs where bone scanning shows a high incidence of pseudofractures.

In rickets, the X-ray changes are most marked in the metaphysis which shows cupping and a poorly defined border. The growth plate (zone of provisional calcification) is widened. These changes are often most apparent at the wrists and knees in young children, and in the pelvis in older children. Deformity, especially bowing of long bones, may occur due to bone softening.

Bone Infection

Acute bone infection may arise by haematogenous spread or by direct spread from an infected wound or fracture or from pyogenic arthritis. Haematogenous spread may occur at any age but is more common in children. The original site of the infection may be relatively trivial; the patient presents with bone pain, usually with accompanying fever. Clinical differentiation between osteomyelitis and overlying soft-tissue infection may be difficult. This is especially true of diabetics who represent a high risk group.

Radiological changes are often delayed in acute osteomyelitis. For the first 10–14 days, usually the only radiological change is some soft-tissue swelling and loss of normal intermuscular planes. Subsequently, a local loss of bone density and periosteal reaction may appear. If the infection is not controlled, progressive bone destruction occurs. Vertebral osteomyelitis may affect either the margin or central portion of the vertebral body and may be associated with vertebral disc destruction or with paravertebral abscesses which may cause pressure on local structures. In children, there may be marked destruction of the intervertebral disc with minimal bone involvement — so-called 'discitis'.

Scintigraphic changes occur early in acute osteomyelitis but are often subtle at first. The earliest bone scan change consists of a focus of decreased uptake followed by a hot spot. Bone scanning has been found to be sensitive in detecting acute osteomyelitis in all age groups except the neonate in whom false-negative studies are not uncommon. Differentiation of scan changes of acute osteomyelitis from cellulitis can be difficult on occasions but can often be resolved by three-phase bone scanning. Both cellulitis and osteomyelitis can cause increased uptake on all three phases of the study. In osteomyelitis, however, the increase is more marked on the delayed than on the flow or blood pool images. In cellulitis the delayed images do not show more intense tracer uptake compared to the early phases.

Recrudescence of bone infection at a previously affected site is a difficult diagnostic problem. X-ray may show some new subperiosteal reaction but this is frequently absent. Bone scintigraphy is valueless in distinguishing inactive but healing osteomyelitis from recurrent infection. Increased uptake of indium-111-labelled leucocytes suggests the presence of infection. Gallium-67 uptake may be increased in healing osteomyelitis but ^{67}Ga uptake which is greater than that of the bone scan agent, favours recrudescence of infection.

Patients with orthopaedic appliances, such as artificial joints, may develop osteomyelitis but, clinically, it may be very difficult to differentiate this from loosening of the prosthesis. The presence of infection is suggested by an irregular area of bone resorption adjacent to the prosthesis and subperiosteal new bone formation. Arthrography can be useful by showing periprosthetic loosening or by outlining the sinus tract. Standard bone scintigraphy is of debatable value. Diffusely increased uptake around the prosthesis has been claimed to be indicative of infection by some workers, but others have found the scintigraphic changes of infection and loosening to be indistinguishable. Gallium-67 uptake which is incongruent with the bone scan changes appears to be a more reliable sign of infection, but is not completely specific. Increased uptake of indium-111-labelled leucocytes is specific for periprosthetic infection but false-negatives do occur, probably because of low-grade infection in many cases.

Bone infection by tuberculosis is due to haematogenous spread from either a primary infection or from a postprimary focus. Common sites are the greater trochanter of the femur, and the spine. Tuberculous osteitis is characterized by considerable bone destruction and the production of pus which may later calcify. The metaphysis is the common area to be involved and articular cartilage or intravertebral disc destruction are often seen. The lesion does not have much surrounding sclerosis usually and the subperiosteal reaction is often slight. In the case of spinal lesions, the anterior aspect of the vertebral body is most often affected producing a local kyphosis or gibbus. A paravertebral abscess may occur, and, in the lumbar spine, take the form of tracking down the psoas sheath, producing lateral bulging of the psoas outlines on X-ray. Computed tomography is valuable in demonstrating the extent of any soft-tissue mass in vertebral TB. Both the bone scan and the gallium-67 scan will show locally increased uptake in bony tuberculosis while indium-111-labelled leucocyte images are normal. Scintigraphy is rarely required in bony TB but may be useful in indicating a site for biopsy in suspected cases with a negative or equivocal X-ray.

Arthritis

Arthritis is a common condition, especially in older individuals, and imaging plays a significant role in confirming the presence of joint abnormalities, in the differential diagnosis of joint problems and in monitoring their progress. Plain radiographs are most frequently used and are sufficient in most situations. Computed tomography is occasionally of assistance, especially in the spine and sacroiliac (SI) joints; magnetic resonance imaging may be of considerable value here in the future. Ultrasound is rarely employed but may be useful in the diagnosis of Baker's cyst rupture (see Chapter 3), in the study of articular cartilage loss and in confirming the presence of a joint effusion. Bone scan changes occur early in arthritis but bone scanning is not routinely employed in arthritis. Occasionally, it can be valuable in demonstrating changes in a symptomatic joint which has no clinical or radiological abnormality. The main use for scintigraphy in arthritis is in the search for alternative causes of musculoskeletal pain, such as infection, tumour or Paget's disease.

The differential diagnosis of arthritis is complex and, in many cases, relies upon clinical features and specific blood tests. Radiology may help in the process by demonstrating particular radiographic changes within joints or the distribution of the affected joints.

Osteoarthritis

Osteoarthritis is the most common form of joint disorder and occurs due to degenerative changes. It may affect any joint but, in the absence of predisposing factors such as trauma or destructive arthritis, is seen most commonly in the large joints of the lower limb, the distal interphalangeal joints of the hands, the metacarpophalangeal joint of the thumb, and in the spine.

Radiologically, osteoarthritis is characterized by new bone formation and, later, by narrowing of the joint space. The commonest form of new bone formation in osteoarthritis is the appearance of spurs of compact bone, known as osteophytes, on joint margins. Osteophytes may cause marked irregularity of the joint surface and may also become detached forming intra-articular loose bodies. In the distal interphalangeal joints of the hands new bone formation leads to the production of Heberden's nodes which can be palpated clinically. Smaller flecks of calcification (ossicles) may be seen in the soft tissues around the interphalangeal joints.

Loss of articular cartilage occurs early in the course of osteoarthritis but is often a late radiographic change with loss of joint space. Subarticular ('pseudo') cysts are commonly seen in affected hips and shoulders, while subperiosteal new bone formation may be seen on the inferior surface of the femoral neck. Osteoporosis is not a feature of osteoarthritis, although patients with osteoporosis are at increased risk of the degenerative changes of osteoarthritis. Osteoarthritis does not lead to bony fusion (ankylosis) of joints and occurrence of this change indicates a destructive arthropathy.

In the spine, the posterior interfacetal (or apophyseal) joints are true synovial joints and may develop osteoarthritis. The intervertebral disc joints do not have synovium but frequently develop changes similar to osteoarthritis. The primary abnormality in the intervertebral joints is usually loss of disc substance producing a decrease in disc height and osteophyte formation; the combined changes are known as spondylosis. Spondylosis may occur at any level in the spine but is especially common in the cervical spine. Osteophytes may encroach on the spinal foramina producing nerve root compression. Reduction in the diameter of the spinal canal diameter by osteophytes or prolapsed intervertebral discs causes spinal cord compression. This occurs most commonly in the cervical and lumbar regions. This subject and the role of imaging in its investigation is described in Chapter 9.

Inflammatory Arthritis

Destructive or inflammatory arthritis may be caused by a variety of pathological processes, notably rheumatoid arthritis and other collagen diseases. Destructive articular changes are also well recognized in psoriasis, various bowel abnormalities and in gout.

In patients presenting with acute polyarthritis, X-ray changes are often slight and confined to the soft tissues in the early stages. When obtaining radiographs, a full body joint survey is rarely required and X-ray examination should be confined to symptomatic joints and asymptomatic joints with a high risk of involvement, such as the small joints of the hands and the feet.

Rheumatoid arthritis may occur in any synovial joint but most commonly produces symmetrical involvement of the proximal interphalangeal and metacarpophalangeal joints in the hands, the wrists and the metatarsophalangeal joints of the feet. It is rare to find X-ray changes in joints which have not caused symptoms either at presentation or in the past. The diagnosis of rheumatoid arthritis is made by a combination of the clinical features, blood findings — especially the presence of rheumatoid factor in 70 per cent of those affected — and characteristic radiographic changes.

The earliest X-ray changes consist of soft-tissue swelling which is seen in the region of clinically affected joints (Fig. 8.10). Imaging does not have any advantage over clinical examination in showing soft-tissue swelling. An early bone change is osteoporosis which starts in the juxta-articular bone. More generalized osteoporosis can occur in patients with long-standing severe rheumatoid arthritis, especially if they are immobile or receive prolonged steroid therapy. Destructive changes are manifest as erosions — the most characteristic radiographic change in rheumatoid arthritis. They are first seen on the joint margins and early in the course of the disease special projections may be necessary to show them. With progressive disease there is more marked bone destruction and X-rays at regular intervals can provide a useful index of disease progression. Arthritis mutilans, in which the bone ends are resorbed, is regarded as a particularly severe variant of rheumatoid arthritis (Fig. 8.11).

Changes in the joint space in rheumatoid arthritis vary according to the

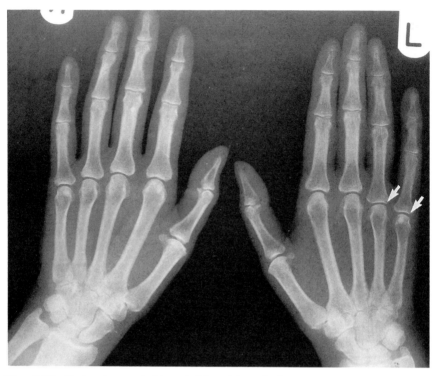

Fig. 8.10 Early rheumatoid arthritis with soft-tissue swelling around the interphalangeal and metacarpophalangeal joints and some periarticular osteoporosis. Some early erosions are seen, e.g. at the proximal end of the left third and fourth proximal phalanges (arrows).

stage of the disease. Initially, the joints are widened due to the presence of an effusion, but later on there is cartilage destruction and loss of joint space. The onset of secondary osteoarthritic changes is almost inevitable in any affected weight-bearing joint. In the end-stages of the disease there is bony ankylosis of affected joints.

The combination of bone destruction, loss of joint space and soft-tissue changes leads to joint deformity. Ulnar deviation of the phalanges at the metacarpophalangeal joint, subluxation of the interphalangeal joints to produce bouttonière, swan-neck and Z-shaped deformity of the fingers and subluxation of the metatarsophalangeal joints can all be demonstrated radiologically but are easily diagnosed on clinical examination. Atlantoaxial subluxation in the cervical spine is very difficult to detect clinically and is potentially lethal. Vertical or lateral subluxation of the atlantoaxial joint may occur but, more commonly, the problem is increased separation between the odontoid peg of the axis and the posterior border of the anterior arch of the atlas. This abnormality may lead to cervical cord or medulla oblongata compression by the odontoid peg, which can produce

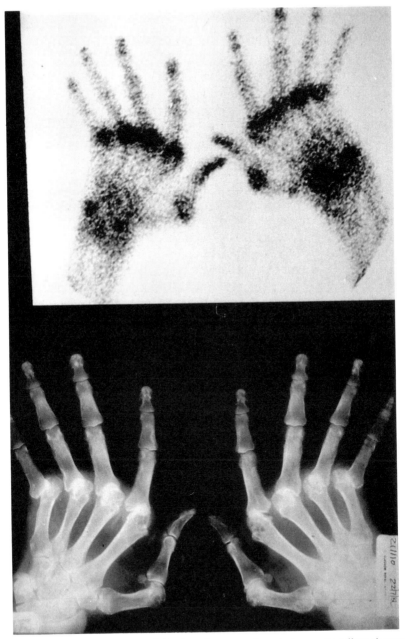

Fig. 8.11   Arthritis mutilans of the hands secondary to long-standing rheumatoid arthritis. The bone scan shows increased tracer uptake in some affected joints.

severe neurological deficit or even sudden death. It can be demonstrated in most cases by a lateral X-ray taken with the neck in flexion. In doubtful cases computed tomography may clarify the state and position of the odontoid peg.

Rheumatoid patients are especially at risk of cervical cord compression during general anaesthesia because of loss of muscle tone and muscle reflexes. Any patient with rheumatoid arthritis should have a flexion and extension cervical spine X-ray prior to a general anaesthetic so that the anaesthetist can be alerted of any potentially dangerous abnormality.

Extra-articular manifestations may occur in rheumatoid arthritis. Imaging methods can be used to demonstrate some of them, including pleural effusions, pulmonary nodules, pulmonary fibrosis, pericardial effusion and splenomegaly (known as Felty's syndrome if there is associated leucopenia).

Juvenile rheumatoid arthritis (Still's disease) may show radiographic changes identical to those seen in adults but periosteal changes are commoner in children. Spinal, sacroiliac and large joint involvement are more often seen in Still's disease. Epiphyseal damage or hyperaemia can produce either growth retardation or acceleration, respectively.

Arthritis is a common feature of systemic lupus erythematosus but, even with severe joint changes, erosions are rarely seen. Polyarthritis is seen in approximately 25 per cent of patients with scleroderma. Erosive changes are rare. Bone resorption occurs, especially in the distal phalanges, though other sites may be involved. Subcutaneous calcification in the fingers can occur and may be part of the CRST syndrome (see Chapter 3).

Joint changes are the major feature of ankylosing spondylitis. The principal pathological abnormalities occur at the bony insertions of tendons, ligaments and joint capsules. The most characteristic finding is sacroiliitis and the diagnosis cannot be made in its absence. Sacroiliac joint (SI) abnormalities are usually bilateral and symmetrical, though on occasions one SI joint may be more severely affected. Erosion of the joint margin causes initial widening and loss of the joint space, especially in its lower half. Detection of early forms of sacroiliitis on plain X-ray is difficult, but quantitative analysis of bone scintigraphy may help. Later, there is bone sclerosis around the joint followed by loss of joint space and bony ankylosis. Spinal involvement (spondylitis) is also an important component of ankylosing spondylitis, with marked new bone formation producing squaring of the vertebral bodies and the appearance of bone spurs. Calcification of the spinal ligaments is a later feature, resulting in the end-stage picture of bamboo spine (Fig. 8.12). In addition to loss of mobility, the rigid spine is prone to fractures, which may be multiple. Peripheral joint involvement, especially of the hip, is well recognized in ankylosing spondylitis and leads to X-ray changes identical to rheumatoid arthritis.

A picture resembling ankylosing spondylitis is seen in some patients with inflammatory bowel disease and in Reiter's disease. Frequently, patients

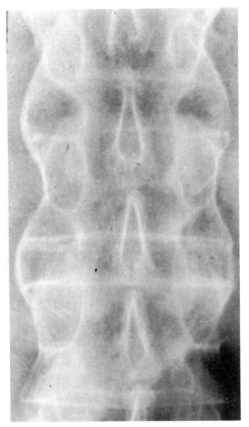

Fig. 8.12 Calcification of the lateral ligaments of the thoracic spine in ankylosing spondylitis — 'bamboo spine'.

with Reiter's disease also show 'spurring' of the calcaneum in which there is new bone formation involving the plantar surface of the bone in particular.

About 5 per cent of patients with psoriasis develop arthritis. Some cases are similar to rheumatoid arthritis, others involve predominantly the sacroiliac joints or spine. Distal interphalangeal joint involvement is more typical of psoriatic arthritis than rheumatoid arthritis and can be associated with absorption of the terminal phalanges.

A diagnosis of gout can be proven absolutely only by showing the presence of urate crystals in the synovial fluid of an affected joint, although certain X-ray changes may support the diagnosis. Punched-out erosions around the joint are seen, cartilaginous destruction tends to occur late and osteoporosis is rare. Tophi are soft-tissue depositions of urate; they may calcify but more often appear as eccentric soft-tissue swellings.

Pyogenic Arthritis

Infective arthritis may occur by either direct joint infection due to trauma, or spread from adjacent osteitis. Blood-borne infection may also occur and there is an increased risk in patients with pre-existing inflammatory arthritis, e.g. rheumatoid arthritis. Pyogenic arthritis is usually suspected on clinical grounds and confirmed by joint aspiration. Rapid and progressive cartilaginous destruction, with erosions and destruction of subarticular bone occur if the diagnosis is not made promptly. Indium-111-labelled white cell imaging can sometimes be helpful in detecting infection in a joint which is affected by a chronic arthropathy. Gallium-67 uptake does not differentiate inflammatory arthritis from infection.

Hypertrophic Pulmonary Osteoarthropathy

Patients with bronchial carcinoma and, less commonly, other forms of lung disease, may develop hypertrophic pulmonary osteoarthropathy (HPOA). In this condition, there is an inflammatory polyarthritis, most often affecting the wrists and ankles. In addition to soft-tissue swelling, X-rays also show subperiosteal new bone formation, usually most marked in the distal areas of the tibial and forearm bones. Cure of the underlying lung condition may result in resolution of the HPOA.

Haemoglobinopathies

Crises in sickle cell disease often produce polyarthropathy probably due to sub-chondral infarction. Skeletal lesions may develop due to thrombosis of nutrient vessels and bone infarction. The resultant aseptic necrosis is seen most commonly in the head of the femur but may occur also in the humeral head, the patella and vertebral bodies. Similar bone infarcts have been described in severe thalassaemia.

Sickle cell disease, and other haemolytic anaemias, may cause marrow extension with widening of the medullary cavities and thinning of the arteries. Thickening of the vault of the skull may also occur in the congenital haemolytic anaemias.

Further Reading

Fogelman, I. (Ed.) (1987). *Bone Scanning in Clinical Practice*. Springer–Verlag, London and Berlin.

Griffiths, H.J. (1987). *Basic Bone Radiology*. Appleton and Lange, Connecticut and California.

McKillop, J.H. (1986). Radionuclide bone imaging for staging and follow-up of secondary malignancy. *Clinics in Oncology* **5**, 125. W.B. Saunders & Co., Eastbourne and Philadelphia.

Park, W.M. and Hughes, S.P.F. (1987). *Orthopaedic Radiology*. Blackwell Scientific Publications, Oxford.

Scott, W.W., Magid, D. and Fishman, E.K. (Eds.) (1987). *Computed Tomography of the Musculoskeletal System*. Churchill Livingstone, Edinburgh.

9

THE CENTRAL NERVOUS SYSTEM

Imaging the Skull and Intracranial Structures

Skull Radiology and Tomography

A full examination of the skull consists of five standard projections, generally. These include a lateral, two anteroposterior and one posteroanterior projections, and a submentovertical (base of skull) view. Standard views can be supplemented by coned views, specialized views, e.g. optic canals, internal auditory canals and tomography where necessary. CT has largely replaced the need for the more specialized projections and conventional tomography is now requested somewhat infrequently when compared with the pre-CT era. In practice, a modified skull series is performed, tailored to suit the particular clinical problem.

Probably the commonest reason for a skull X-ray request is trauma. The most important signs are:

1. The position of a calcified pineal and any evidence of shift from the midline
2. The presence of air within the skull vault
3. The presence of a depressed fracture
4. The status of the craniovertebral junction.

The above radiological features must be assessed carefully by any clinician or casualty officer who is assessing a patient with head trauma. Whether a simple linear skull fracture is identified or not may be of minor consequence to the patient when compared with the seriousness of missing pineal shift or the presence of air in the head. A pineal shift of 3 mm or more is significant and further investigation is urgently required. The imaging of head injury is discussed further on pp. 280–83.

Other than trauma, skull radiographs may be helpful in a variety of clinical situations. Metastases and multiple myeloma deposits may be demonstrated, and metabolic disorders such as hyperparathyroidism will give characteristic skull appearances. Intracranial calcification may reflect a tumour, a vascular malformation, or old infection such as tuberculous meningitis or toxoplasmosis. Haematological disorders such as thalassaemia, leukaemia, and histiocytosis can all give skull vault changes as can Paget's disease and acromegaly.

A full discussion on the uses of specialized skull projections are beyond the scope of this text. Suffice to say that these can demonstrate in some detail the various cranial foramina and canals. Changes in these structures may give direct and indirect information concerning intracranial pathology.

Computed Tomography (CT)

Neuroradiological practice has been transformed by the development of CT. Unpleasant and invasive procedures such as air encephalography have all but become extinct and the need for diagnostic angiography has fallen significantly. CT will give cross-sectional images from the skull base to the vertex demonstrating the ventricular system, the white and grey matter the skull vault and any intracranial calcification. Masses will be demonstrated and their effects on adjacent structures assessed. A standard head CT examination usually involves 10 mm slices from base to vertex but modified examination techniques are used when the pituitary fossa, posterior fossa, middle ear or orbits require more detailed study. By using modern computer technology, reconstructions in the coronal and sagittal or oblique planes can be produced using data acquired during axial scanning, (Fig. 9.1a and b). Scanning can also be performed in the direct coronal plane and, occasionally, depending on the type of CT machine, in the direct sagittal plane. However, the ability of modern scanners to provide re-formatted images of high quality has meant that direct scanning, other than in the axial plane, is rarely required.

A variety of contrast media can be used to gain further information during a head CT study. Most often, standard intravenous radiographic contrast is appropriate, and contrast enhancement will give a clearer demonstration of vascular lesions, e.g. arteriovenous malformations and aneurysms, or pathology which disturbs the blood-brain barrier, e.g. tumours and infective processes. Investigations of dementia, hydrocephalus, or acute head injury generally do not require routine intravenous contrast. Two other types of contrast may be used. First, in assessing pathology around the cranio-vertebral junction and at the cerebellopontine angle, CT scans after intrathecal contrast may prove helpful. Contrast is introduced into the thecal canal by lumbar puncture. Contrast can then be run in a cephalad direction to demonstrate structures around the skull base and also the basal cisterns, if required. Second, the introduction of a small volume of filtered air or carbon dioxide, again using a lumbar puncture approach, may help in the diagnosis of a small mass in the cerebellopontine angle, e.g. an acoustic

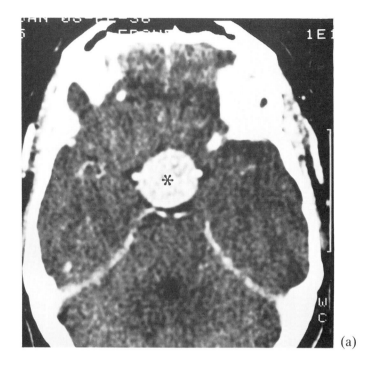

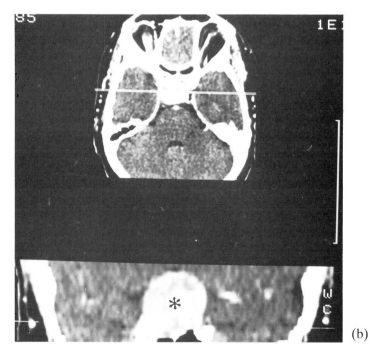

Fig. 9.1 Pituitary neoplasm. (a) CT axial section showing a large pituitary tumour (asterisk). (b) Coronal reconstruction through the pituitary fossa demonstrating the superior extent of the pituitary tumour.

neuroma. In the normal patient, air can be manoeuvred into the internal auditory meatus and the cisterns around the cerebellopontine angle. Tumours are demonstrated by showing either a blockage of the internal auditory canal or their projection into the cerebello-pontine cistern, when the tumour mass will be outlined by air.

Magnetic Resonance Imaging (MRI)

The technique of magnetic resonance imaging (MRI) has been discussed in some detail in Chapter 1. At the time of writing, this technique has had its greatest impact in the investigation of patients with neurological and neurosurgical disorders. When compared with CT, MRI provides additional information resulting from its superior display of soft-tissue density. As well as demonstrating mass lesions and their effect on adjacent tissue, MRI can assess the tissue characteristics of the mass itself plus some assessment of spread into adjacent structures. Another advantage of MRI over CT is the absence of image degradation due to streak artefact production which is a recurring problem with CT scanning, particularly of the posterior fossa and the base of the skull. MRI is free of such difficulties and therefore gives a better anatomical display of the structures around the brain stem and craniovertebral angle, particularly when direct sagittal scanning is used.

Ultrasound

One of the earliest uses of ultrasound in radiological practice involved the detection of a midline shift. The technique was used in the pre-CT era to demonstrate the effects on midline structures of mass lesions in and around the brain. Like much of early ultrasound, there was considerable operator dependence which reduced the credibility of the examination. Ultrasound in neurological practice has come to the fore with its use in neonatal imaging. Real-time high resolution ultrasound provides a speedy, simple, and harmless means of assessing the intracranial contents of a neonate. The anterior fontanelle provides access for the ultrasound beam into the neonatal cranial vault. The ventricular system can be assessed and intracranial haemorrhage be it ventricular, intracerebral, or intracerebellar can be diagnosed. Because of the relative simplicity of the technique, portable examinations have proved invaluable on special care baby units and the like. Ultrasound is particularly useful for follow-up examinations in order to monitor progression or deterioration. Whereas ultrasound is very useful in the detection of neonatal haemorrhage and hydrocephalus, CT is still required for the assessment of much of neonatal intracranial pathology. Ultrasound will give no information about diffuse cerebral pathology such as white matter disease.

Angiography

Although the number of angiograms performed has fallen with the widespread availability of CT, angiography retains an important place in neuroradiological practice. The arrival of digital angiography has added the further dimension of outpatient procedures which can be performed either intravenously through a central catheter or intrarterially using small calibre high flow catheters. Digital angiography is particularly useful in the assessment of vessels in the neck. An initial digital study can be performed on patients being investigated for transient ischaemic attacks without recourse to arch aortography or selective conventional catheterization of each neck vessel. Currently, opinion is divided on the role of digital vascular imaging (DVI) in neuroradiological practice. Some centres require conventional angiographic studies before a patient is considered for surgical treatment. DVI is appropriate in certain clinical situations, e.g. venous sinus thrombosis and conventional studies may then be avoided. Moreover, if for technical reasons arterial access proves impossible, then a digital venous study may provide sufficient information to allow a management decision to be made.

While traditionally direct carotid or vertebral arterial puncture was used, virtually all angiography in modern neuroradiological practice involves a puncture of a femoral artery followed by pre-shaped catheters being advanced into the arch of the aorta and manipulated into the required vessel for study. Femoral puncture is safer and has less associated morbidity than either direct puncture or angiography via the brachial approach. A femoral puncture also gives the operator much more versatility than with the direct puncture approach.

Indications for angiography include the following:

1. Demonstration of the vascular supply to a neoplasm or vascular malformation previously diagnosed on CT prior to surgery (Fig. 9.2)
2. Demonstration of the source of haemorrhage in a patient presenting with a subarachnoid haemorrhage, e.g. an aneurysm or arteriovenous malformation
3. Assistance in the differential diagnosis of certain intracranial masses, e.g. meningioma versus glioma
4. Assessment of the neck vessels in a patient presenting with transient ischaemic attacks.

Phlebography

The venous drainage of the intracranial or neck circulation is usually demonstrated on the late films of an arterial study. Requirements for a detailed knowledge of the significance of subtle changes in venous distribution or venous filling has been diminished following the arrival of CT. One of the few indications for direct phlebography is in the examination of pathology in and around the orbits. Orbital venography may give useful

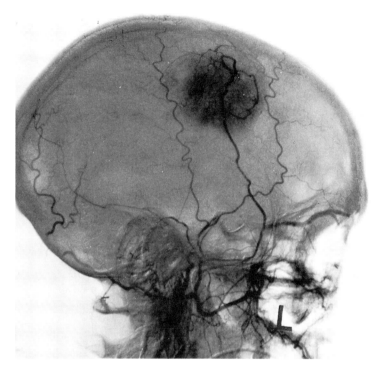

Fig. 9.2 Meningioma. External carotid angiogram. Characteristic tumour 'blush' of a meningioma.

information although once again, CT is the method of choice for orbital imaging.

Air Encephalography and Air Ventriculography

As these investigations are now only of historical interest, they will not be discussed further.

Scintigraphy

The use of scintigraphy has fallen from its past position as one of the first investigations of a possible space-occupying lesion or cerebrovascular accident, to that of acting largely as a backup for CT. Its current role is, mainly, to relieve the pressure of work on a CT unit or to provide a screening service when CT is not immediately available.

Following intravenous injection, a radiopharmaceutical will accumulate at sites where there is blood-brain barrier breakdown. The exact mechanisms are complex and not fully understood but accumulation of isotope will occur within neoplasms, inflammatory disorders either focal or diffuse, and intracranial collections, e.g. a subdural haematoma.

Several radiopharmaceuticals are used. The demonstration of an intracranial abnormality depends on some local disturbance of the blood-brain barrier which allows an increased concentration of activity at the site of the abnormality when contrasted with normal brain substance which does not allow the uptake and concentration of isotope. The commonest agents for assessing intracranial pathology and cerebral blood flow are $^{99}Tc^m$-pertechnetate or $^{99}Tc^m$-DTPA, with images usually being obtained 1 hour following injection. Equivocal abnormalities can sometimes be clarified by obtaining images 3–4 hours after the injection of $^{99}Tc^m$-glucoheptonate. Dynamic scintigraphic studies can assess cerebral blood flow patterns. By observing the early vascular phase of a brain scintiscan, an effective radionuclide cerebral angiogram is obtained. This allows assessment of blood flow through the neck vessels and also within the brain substance, and any asymmetry between hemispheres can be identified.

One of the major limitations of brain scintigraphy is its lack of specificity. Any process which disrupts the blood-brain barrier will appear as an abnormal accumulation of activity. Differentiation between neoplasm, infection, or even infarction may not be possible. Resolution with current technology is limited to masses measuring 1 cm or more. Iodine-labelled amphetamine derivatives and $^{99}Tc^m$-HMPAO are currently being evaluated for their detection of areas of impaired brain metabolism or perfusion in cerebrovascular disease and dementia.

Radionuclide cisternography

Indium-111-DTPA is the commonest radiopharmaceutical used to assess CSF flow. The tracer is introduced into the CSF circulation generally by a standard lumbar puncture approach, but occasionally directly into the cisterna magna, or directly into a ventricle at surgery. The uses of radionuclide cisternography include:

1.   Assessment of hydrocephalus
2.   Assessment of shunt patency
3.   Detection and localization of the source of CSF rhinorrhoea.

Following the introduction of a suitable radiopharmaceutical, scans are obtained at intervals up to 48 hours. A normal scintigraphic pattern will demonstrate activity within the basal cisterns within 1–3 hours, within the sylvian fissures and hemispheric fissure by 3–6 hours, and at 24 hours, activity will surround both hemispheres. In the normal patient, at no time will activity be identified within the ventricular system. When ventricular reflux occurs, this generally indicates the presence of a communicating hydrocephalus (see p. 278). Where CT is available, the indications for radionuclide cisternography are rather limited, but the technique may be of some value in differentiating between a normal pressure hydrocephalus and cerebral atrophy. Normal pressure hydrocephalus is one of the few treatable causes of presenile dementia, its diagnosis is of great importance. Scintigraphically, a patient with normal pressure hydrocephalus tends to show ventricular reflux. In cerebral atrophy, ventricular reflux is absent.

Following the insertion of a ventriculocardiac or ventriculoperitoneal shunt for the relief of hydrocephalus, it is important to demonstrate whether the shunt is patent or compromised should the clinical picture alter. This can be assessed scintigraphically by injecting isotope into the shunt reservoir and assessing its rate of clearance.

CSF rhinorrhoea is often a difficult diagnostic problem. The majority of cases are secondary to head trauma but, occasionally, CSF rhinorrhoea may occur spontaneously or following surgery. The site of a leak may be detected by assessing the presence of any radioactivity within nasal pledgets which are placed within the nasal cavity prior to injection of isotope into the lumbar canal.

When surgery is being considered, high resolution CT with or without intrathecal contrast usually gives much more specific surgical information.

Imaging the Spinal Cord and Related Structures

Spinal Radiographs

Any assessment of spinal pathology should include plain radiographs prior to consideration of any more invasive techniques. At least two projections are performed and the overall status of bony mineralization, focal or multiple areas of bone destruction, or the presence of any developmental anomalies can be assessed. Apart from the bony structures, the disc spaces and paravertebral regions will also be demonstrated. When metastatic disease is suggested, i.e. destruction of one or more pedicles, or if an inflammatory process is likely, i.e. evidence of a discitis, then a chest X-ray should be performed to ensure that a bronchogenic carcinoma or active pulmonary tubercle is not responsible. Conventional tomography is often helpful in assessing spinal pathology when the plain films are non-diagnostic. This applies not uncommonly in cases of spinal trauma where a fracture or disruption of a facet joint may be appreciated on tomography. Similarly, defects of the pars interarticularis or the morphology of a spinal bone tumour, may be more fully assessed using tomographic techniques.

Plain radiographs may show a number of developmental variations. One of the commonest is spina bifida occulta, but various transitional vertebrae in the lumbar, dorsal or cervical spine, and the occasional intervertebral fusion in the cervical spine represent not uncommon anomalies — sometimes with no significant clinical relevance.

Computed Tomography (CT)

The use of CT is becoming more widespread in the assessment of disc disease and some centres now prefer CT to myelography as the initial investigation of the patient with back pain. Typically, 3 mm axial slices are obtained at the level of the suspected lesion. Modern computer technology allows high quality reconstructions in the coronal and sagittal planes, but to obtain images of diagnostic quality, 3 mm or 1·5 mm slices are required. The balance between useful additional information and increased radiation dose then requires careful consideration, particularly in the younger patient.

CT can also be performed after the introduction of intrathecal contrast. Generally, the CT study follows a conventional myelogram. A delay between examinations of several hours is often required to allow dilution of the radiographic contrast. Intrathecal contrast allows greater differentiation between the cord itself, the CSF, and the epidural structures. Lumbar and sacral nerve roots will usually be demonstrated, and lateral disc lesions which are not shown on myelography will be identified. When a contrast CT is performed to demonstrate a syrinx or other spinal cystic structure, delayed scans are usually required. These may involve scanning 24 hours post-myelography before the syrinx opacifies with contrast.

Spinal CT has proved particularly valuable in the assessment of the trauma patient. Spinal fractures will be elegantly demonstrated but, in addition, any bony fragments compromising the neural canal or damage to the paraspinal and retroperitoneal tissues will be demonstrated at the same examination.

Some developmental disorders are best demonstrated after intrathecal contrast. One such example is diastematomyelia. As well as showing the split cord and the diastem, associated abnormalities such as lipomata and tethering of the cord will be displayed.

Magnetic Resonance Imaging (MRI)

Spinal cord imaging has also been revolutionized by the use of MRI. Prior to this technique being available, most examinations of the spinal cord required some form of contrast media to separate the spinal cord elements from the surrounding bony canal. With MRI, the spinal cord can be demonstrated in considerable detail without the need for any contrast material. The availability of direct sagittal and coronal scanning results in the production of images containing high anatomical detail, compared with the inferior resolution of reconstructed CT images.

Scintigraphy

The most sensitive method of demonstrating spinal metastatic disease is by a bone scintigram. Early images taken soon after injection will give some assessment as to the vascularity of a spinal lesion. The demonstration of a hot spot on a bone scintigram is not specific for metastatic disease. Degenerative changes will also produce focal areas of increased activity but, generally, the distribution of the abnormalities will suggest whether degenerative or neoplastic disease is present. When doubt remains, a combination of plain radiographs and scintigraphy may resolve any difficulty.

Gallium or labelled white cell scanning may be helpful in the detection of intraspinal or paraspinal abscess formation.

Myelography

The development of water-soluble agents suitable for intrathecal injection has significantly diminished the morbidity associated with myelography. Myelography involves an intrathecal injection of radiographic contrast media by either a lumbar or cervical puncture. The lumbar approach is the more common. Under fluoroscopic control, contrast can then be run in a cephalad direction up into the neck. The whole spinal cord can, therefore, be examined from the same lumbar puncture. However, dilution of contrast with CSF will occur which inevitably leads to some reduction in the quality of the cervical examination. When a detailed study of the cervical region is required, a lateral cervical puncture is often more appropriate. Careful consideration is required as to whether a cervical puncture is appropriate in some clinical situations. When a high spinal block is likely, then it is best to puncture via the lumbar route in the first instance. Should a block be confirmed, then contrast will usually be required from the other direction if the full extent of the lesion was not shown at the initial puncture. However, in this situation, care will be taken to ensure that the minimum volume of contrast is used thereby reducing the likelihood of complications due to intracranial spillover. Similarly, if pathology is suspected at the craniovertebral junction, e.g. cerebellar tonsilar ectopia, a cervical puncture is inappropriate as the initial technique and again contrast should be run from the lumbar region.

As has been stated already, the complications associated with myelography have fallen significantly since water-soluble contrast has replaced the oil-based media. At the time of writing, there have been no proven cases of arachnoiditis associated with water-soluble contrast. This particular complication was a frequent sequel to oil-based contrast media. The majority of complications associated with myelography are associated with the lumbar puncture itself. Lumbar puncture headache is due to a combination of factors, but by using a small gauge needle, postlumbar puncture leakage will be minimized and therefore severe headache prevented. Early water-soluble intrathecal contrast agents had a definite epileptogenic potential. The newer agents are safer but caution is required when a myelogram is performed in a patient either with epilepsy or on drugs which lower the threshold to epilepsy.

Discography

Discography remains a further imaging option in the diagnosis or confirmation of degenerative disc disease. If a patient's signs and symptoms are reproduced by injecting contrast or saline into a particular disc and not its immediate neighbours, then it is reasonable to presume that this disc level is the one responsible for the patient's presenting complaint. Surgery may be avoided in some patients by the injection of chemolytic agents into the symptomatic disc.

Spinal Angiography

Spinal angiography is often a prolonged and technically demanding investigation usually performed under general anaesthesia. It is used in the investigation of spinal arteriovenous malformations or in the assessment of vascular tumours of the spinal cord. Contrast is injected into lumbar, intercostal, and often vertebral arteries depending on the level under investigation. When a vascular lesion is shown, the therapeutic option of embolization may be appropriate should surgery be considered unfavourable.

Phlebography

Spinal phlebography can be applied to the assessment of degenerative disc disease. Bulging discs may cause compression or lack of filling of the adjacent venous system and the technique has occasionally been useful when myelography gave equivocal appearances. However, CT has largely replaced any need for phlebography in modern radiological practice.

Intracranial Pathology

Intracranial Calcification

There are numerous causes of intracranial calcification both physiological and pathological. Physiological causes include calcification of the pineal gland, the choroid plexus, the habenular commissure, and the arachnoid granulations. As a rough rule of thumb, the percentage of the population with pineal calcification on a skull radiograph is proportional to their age, i.e. 70 per cent of 70 year olds will show pineal calcification, whereas it will be present in only 20 per cent of 20 year olds. Before pronouncing the pineal to be central or displaced, it must be identified on both a frontal and lateral view. This avoids any confusion with a calcified choroid plexus or calcification within the basal ganglia. Pineal calcification is shown in virtually 100 per cent of adult patients on CT.

Calcification of the dura and related ligamentous structures may be demonstrated on a skull radiograph and is generally of no pathological consequence. The falx cerebrae, diaphragma sella, petroclinoid and inter-clinoid ligaments can all demonstrate physiological calcification. When excessive, it may be pathological. Prominent dural calcification is seen in some patients on long-term renal dialysis and following intrathecal chemotherapy.

Basal ganglia calcification can occur in a wide range of conditions. Metabolic causes include hypoparathyroidism, pseudohypoparathyroidism and, rarely, hyperparathyroidism. It can be idiopathic or familial and may result from various poisons, e.g. lead or carbon monoxide.

Calcification around the pituitary fossa can be physiological, e.g. in the petroclinoid ligament or carotid syphon in the elderly, or can reflect past infection, e.g. tuberculous meningitis. It may, however, indicate more serious pathology including the presence of a tumour, e.g. craniopharyn-gioma, meningioma, pituitary adenoma or chordoma, and other supporting signs, such as those of raised intracranial pressure, should be searched for. Sometimes calcification in an aneurysm may mimic a parasellar tumour but generally the ring type calcification is characteristic of a vascular aetiology.

Skull radiographs are often requested to 'exclude an intracranial tumour'. As only 5–10 per cent of gliomas and 10–15 per cent of meningiomas actually calcify, it is only the minority of brain tumours which will be demonstrated on a skull radiograph. However, indirect signs such as erosion of the dorsum sellae indicating raised intracranial pressure, a shift of the pineal, or changes in the skull vault, e.g. thickening or thinning of bone may indicate the presence of a space-occupying lesion.

Vascular malformations calcify in only about 15–30 per cent of cases. The 'tramline' calcification which occurs in Sturge–Weber syndrome is character-istic but more often the type of calcification which occurs in a vascular malformation is indistinguishable from that seen in an intracranial neo-plasm.

Multiple areas of intracranial calcification may result from past intracra-nial infection, e.g. toxoplasmosis, cytomegalovirus, and cysticercosis. Mul-tiple foci of calcification may reflect tubers in tuberose sclerosis where the distribution is typically periventricular and other skeletal stigmata of the disease may be identified. Widespread dural calcification may occur after some forms of chemotherapy, e.g. intrathecal methotrexate for leukaemia. When multiple areas of focal calcification are shown, then metastases will not account for the appearances.

Intracranial Tumours

As has already been stated, the not infrequent request for skull radiography on the basis of '? intracranial tumour' yields positive findings rather uncommonly, especially in the early stages of illness. A lateral skull radiograph will detect most abnormalities if they are present. Calcification occurs in only the minority of intracranial tumours (see p. 271); exceptions include a craniopharyngioma in a child — but indirect evidence of a tumour may be demonstrated by identifying signs of raised intracranial pressure. In adults, raised intracranial pressure (ICP) results in changes at the pituitary fossa ranging from early erosion of the lamina dura at the base of the dorsum sellae, to truncation of the dorsum, and deossification and enlargement of the sella turcica. Changes in children are age dependent. In infancy, the cranium enlarges, convolutional markings on the skull vault become more prominent, and the sutures show widening and increased interdigitations. Suture diastasis is a feature of raised ICP in older children who may also show changes at the pituitary fossa.

Computed tomography (CT)

Intracranial tumours are either primary or secondary. It should be remembered that a metastasis is the commonest cause of a mass in the cerebellum. Gliomas represent the commonest primary intracranial tumours and these show varying degrees of malignancy. CT typically shows a mass of mixed attenuation surrounded by oedema and which displays a mass effect. Contrast enhancement almost always occurs with malignant tumours and a variety of patterns of enhancement may be shown, (Fig. 9.3).

Some gliomas, typically grade I astrocytomas, may be difficult to detect even with CT. Brain stem gliomas may also be difficult to demonstrate as,

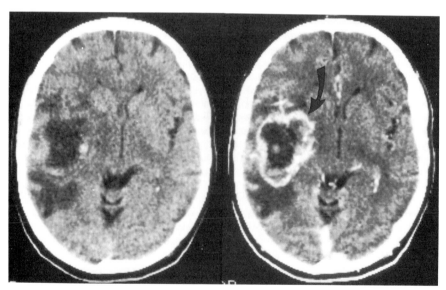

Fig. 9.3 Intracranial glioma. Pre- and postcontrast CT images. The tumour is shown as an area of low attenuation, which after intravenous contrast, shows peripheral enhancement (curved arrow). The low density areas adjacent to enhancing tumour are due to oedema.

characteristically, they do not enhance and may produce only subtle changes at the basal cisterns or on the configuration of the fourth ventricle. The position of certain gliomas may help the radiological diagnosis. For example, ependymomas are characteristically para- or intraventricular, and oligodendrogliomas are frequently calcified and, typically, situated within the frontal lobes.

Scintigraphy

On scintigraphy, intracranial neoplasms appear as a focal area of increased uptake of isotope. Occasionally, gliomas may outgrow their vascular supply which results in a 'doughnut' appearance on scintigraphy. Central necrosis within the tumour mass shows as an area of diminished activity surrounded by a rim of increased activity within the remainder of the viable tumour mass. Unfortunately, the scintigraphic appearances are not specific as they can be seen in a variety of other pathologies including abscesses. Scintigraphy is less sensitive than CT for the detection of an intracranial neoplasm. However, if multiple areas of abnormal activity are shown, then metastatic disease is a likely diagnosis. When a solitary focus is demonstrated, the differentiation between a primary or secondary tumour is not possible.

Magnetic resonance imaging (MRI)

Magnetic resonance imaging, like CT, will demonstrate the tumour mass and its effect on surrounding structures including the ventricular systems. MRI will not demonstrate foci of calcification, but is considered superior to CT in its assessment of the extent or degree of invasion of the tumour into adjacent tissues. Like CT, however, it may be difficult to differentiate surrounding oedema from actual tumour mass. Recent developments in the use of paramagnetic substances such as gadolinium result in improved density differentiation between tumour and oedema similar to the effects of conventional intravenous contrast on CT scanning. A major advantage of MRI in the assessment of intracranial tumours is in its ability to accurately demonstrate the position, volume, and relationships of the lesion using multiplanar scanning.

Intracranial Infection

Abscess

As approximately 50 per cent of cerebral abscesses arise via direct extension from adjacent structures such as the middle ear or a paranasal sinus, skull radiographs may be useful in demonstrating the site of origin of an infection. Fractures, foreign bodies, and osteomyelitis may be shown. The chest is a common origin for haematogenous spread, therefore a chest radiograph should be obtained in all patients with a suspected intracranial infection. Echocardiography may reveal a further focus of infection.

Whenever an intracranial abscess is considered possible on clinical grounds, there should be no delay in arranging appropriate radiological investigation. CT is the procedure of choice and, characteristically, an abscess will appear as a mass of low attenuation, showing, after contrast, a peripheral, strongly enhancing 'ring' of uniform thickness and surrounded by oedema. This ring or 'doughnut' appearance results from enhancement in the wall of the abscess contrasting with the central area of pus or necrotic debris. The position of such a lesion, e.g. adjacent to the mastoids, when coupled with an appropriate clinical history, may suggest extension of infection from an extracranial source. Abscesses resulting from haematogenous spread can occur at any site within the brain.

Scintigraphy is very accurate in detecting intracranial abscesses with most reported series showing successful detection in almost 100 per cent of cases. Like CT however, the scintigraphic appearances are not pathognomonic.

Meningitis

Imaging has no place in the assessment of uncomplicated meningitis. When complications occur and clinical examination suggests possible abscess formation or venous thrombosis, CT should be performed.

Encephalitis

Encephalitis produces a diffuse or occasionally focal cerebral abnormality affecting predominantly the white matter areas. Scintigraphy generally demonstrates a diffuse increase in activity, often over both convexities, and therefore can differentiate between an encephalitis and a focal intracranial abscess. CT will show reduced attenuation in the white matter territory. The extent of changes on CT varies with the cause and severity of illness. In children, herpes simplex encephalitis results in severe generalized oedema compressing the ventricular system, and often resulting in rapid death. In adults, herpes simplex encephalitis may be focal and typically involves the frontal and anterior temporal regions. Focal haemorrhage may be identified and when compared with other forms of encephalitis, contrast enhancement may be a prominent feature.

Cerebral Vascular Disorders

Cerebrovascular accident or 'stroke'

Cerebral vascular accidents can be broadly grouped into those due to haemorrhage and those secondary to infarction. Infarction is responsible for the majority of 'strokes', haemorrhage being the precipitating event in only 10 per cent.

Infarction

The clinical consequences of cerebral ischaemia vary with the duration of the insult. Brief compromise to the circulation will result in transient ischaemic attacks, with symptoms and signs resolving within 24 hours. Longer periods of ischaemia may still be reversible but progressive neurological deficit is more likely to result in a completed stroke. The occasions when a skull radiograph gives any useful information in the assessment of the patient with a CVA are so few that conventional radiography can be dismissed for the purposes of discussion. CT and, occasionally, scintigraphy are the most useful techniques currently available. It is well recognized that CT is often normal in the early stages of infarction and may be difficult to interpret when there is a history of previous infarct. Tissue hypoxia and associated oedema result in an area of low attentuation within the distribution of the compromised vessel. These are most commonly recognized in middle and posterior cerebral artery territories and the majority are demonstrated on CT by 48 hours (Fig. 9.4). Oedema may result in mass effect which is usually identified by effacement of sulci and ventricular compression on the side of the infarct. The use of intravenous contrast is controversial and, at the time of writing, is considered both unnecessary and possibly harmful. Although the majority of recent infarcts enhance, little additional information will be achieved to justify the theoretical possibility of further damage to the cerebral substance.

Scintigraphy may be helpful in the early stages of an infarct. Images acquired during the dynamic phase of a brain scan may show diminished flow to the affected part of the cortex before the static images become

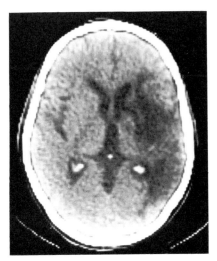

Fig. 9.4 Middle cerebral infarct. Non-contrast CT study showing extensive areas of low attenuation within the middle cerebral artery territory characteristic of infarction. (By courtesy of Dr R J Bartlett.)

abnormal. Scintigraphy may also help when there is difficulty in deciding whether an infarction shown on CT represents an acute or a previous insult.

Because of its superior ability to characterize tissue densities, magnetic resonance imaging is considered better than CT in the demonstration of the extent of an infarct. Infarcts at certain sites, e.g. the brain stem, may be very difficult to show on CT but are demonstrated relatively simply by magnetic resonance imaging.

Angiography has a very limited role in cerebral ischaemic disease with the exception of displaying the neck vessels in patients with transient ischaemic attacks. As has been previously discussed, digital vascular imaging (DVI) is an appropriate means of investigating these patients and surgically relevant lesions around the carotid bifurcation can be assessed. Atheromatous plaques and ulcers, as well as the typical features of fibromuscular hyperplasia, will be demonstrated. Before surgery is considered, angiography must demonstrate the status of the ipsilateral system distal to the carotid bifurcation, and the contribution of flow from the contralateral carotid and the ipsilateral external carotid supplies.

Intracranial haemorrhage

In the majority of patients with intracranial haemorrhage, skull radiographs are of limited value, but they may be helpful where there is a history of trauma and will occasionally reveal calcification in an arteriovenous malformation. As with cerebral infarction, CT is the most appropriate means of diagnosis. The appearance of blood on a CT scan varies with its age. Fresh blood will give an area of high attenuation but over a period of time, the haematoma will gradually reduce in density until finally it appears as a focal area of low attenuation. CT will also demonstrate the extent of any intraventricular bleeding.

The use of head ultrasound in neonates to detect intracranial haemorrhage has already been discussed (see p. 263).

Subarachnoid haemorrhage

Subarachnoid haemorrhage is usually due to a bleed from an intracranial aneurysm or arteriovenous malformation (AVM) and can also accompany severe head trauma. Aneurysms and AVMs may be demonstrated on the plain skull radiograph if calcification is present. Radionuclide flow studies may demonstrate a vascular malformation as a localized area of increased flow. There may, however, be difficulty in differentiating such lesions from a vascular tumour or a meningioma. CT is generally required to show both the extent of the bleed and the possible site of origin (Fig. 9.5). Fresh blood within the subarachnoid space results in increased density of the CSF and its detection depends on the severity of the haemorrhage. Most bleeds will be demonstrated if scans are performed within 7 days. The distribution of blood

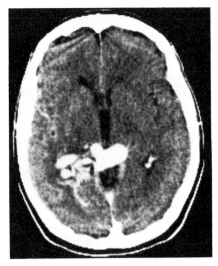

Fig. 9.5 Arteriovenous malformation. Postcontrast CT study in a patient who recently presented with a subarachnoid haemorrhage reveals a large arteriovenous malformation.

may suggest the artery of origin; blood in the septum pellucidum or frontal lobe indicating an anterior communicating artery aneurysm as the likely source; blood in the temporal lobe or sylvian fissure being typical of a middle cerebral artery bleed. Multiple aneurysms occur in 20 per cent of patients (Fig. 9.6), and CT may be able to suggest which is the most likely source of haemorrhage. Associated hydrocephalus and areas of reactive ischaemia can also be identified. Haemorrhage may cause secondary spasm of adjacent vessels which, not uncommonly, can lead to ischaemic infarction. Although CT will demonstrate the severity of a subarachnoid haemorrhage, angiography is required before surgery is contemplated. Despite careful examination, angiography will fail to demonstrate a cause in as many as 10 per cent of cases.

Interventional radiological techniques are being applied to the treatment of arteriovenous malformations (AVMs). After angiography has carefully demonstrated the feeding vessels and the general status of the vascular tree, including the extent of supply from internal carotid and vertebral arteries, embolization may be an appropriate means of treatment. The risks associated with embolization are substantial and meticulous technique is required to prevent irreversible damage to vital brain tissue.

Cerebral Atrophy and Hydrocephalus

As most of the causes of cerebral atrophy are irreversible and untreatable, the most important use of imaging is in the differentiation of atrophy from a

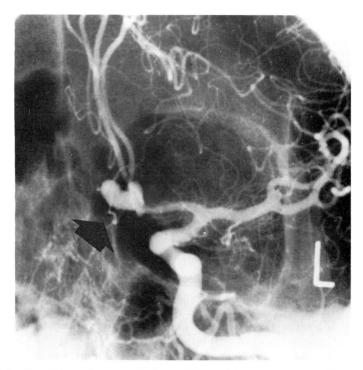

Fig. 9.6 Carotid angiogram. Patient with multiple aneurysms. Two middle cerebral artery aneurysms demonstrated (arrow) on this left carotid arterial injection. (By courtesy of Dr R J Bartlett.)

potentially treatable hydrocephalus. The use of radionuclide cisternography in this situation has already been discussed (see p. 266).

Atrophy results in brain shrinkage with a subsequent increase in the volume of CSF. CT will demonstrate enlargement of the ventricular system with prominence of the sylvian fissures, cerebral sulci and basal cisterns. Atrophy which occurs as part of the normal ageing process tends to affect the brain diffusely and symmetrically. Other causes of atrophy may characteristically affect particular parts of the brain substance. Wilson's disease affects the basal ganglia and brain stem predominantly, whereas the atrophy associated with chronic alcoholism tends to involve the cerebellum specifically. Dementia may also occur in patients with small vessel disease and CT may show multiple areas of low attenuation, characteristically more marked around the frontal horns of the lateral ventricles.

Failure of resorption or overproduction of CSF from whatever cause leads to hydrocephalus. Where there is some compromise to the flow of CSF from the ventricles to the subarachnoid space, then non-communicating or obstructive hydrocephalus is present. Causes are many and include neoplasms, infections, and congenital abnormalities such as aqueduct stenosis. Communicating hydrocephalus occurs where CSF can freely communicate between the ventricle and subarachnoid space. In this situation, pathology lies distal to the exit foramina of the fourth ventricle and may affect the basal cisterns, the cerebral convexities, or the arachnoid villi. The commonest causes of communicating hydrocephalus include subarachnoid haemorrhage, infection such as meningitis and trauma.

The use of scintigraphy in the assessment of CSF flow has already been described. CT is now the technique of choice for assessing hydrocephalus and, more importantly, its differentiation from cerebral atrophy. In communicating hydrocephalus, ventricular dilatation is global and prominent with the possible exception of a normal-sized fourth ventricle. There is usually no associated enlargement of the cerebral sulci. Conversely, in cerebral atrophy, there is, typically, additional enlargement of the cerebral sulci whereas the temporal horns of the lateral ventricles are usually of normal size. Temporal horn enlargement is usually a feature of hydrocephalus and not cerebral atrophy.

In neonates, hydrocephalus is usually a consequence of intracranial haemorrhage leading to blockage of the outflow tract. Ultrasound will demonstrate any ventricular dilatation and will provide a means of regular follow-up to detect the exceptional case which may require ventricular shunting. Congenital causes of hydrocephalus such as aqueduct stenosis or Dandy–Walker syndrome can be suggested by ultrasound before the need for more complex neuroradiological investigation.

Demyelination

Conventional radiology has little role to play in the imaging of patients with multiple sclerosis (MS). With improvements in CT technology, particularly when thin sections are used, CT can show a variety of appearances in patients with multiple sclerosis. Although many patients with the disease will have a normal CT scan, areas of low attenuation or 'plaques', typically situated in the periventricular white matter regions, may be demonstrated during acute phases of the illness. These plaques are located predominantly near the anterior and posterior horns of the lateral ventricles, may enhance after contrast and show variable resolution after the acute episode has passed. Plaques which persist during the remission phases characteristically do not enhance. The number of plaques demonstrated varies, but between one and eight plaques are typical. Postmortem studies have shown that both CT and clinical examination are relatively insensitive in demonstrating the extent of multiple sclerosis lesions in the brain. Patients with chronic MS may show areas of periventricular and cortical atrophy.

As the majority of plaques are small, the limited resolution of scintigraphy may be inadequate for their demonstration. Scintigraphy is invariably normal during inactive phases of the disease. During the acute phase however, breakdown of the blood-brain barrier which results in contrast enhancement on CT, may produce focal areas of abnormal activity on scintigraphy.

Magnetic resonance imaging (MRI) is currently the technique of choice in differentiating between white and grey matter disease. MR imaging shows considerably greater sensitivity than CT and is considered the best imaging option in the study of multiple sclerosis. MRI will detect smaller plaques than CT and recently differences in the appearance of MS lesions in younger and older populations have been appreciated. MR imaging can also detect lesions in the cervical cord and the optic nerves.

Imaging of Head Trauma

Skull radiographs

There has been much debate concerning the value of skull radiographs in the patient presenting with head trauma. Various bodies, including the Royal College of Radiologists, have suggested that an automatic request for skull films in every case of head trauma is no longer necessary. Each case should be assessed on its own merit and management tailored accordingly. There are certain situations when skull radiographs are mandatory and include any degree of the following:

1. Documented loss of consciousness
2. Neurological deficit on clinical examination
3. Loss of blood or CSF from the nose or ear
4. Suggestion of a penetrating injury
5. Head injury combined with alcohol intoxication.

As has already been discussed, the detection of a simple linear skull fracture, although more likely to be associated with underlying intracranial trauma than in the patient with head trauma and no fracture, is of less importance than the detection of a depressed fracture, a shift of the pineal gland, or the presence of air within the skull vault. Depressed fractures may occasionally show only subtle radiographic features on standard projections. These may take the form of a linear area of sclerosis or increased density on the lateral view of a minor buckling of the skull table (Figs. 9.7a & b). When skull trauma is being assessed, it is valuable to 'bright light' the films to demonstrate the soft tissues. Any localized swelling will indicate the site of injury. Unexplained areas of sclerosis or irregularities of the vault should lead to a request for oblique or tangential views to exclude a depressed fracture. Depressed fractures are relatively simply demonstrated by CT. Lateral skull radiographs are taken in the 'brow-up' position which means that any intracranial air will lie anteriorly and will often be associated with an air/fluid level. Similarly, an air/fluid level in the sphenoid sinus or maxillary antrum may be the only indication of a base of skull fracture. Careful note should be made of the position of the pineal gland. Pineal

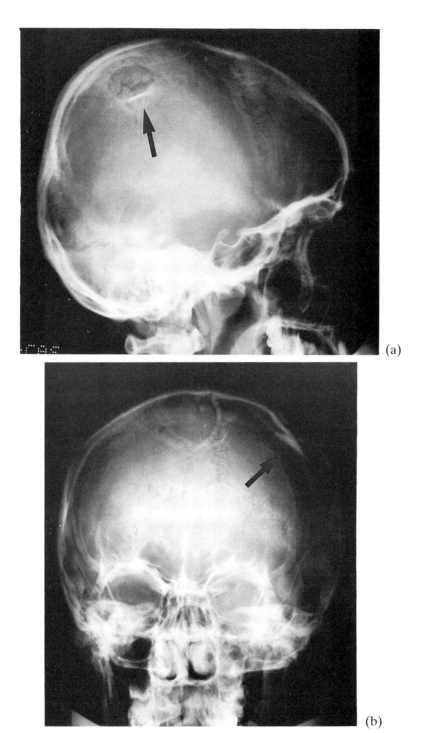

(a)

(b)

Fig. 9.7 Depressed skull fracture. Lateral (a) and frontal (b) projections showing classical depressed skull fracture. On the lateral view, the depressed fracture is indicated by the linear area of sclerosis (arrow). (By courtesy of Dr R J Bartlett.)

displacement generally indicates intracranial haemorrhage which may be either on the side of the fracture or on the opposite side. When serious trauma has occurred, e.g. following a road traffic accident, it is prudent to obtain at least two views of the cervical spine. Serious head trauma is often associated with cervical trauma. If neurological deterioration occurs, or if a patient is unconscious at presentation, then CT is required as a matter of some urgency.

Computed tomography (CT)

CT is the definitive means of imaging post-traumatic haemorrhage and it will show whether blood is located within the intracerebral substance, the subarachnoid space, or whether it lies subdurally or extradurally. Other sequelae of trauma such as cerebral contusion or generalized focal brain swelling will also be displayed. As has already been discussed, the appearance of intracranial haemorrhage on CT is age-related. Acute haemorrhage appears as a high density mass, but after 2 or 3 weeks the clot will become less dense than adjacent brain. Therefore, it follows that at some stage a collection of blood will be isodense with brain and its detection will subsequently depend on any effect on adjacent structures. For example, cerebral sulci will be effaced and midline structures may be displaced. Bilateral isodense subdural collections may be difficult to identify. In this situation, the ventricles may be distorted or compressed rather than displaced. Contrast enhancement may help make the diagnosis by identifying cerebral cortical vessels and noting any separation from the skull vault. An acute extradural haematoma typically displays a convex medial border (Fig. 9.8), whereas an acute subdural collection shows a concave medial border. Extradural haematomas are often associated with fracture resulting in damage to the middle meningeal artery. Extradural collections occur most commonly in the temporoparietal region. Subdural haematomas are less often associated with fracture as, most frequently, the haemorrhage results from a rupture of subdural veins.

Angiography and scintigraphy have little place in the initial assessment of acute head injury. Angiography is still required when the possibility of vascular damage following a penetrating injury exists, or when assessment of a post-traumatic arteriovenous fistula is required.

In the trauma patient, any injury to the skull vault or extracranial soft tissues may result in an abnormal accumulation of isotope which can cause difficulties in interpretation. Although it generally takes some 7–10 days for static brain scans to show changes associated with a subdural collection, this activity will persist for several months. Therefore, scintigraphy may be a useful investigation for the exclusion of a subdural collection without resort to more expensive investigations. It may be particularly valuable in the elderly patient who presents with indeterminate neurological signs and a

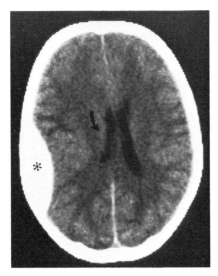

Fig. 9.8 Acute extradural haemorrhage. High density mass (asterisk) representing acute haemorrhage and showing a characteristic convex medial border. Mass effect is demonstrated by displacement of the adjacent lateral ventricle (curved arrow). (By courtesy of Dr R J Bartlett.)

history of a recent fall. At the time of writing, it would appear that magnetic resonance imaging demonstrates post-traumatic intracranial haemorrhage as well as CT scanning.

Spinal Pathology

Degenerative Spinal Disease

Conventional radiographs are often all that is required in the assessment of the patient with degenerative spinal disease. Standard views can be supplemented by oblique projections to assess the facet joints, intervertebral foramina, and pars interarticularis, and tomography may be of value in the assessment of spondylolisthesis. If surgery is being considered, myelography or CT will be required, particularly in cases of cervical spondylosis. Myelography typically shows multiple extradural irregularities but sites of critical cord compression or block will also be demonstrated. CT will also show the effects of hypertrophy of the ligamentum flavum and hypertrophic changes in the posterior joints. Gas in a degenerate disc or a herniated portion of a disc is shown clearly by CT.

Lumbar Disc Herniation

Most patients with low back pain can be treated conservatively and only about 10 per cent will require surgical intervention. The success of surgical management depends on strict correlation between 'abnormalities' shown by imaging and the patient's symptoms and neurological deficit. It has been shown that asymptomatic disc herniation may be present in over 30 per cent of patients undergoing myelography for reasons other than low back pain. As, increasingly, CT without myelography becomes the initial investigation of choice in a back pain patient, any abnormality shown must agree with the clinical deficit before surgical intervention is considered, particularly if abnormalities are displayed at several levels.

Both myelography and CT aim to demonstrate unequivocal lumbar disc herniation which is compressing the clinically affected nerve root. Myelographic evidence of a disc herniation is indirect in that an extradural compression on the contrast column is shown (Figs 9.9 a and b). CT however will demonstrate the actual herniated disc itself. Management difficulties arise when a patient presents with persistent back pain or leg pain but no neurological deficit. In this situation, the clinician must decide whether intervention on radiological grounds alone is indicated. Clinical assessment also influences the choice of imaging technique used in patients with low back pain. In patients with back pain and an appropriate neurological deficit, a CT scan in the first instance may well be all that is required. As myelography is an invasive procedure and has definite, albeit minor, associated morbidity, a CT abnormality which is consistent with the clinical assessment will preclude the need for supplementary myelography. Often, in any patient with back or leg pain without neurological deficit, CT and myelography will be required before surgery is contemplated. CT in the first instance will help establish which patients require further assessment and which do not. As well as displaying the lumbar nerve roots, myelography will allow assessment of the conus medullaris to exclude a tumour at that site as the cause of indeterminate back pain.

In centres where myelography is the initial investigation of back pain patients, clinicians should remember that a false-negative examination may occur in two well-recognized situations. First, myelography may fail to demonstrate lateral disc herniations particularly when they occur distal to the neural foramen. Contrast-filling of nerve roots does not extend this far and any nerve root compression at this site will be missed at myelography. Lateral disc fragments may accompany central disc herniation. As myelography will generally show only the central lesion, a surgeon acting on this information alone, may plan an inappropriate operation. Such lateral disc herniations or fragments, are easily demonstrated by CT. Second, at the L5/S1 level, the spinal canal becomes widened and the thecal sac lies posteriorly which may result in incomplete contrast-filling of a nerve root sleeve. This may result in an absence of the characteristic myelographic features of root compression by a herniated disc. Again, CT may be required to demonstrate herniations at this level.

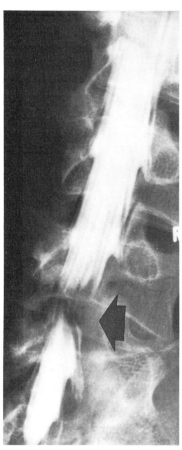 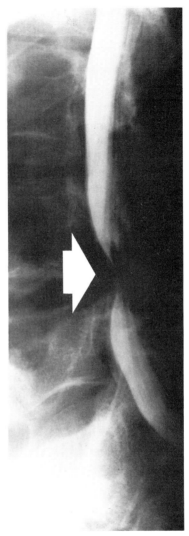

(a)   (b)

Fig. 9.9 Water-soluble lumbar myelogram. (a) Oblique and (b) lateral projections. Extradural impression on the contrast column characteristic of a right posterolateral lumbar disc herniation. The 5th lumbar nerve root sleeve is compromised and fails to fill with contrast.

Possible disc herniation — which investigation to choose?

Whether CT or myelography is used as the primary investigation of low back pain will vary according to availability and from department to department. Whichever technique is used, if the CT or myelographic appearances correlate with the clinical findings, then no further investigations are required. When radiological findings are equivocal or are at variance with the clinical impression, then the alternative imaging technique should be performed to clarify the situation. Cervical spine CT is less successful in the detection of disc herniation than in the lumbar region and cervical myelography is considered the investigation of choice. The relative lack of epidural fat in the cervical region combined with technical problems due to scanning artefacts from the shoulders, may make the CT examination difficult to interpret. Magnetic resonance imaging may provide a further means of evaluating disc disease. Early experience appears promising as degenerate discs exhibit changes in signal intensity when compared to healthy disc material.

Difficulties association with the diagnosis of lateral disc herniation have already been discussed. Lumbar phlebography may help in both the diagnosis of lateral discs and also of disc herniation at the L5/S1 level. Bulging of disc material is identified by amputation, displacement, or a focal non-filling of the epidural veins.

Persistent or Recurrent Back Pain in Patients with Previous Surgery

Myelography in such patients is often unhelpful as postoperative scarring prevents an adequate demonstration of the nerve root sleeve. CT will demonstrate recurrent disc herniation or scar tissue compressing the nerve root. Occasionally, it may be difficult to differentiate scar tissue from herniated disc material. Intravenous contrast may help as scar tissue can enhance thereby allowing differentiation from disc material. The place of magnetic resonance imaging in the diagnosis of postoperative scarring is as yet not fully established but may well prove helpful. Some surgeons consider that if CT demonstrates nerve root compression, surgery is indicated regardless of whether scar tissue or herniated disc is responsible. As well as proving superior to myelography in the demonstration of recurrent disc herniation and extradural scarring, CT will also provide a better assessment of postoperative bony stenosis and disturbances of the facet joints which may account for symptoms. Myelography remains superior to CT in the assessment of postoperative adhesive arachnoiditis.

Spinal Stenosis

Lumbar spinal stenosis can be congenital where the AP diameter of the canal is reduced *de novo*, or acquired where degenerative changes in the facet joints and thickening of the ligamentum flavum result in a similar narrowing of the spinal canal. Before CT, the diagnosis of spinal stenosis relied on sagittal and transverse measurements of the spinal canal from a conventional spinal radiograph. Myelography gives additional information regarding the thecal sac and, characteristically, shows either partial or complete obstruction to flow opposite disc spaces, crowding of nerve roots, and wasting of the contrast column at multiple levels. However, it is generally agreed that plain films and myelography are inadequate to assess fully the cross-sectional configuration of the spinal canal. CT provides an imaging technique that can assess both the bony and soft-tissue structures of the spinal canal and give a cross-sectional display of the spinal canal dimensions. Although a normal range for AP lumbar canal measurements has been established, absolute values are not generally regarded as essential for diagnosis. A careful assessment of the soft-tissue characteristics, e.g the thickness of the ligamentum flavum or the presence or absence of epidural fat at the level of the intervertebral disc, are as important as specific canal measurements. CT can also help in differentiating between a congenital stenosis and an acquired stenosis secondary to degenerative change. The contributions of marginal vertebral osteophytes and hypertrophy of the facet joints to the stenosis can be appreciated.

Mass Lesions of the Spinal Cord: Cord Compression

When clinical findings indicate mechanical compromise to the spinal cord, radiological investigation must be tailored to provide a diagnosis which includes the level of compression as a matter of some urgency. If cord compression is suspected, myelography should be performed only with the close co-operation of the radiologist and neurosurgeon as the investigation may lead to a deterioration in the patient's neurological status. As with any contrast examination, plain radiographs will give useful information prior to the more invasive study. Focal metastases or infection leading to bony destruction, reduction in a disc space, or a paravertebral mass, will give some indication as to the likely level of cord compromise. Calcification may indicate a meningioma or a dorsal disc herniation, and spinal cord expansion may accompany an intramedullary neoplasm.

Myelography is the commonest intitial approach to a possible mass lesion compromising the spinal cord. Characteristic myelographic appearances will depend on the site of the offending lesion. Pathology which results in extradural compression, can be differentiated from lesions which are intradural, which can be further divided into abnormalities of intra- or extramedullary origin (Figs. 9.10 a and b).

When a spinal block is demonstrated by myelography, a subsequent CT scan may define the extent of the block without the need for repuncture above the level of obstruction. Although a block may appear total on myelographic criteria, CT can detect very low levels of contrast which bypass the lesion, thereby defining the upper or lower margins. The ability of CT to detect low levels of contrast can also be of value in demonstrating the entry of intrathecal contrast into a syrinx following myelography. This finding is not specific as similar staining of the centre of the cord can occur with a variety of other pathologies. Myelographic differentiation between a

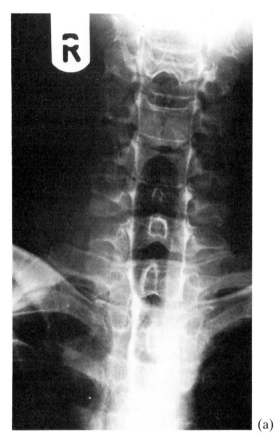

(a)

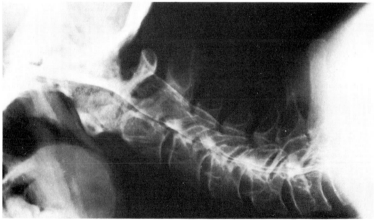

(b)

Fig. 9.10 Intradural intramedullary mass. Water-soluble cervical myelogram. (a) AP and (b) prone lateral projections demonstrating an extensive intramedullary mass. Myelography is not able to determine the underlying pathology. Further investigation showed the lesion to be an extensive haemangioblastoma.

syrinx and an intramedullary tumour may be difficult and CT can usually resolve any diagnostic dilemma. CT can demonstrate levels of calcification which will not be detected by a conventional radiograph. With the exception of a lipoma or a syrinx, CT is of little help in differentiating between one solid tumour and another.

Although secondary malignancy results most commonly in extradural compression, occasionally, systemic metastases will appear as intradural abnormalities on myelography. These may produce multiple irregular nodular filling defects or generalized nerve root thickening. Similar appearances may occur in lymphoma and leukaemia. Certain intracranial tumours metastasize to the subarachnoid space, e.g. medulloblastoma.

Spinal Infection

The radiological appearances of spinal infection largely depend on the responsible organism. Pyogenic and tuberculous infections have certain similarities but also demonstrate characteristics which are more specific to one pathology than the other. Pyogenic infection of the spine is associated with urinary tract infection in both men and women but is also a recognized complication of urinary outflow obstruction secondary to prostatic hypertrophy. Spinal tuberculosis is less common nowadays, but the spine remains a more frequent site of tuberculous infection than the appendicular skeleton.

In spinal tuberculosis, plain radiography will often show vertebral body destruction, characteristically in the anterior third and often associated with a paravertebral mass. Tuberculous infection spreads up and down the spine via paraspinal ligaments but there may be little evidence of this spread on the plain radiograph. Bone scintigraphy and CT will often detect early changes which may not be appreciated on plain radiographs. CT is particularly useful for assessing paraspinal progression and any extension into the spinal canal. Intervertebral disc infection tends to occur when adjacent vertebral bodies are involved. Spinal tuberculosis generally leads to vertebral body destruction with relatively little reactive bone formation. Calcification may be recognized in the paraspinal tissues, particularly by CT.

Pyogenic spinal infection is usually centred on the intervertebral disc. Early changes show a reduction in the disc space height. Thereafter, reactive sclerosis of the vertebral body endplates develops and may be associated with periosteal reaction and eventual bony buttressing. Only occasionally will infection breach the cartilaginous endplates and lead to vertebral body destruction. Paravertebral masses are less often seen with pyogenic infection. When present, they are typically small.

Plain radiographs can be supplemented by conventional tomography which may give more information concerning vertebral body or disc space infection, particularly in the early stages of the disease. Myelography will be indicated if there is any neurological deficit to confirm or exclude spinal block. When myelography is performed, care must be taken to ensure that any puncture is well clear of the suspected level of infection to prevent subarachnoid spread.

Magnetic resonance imaging may prove to be the most sensitive technique available in the detection of early spinal infection. Infected discs and the adjacent vertebral bodies show decreased signal intensity at an early stage of the disease. Certainly at the time of writing, MRI would appear more sensitive in detecting early infection than CT, which again is more sensitive than conventional radiography. If the cause of vertebral body destruction or an associated paraspinal mass is uncertain and differentiation between an inflammatory and a neoplastic process cannot be made, percutaneous biopsy or aspiration under fluoroscopic control should be performed.

Developmental Abnormalities of the Spinal Cord

Although plain radiographs will demonstrate spinal dysraphic changes and may show, for example, the bony spur which dissects the cord in diastematomyelia, contrast studies are usually required to demonstrate the full extent of any developmental abnormality. Myelography remains the investigation of choice as the entire cord can be examined at the same examination. Tethering of the cord, splitting of the cord, and any congenital neoplasms such as lipomas or dermoids will be demonstrated. CT will give similar information but intrathecal contrast is usually required and the radiation doses may well be higher depending on the number of CT sections required (Figs. 9.11 A and B).

Intrathecal contrast is also required for the assessment of developmental disorders around the craniovertebral junction. Myelography or CT following myelography will demonstrate cerebellar tonsillar ectopia, but CT is the better technique for detecting an associated syrinx. The possibility of an Arnold–Chiari type malformation with or without an associated syrinx may be suggested on plain radiographs of the cervical spine. Not uncommonly, atlanto-occipital fusion or other intervertebral fusions are present.

Magnetic resonance imaging is currently the best imaging technique for the assessment of the craniovertebral junction and spinal cord. Direct sagittal scanning will demonstrate the cervicomedullary junction and the entire length of the spinal cord without the need for either intrathecal contrast or ionizing radiation (Fig. 9.12). Early experience suggests that magnetic resonance imaging is very accurate in detecting intramedullary cystic and solid tumours and is probably the technique of choice for the diagnosis of syringomyelia. In spinal dysraphism, MRI will demonstrate meningoceles, split and tethered cords and associated tumours. Any bony defects of the spinal canal will not be shown.

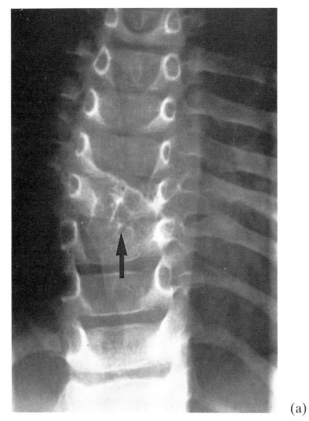

(a)

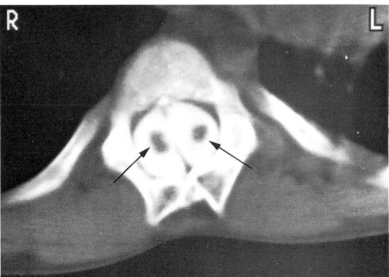

(b)

Fig. 9.11 Diastematomyelia. (a) Dorsal spinal radiograph showing widening of the dorsal canal and characteristic bony spur (arrow). (b) CT myelogram at level of diastem which clearly shows the split nature of the cord (arrows).

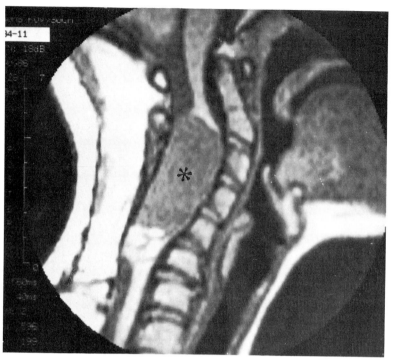

Fig. 9.12 Cervical intra-medullary dermoid. MRI direct sagittal scan. Short spin-echo sequence. Asterisk marks centre of the lesion. Areas of high signal at its inferior limit indicate fatty components. (By courtesy of Dr R J Bartlett.)

Imaging of Spinal Trauma

The majority of spinal fractures will be adequately demonstrated by conventional radiographic techniques. Any spinal examination will be limited initially by necessary restrictions on the patient's movement until the nature of the injury has been elucidated. Basic projections will usually determine if subsequent patient movement is appropriate or not. Angling the X-ray tube and using tomographic techniques will often compensate for restrictions of patient positioning due to unstable injuries of the spine.

A detailed discussion on spinal trauma is beyond the scope of this text but certain basic principles should be emphasized with regard to imaging in these patients. All dislocations and shear fracture/dislocations are inherently unstable as the major brunt of the trauma falls on the components of the posterior arch. Neurological deficit may be severe as coexistant damage to the spinal cord or nerve roots will be present. CT scanning is often helpful to demonstrate in more detail the extent of the bony injury and to show the presence of any avulsed bony fragments within the neural canal. When a fracture is demonstrated in a patient with ankylosing spondylitis, the incidence of multiple spinal fractures is high so the whole spinal column should be radiographed to prevent missing a significant injury elsewhere. All spinal fractures in patients with established ankylosing spondylitis are unstable.

As well as bony injury, adjacent soft-tissue structures may be damaged during trauma which results in a spinal fracture. Soft-tissue injury is particularly common in the hyperflexion, lap strap or Chance fracture. The fracture itself is generally stable but the significant trauma is usually to the adjacent soft tissues, e.g. the bowel — especially at points of relative fixation such as the duodenojejunal junction, and the ileocaecal junction — solid organs such as the pancreas, or adjacent vascular structures such as the abdominal aorta. CT is the single most appropriate examination in such patients. Some indication of a soft-tissue injury may be shown on initial plain abdominal radiographs. Effacement of a psoas outline or a soft-tissue mass obscuring a renal outline may be identified.

Contrast studies in spinal trauma are generally reserved for patients who develop progressive neurological deficit, particularly if the plain films are unremarkable. A lumbar approach may be inappropriate in the presence of a spinal fracture and often a lateral cervical puncture may be the best option. Myelography can assess spinal block whether it be due to extrinsic compromise, e.g. traumatic disc herniation, bony fragments, or haematoma formation, or due to an intrinsic haematomyelia. CSF-containing meningoceles may be a consequence of nerve root damage and are frequently seen following injuries to the brachial plexus.

CT may provide sufficient information to avoid a contrast study. Avulsed bony fragments following a spinal fracture, or foreign bodies after a penetrating injury, are more easily demonstrated by CT than conventional radiographs. The relationship of any bony fragment or foreign body to the spinal cord will also be displayed. The diagnosis of intramedullary haemorrhage is difficult: even on CT, often only swelling of the cord is demonstrated. Haemorrhage within the subarachnoid space or within the paraspinal muscle groups will also be demonstrated by CT. CT has also a valuable role in the follow-up of the spinal injuries patient. Post-traumatic syrinxes and meningoceles may be shown and some assessment of bony healing can be made. Progressive neurological deficit occuring some time after an episode of spinal trauma may be due to the development of syrinx. CT following intrathecal contrast is often the best technique for diagnosis as delayed images may show contrast entering a cyst within the cord. Occasionally, direct injection of contrast into the cord can be performed to assess more fully the extent and limits of a syrinx. MRI would appear to be the best method for diagnosing cystic lesions of the cord currently available.

Other Specific Clinical Problems

Disorders of Hearing and Balance

When a patient presents with progressive impairment of hearing, differentiation between nerve deafness and conductive deafness is made by clinical and audiometric examination. Imaging is required to assess the effects of trauma, inflammation or neoplasm on the inner and middle ear structures, the mastoid air cells, and the facial nerve canal. Specialized conventional tomographic projections which demonstrate anatomy ranging from mastoid air cells to the ossicles of the middle ear in great detail are gradually being replaced by high resolution CT scanning. In the patient with symptoms suggestive of an acoustic neuroma, conventional radiographs of the internal auditory meati are performed in the first instance. These may be supplemented by thin section tomography, but usually the patient will proceed directly to CT if there is any suggestion of an abnormality on the conventional views. Scans obtained before contrast typically show a mass which is either isodense or of slightly greater density than adjacent brain. After contrast, acoustic neuromas demonstrate moderate enhancement. In addition, the fourth ventricle is usually displaced and erosion of the internal auditory meatus may be shown if high definition slices are obtained. Intrathecal air or water-soluble contrast may be required to diagnose smaller meatal lesions. This has already been discussed in a previous section.

Glomus jugulare tumours present as hearing disorders: tinnitus, usually, but occasionally deafness. Bone destruction resulting in an enlarged jugular foramen will be shown on conventional radiographs. CT will also demonstrate bony destruction but, in addition, will display any intracranial component. External carotid arteriography is required to assess fully the vascularity of the tumour and to provide a possible means of treatment by therapeutic embolization.

Inflammatory disorders of the ear and mastoid air cells are generally documented adequately by conventional radiographs supplemented by tomography where necessary. Tomography often helps in assessing bone destruction such as may occur when a cholesteatoma complicates chronic inflammation of the middle ear. Apart from evaluating bone destruction, CT will also demonstrate displacement or destruction of the auditory ossicles and will outline the soft-tissue component of the cholesteatoma.

Otosclerosis and Menière's disease are not uncommon causes of impaired hearing in the older population. Radiology is of little value in their assessment but is used to identify other treatable causes of deafness and tinnitus if there is clinical doubt concerning the diagnosis. The skull is a common site for Paget's disease and skull radiographs will demonstrate the characteristic appearances.

Disorders of the Eye and Orbit

Conventional radiographs, ultrasound, CT and angiography can be used to assess disorders of the eye and orbit. As with so much cranial imaging, CT is now the technique of choice in the investigation of orbital pathology. The hypertrophied intraocular muscles of Graves' disease are best displayed by coronal and axial CT scans, and CT also provides a good means of follow-up after treatment.

Orbital cellulitis most often results from the spread of adjacent paranasal sinus infection. Conventional radiographs will show an opaque sinus but CT will outline best the extent of orbital infection, any intracranial extension, or bony destruction.

Visual disturbance may be a symptom of raised intracranial pressure and the previously discussed changes may be identified around the pituitary fossa on a skull radiograph. Pituitary and parasellar masses may also present with visual impairment and, although skull views may be helpful, CT is required to demonstrate in detail the sellar mass and its effect on the adjacent third ventricle.

Vascular lesions, either arterial or venous may lead to visual impairment or proptosis. Vascular malformations may be situated within the orbit or may arise from adjacent structures such as the cavernous sinus. Diagnosis is often made on clinical grounds without the need for imaging. Prior to any surgical intervention or embolization, angiography will be required to document the vascular supply. Orbital varices may be suggested on plain films by the presence of calcified phleboliths. CT with contrast enhancement will show dilated veins, as will ultrasound. Orbital phlebography may be indicated if surgical intervention is being considered.

Neoplasms affecting the orbit and its contents may arise from neural, vascular, lacrimal, or bony tissue. Neural tumours such as optic nerve glioma usually affect children, whereas an optic nerve sheath meningioma is a condition of adults. In both pathologies, CT is the investigation of choice and expansion of the optic nerve with marked contrast enhancement in the case of optic nerve sheath meningioma will be shown. The majority of primary malignant tumours of the orbit are of lymphomatous or sarcomatous origin but, unfortunately, these do not produce characteristic CT appearances. Differentiation between a lymphomatous or granulomatous mass cannot be made. Malignant melanoma is usually of choroidal origin and the ultrasound appearances are characteristic. Ultrasound will detect smaller lesions than can be shown by CT. Metastases to the orbit are not uncommon. In children, neuroblastoma frequently metastasizes to the orbit, as may secondaries from a Ewing's sarcoma. A meningioma of the greater or lesser wings of sphenoid may result in sclerosis of the orbital wall. Differentiation from fibrous dysplasia, chronic frontal sinusitis, or even Paget's disease may be difficult.

Imaging of the Epileptic Patient

Apart from febrile convulsions of childhood, most patients who present with a seizure warrant some investigation. Clinical examination along with an electroencephalogram (EEG) may suffice. The value of skull radiography in detecting intracranial pathology has already been discussed at length and although often requested, the success rate is low. Pathological intracranial calcification may be shown in an arteriovenous malformation or a temporal lobe hamartoma and signs of raised intracranial pressure will be identified. The decision as to whether CT is indicated will depend on the clinical examination and results of the EEG. Most centres would consider CT a necessary investigation for patients presenting with a focal fit and in elderly patients presenting with epilepsy for the first time. If CT is not readily available, scintigraphy may serve as a screening procedure. In the adult, a chest radiograph should always be obtained to ensure that a bronchial carcinoma is not the responsible pathology.

Headache

History and clinical examination will help in determining just how much or how little radiological investigation is required in the patient who presents with non-specific headache. Skull radiographs in patients with tension headaches and migraine are very unlikely to demonstrate any significant abnormality. Any headache characteristic of raised intracranial pressure should be investigated with skull radiography and CT. Sudden onset of a severe headache may result from subarachnoid haemorrhage. The imaging of intracranial haemorrhage has been discussed previously (see p. 276).

The skull radiograph may demonstrate other causes for headache. These include inflammatory conditions of the paranasal sinuses, the mastoid air cells and middle ear. Paget's disease of the skull will produce characteristic changes and it should be remembered that headache may be produced by disorders of the temporomandibular joints, the cervical spine and the teeth.

Further Reading

Bradshaw, J.R. (1985). *Brain CT — an Introduction*. John Wright, Bristol.

Holman, B.L. (Ed.) (1985). Radionuclide imaging of the brain. *Contemporary Issues in Nuclear Medicine* Vol 1. Churchill Livingstone, Edinburgh.

Moseley, I.F. (1987). *Imaging in Neurological Disease*. Churchill Livingstone, Edinburgh.

Rumuck, C.M. and Johnson M.L. (1984). *Perinatal and Infant Brain Imaging — Role of Ultrasound and Computed Tomography*. Year Book Medical Publishers, Chicago, Illinois.

10

ONCOLOGY AND RELATED TOPICS

Imaging Techniques

Plain Films

The major application of plain films is in following the progress of bone and lung lesions. The majority of primary and secondary tumours of the lung will be visible on plain chest films and once the diagnosis has been established these provide a simple and inexpensive method of follow-up for as long as the lesions are visible. Plain films play a major part in the diagnosis and initial management of primary bone tumours and again are adequate for most aspects of routine follow-up. For the diagnosis and follow-up of skeletal metastases, plain films are used as a supplement to bone scintigraphy (see p. 242). The radiographic skeletal survey has been replaced for most purposes by bone scintigraphy but retains an application in a few malignancies affecting the bone marrow, particularly myeloma in adults and childhood leukaemias and neuroblastoma.

Only a minority of abdominal and pelvic tumours will be visible on plain films and appear as a soft-tissue mass, occasionally with calcification (Fig. 10.1). In these cases, the value of plain films in follow-up is apparent, but in others, abdominal films can be reserved for patients with evidence of tumour spread, recurrence, or complications of treatment.

Contrast Radiology

Barium examination and endoscopy are the prime diagnostic methods for detection of tumours in the gastrointestinal tract. The relative merits and disadvantages of the two techniques were discussed in Chapter 4. Both methods suffer the disadvantage of visualizing only the intraluminal part of the tumour. Intravenous urography is the established technique for imaging tumours of the renal parenchyma but, at the time of writing, is being

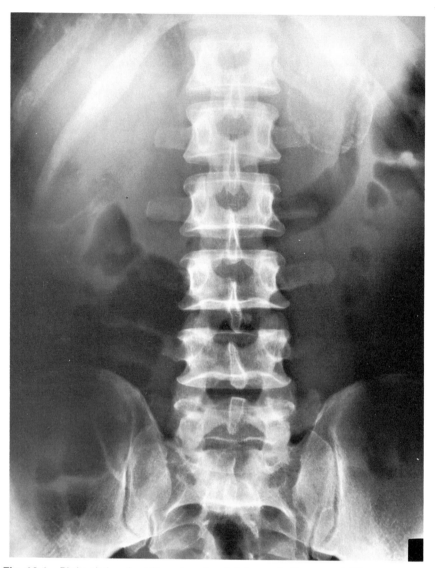

Fig. 10.1 Plain abdominal film showing faint eggshell-type calcification in the left upper quadrant. Diagnosis: adrenal cyst.

increasingly displaced in this role by ultrasound and CT scanning. The IVU still makes a major contribution to the diagnosis of urothelial tumours, particularly in the renal pelvis and ureter. Cystography, particularly double-contrast techniques using air and sterile barium, can give an elegant demonstration of intraluminal bladder tumours but this method is rarely contributory since cystoscopy is almost always used for the investigation of intraluminal tumours.

The role of arteriography in tumour management has been greatly diminished by the introduction of ultrasound and CT. Arterial studies are still useful in the diagnosis of some small tumours — e.g. islet-cell tumours in the pancreas (Fig. 10.2), — and may be used as a second-line test to resolve dubious findings on ultrasound or CT, such as suspected haemangiomas of the liver and dubious mass lesions in the kidney. However, the major role of arteriography is now as a 'road-mapping' technique used pre-operatively in patients being considered for surgical resection. The removal of parenchymal tumours of the kidney, spleen, pancreas and, particularly, the liver is aided by prior knowledge of the source of arterial supply to the tumour, its venous drainage, and its relationship to other major vessels in the vicinity. Similar considerations apply to some head and neck tumours. An indication of the vascularity of tumours can be obtained from their appearance on CT with contrast enhancement. Where a bruit or venous hum is heard by auscultation over the tumour site, it may be wise to carry out arteriography before attempting either percutaneous or open biopsy since it can be difficult to stop the bleeding from hypervascular tumours.

Venography may be of use in patients presenting with either inferior or superior vena cava obstruction, prior to surgery in infants and small children with abdominal masses, and in patients with hypernephroma to exclude tumour thrombus in the renal vein and inferior vena cava. However, these applications can now be replaced to a large extent by ultrasound scanning, CT with contrast enhancement (Fig. 10.3) or radionuclide venography (Fig

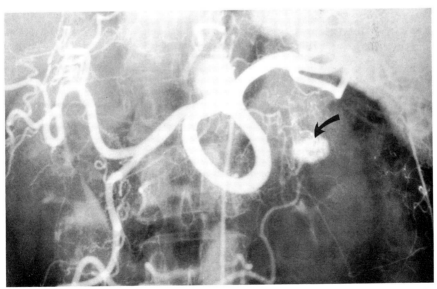

Fig. 10.2 Selective splenic arteriogram in a patient with hypoglycaemic attacks. A small highly vascular tumour is shown (arrow). Diagnosis: benign insulinoma.

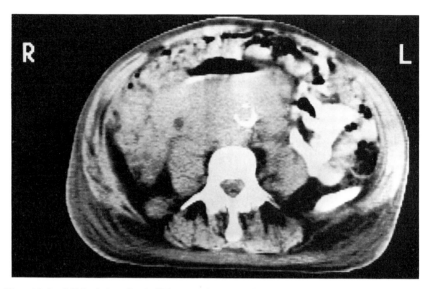

Fig. 10.3 Mid-abdominal CT scan in a patient presenting with gross leg oedema and an abdominal mass. The aorta is lifted off the spine by an encircling mass of tumour density infiltrating the retroperitoneum.

10.4). The introduction of CT has also had a major impact on the use of contrast lymphography.

Lymphography requires careful dissection and cannulation of a lymphatic channel over the dorsum of the foot and the intralymphatic infusion of several millilitres of iodized oil. The technique requires patience and skill on the part of the operator, and is moderately uncomfortable and tedious for the patient. A substantial proportion of the injected oil finds its way into the lungs in the form of microemboli and a temporary diminution in respiratory reserve is usual. Rigors and fever occur infrequently; more severe reactions, rarely. In comparison with other methods of imaging lymph nodes, contrast lymphography has the advangage that it displays not only the size and position of the node, but also gives some detail of its internal structure. Tumours of only a few millimetres in diameter may be recognized as filling defects within a normal-sized node. Lymphography has the further advantage that the contrast medium stays in the lymph nodes for months after the procedure so that response to treatment can be observed by taking sequential plain films. Apart from the disadvantages mentioned above, lymphography has other limitations. The method illustrates the inguinal, external and common iliac, and para-aortic nodes, but does not normally show the lymph nodes of the internal iliac, mesenenteric, coeliac, splenic, hepatic and retrocrural areas. In addition, the presence of large tumour deposits completely replacing or destroying lymph nodes usually causes complete obstruction to lymphatic flow so no contrast will enter these nodes. This is not usually a diagnostic difficulty, since the presence of a non-

Fig. 10.4 Radionuclide venogram in the same patient showing occlusion of the right iliac veins but patency of the vena cava. Diagnosis: metastases from carcinoma of the cervix.

visualized nodal mass can be inferred from the effects on adjacent nodes and lymphatic channels, but it does mean that the largest tumour deposits are not directly visualized by this method.

Ultrasound

The diagnosis of tumours by ultrasound scanning depends upon the recognition of echo patterns which differ from those of the surrounding organ or structures, and on the destruction, distortion or displacement of organs and anatomical landmarks. Because sound waves are scattered in air and absorbed in bone, the applications of ultrasound in the head and thorax are limited. Because both the resolution and the penetration of an ultrasound beam are directly related to the frequency of the sound waves, it is possible to show much better detail when scanning close to the skin surface than when imaging deep structures. Difficulties of technique and interpretation increase with increasing depth, particularly in the upper abdomen. Ultrasonic examination of the gall bladder is technically simpler and its results more reliable than examination of the pancreas, whilst scanning for upper abdominal lymph node disease is more difficult still. However, ultrasound is recommended as the first-line technique for imaging the liver, gall bladder, spleen, pancreas and renal parenchyma. In the pelvis, ultrasound finds application in examination of the uterus, bladder and ovaries. Using a higher frequency transducer (7·5 or 10 MHz) fine detail can be shown in the neck (thyroid and parathyroid tumours), orbit (intraocular tumours) and scrotum (testicular tumours). The combination of ultrasound with endoscopy is providing several new potential applications. At the time of writing, transrectal ultrasound of the prostate is well established as a staging technique, whilst the assessment of rectal, oesophageal and pancreatic tumours by endoscopic ultrasound shows great promise.

Computed Tomography (CT)

CT has applications in every area of oncology and is probably the single most useful technique available. Unlike any other imaging method, CT shows all the different anatomical structures within a body section with equal clarity and resolution. Tumours are recognized by their mass effects and by differences in X-ray attenuation from surrounding tissues. The resolution of CT is the same throughout the image (i.e. it is not reduced at increasing depth as with ultrasound and scintigraphy), but because the scans have a finite thickness, the resolution is much better in the plane of the scan than in perpendicular directions. Although it is possible to make coronal, sagittal or oblique reconstructions from stacks of contiguous axial slices, generally, these images have very poor resolution. The edges of curved structures which pass obliquely through the plane of the scans appear blurred and it is often difficult to distinguish tumour invasion from contact without invasion. As a general rule, tumours in solid organs are more easily seen on scans obtained after intravenous contrast injection. Contrast enhancement, particularly if combined with rapid sequence scanning, will give a good indication of the vascularity of a mass and its relation to nearby major vessels. The digital origin of CT images makes it a simple technical task to measure the diameter or volume of mass lesions so that progress in response to treatment can be quantified. Some tumours respond to treatment by changes in structure rather than in size and these alterations can also be monitored by CT.

Scintigraphy

In the detection of bone metastases, scintigraphy is the technique of choice. The method is simple, free of complications, and provides an opportunity to examine the whole or any part of the skeleton. Since the abnormalities are shown as areas of increased radioactivity (Fig. 10.5), lesions of only a few millimetres in size can be detected provided they are sufficiently active. The uptake of bone-seeking agents is determined by the rate of new bone formation, rather than bone destruction, so that lesions which are purely

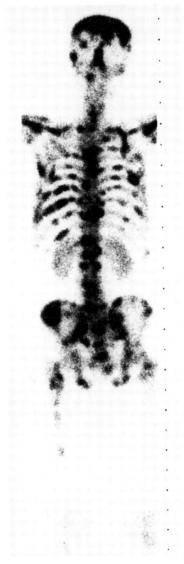

Fig. 10.5 Skeletal scintigraphy in a patient with carcinoma of the prostate showing widespread metastatic disease.

destructive may be missed or appear as photon-deficient ('cold') lesions. However, comparative studies have shown that bone scintigraphy is more sensitive than skeletal radiography in detecting virtually all types of metastatic disease with the few exceptions mentioned earlier. Some scintigraphic patterns of metastatic disease are characteristic, but in other cases the appearances will be non-specific and radiographs of the abnormal areas are needed in order to avoid confusion with benign bone disorders which may coexist. Consecutive bone scans are useful in following the progress of treated bone metastases and in detecting new lesions in patients with progressive disease.

The use of colloid scintigraphy for the detection of liver metastases has been overtaken by CT and ultrasound, to a large extent, as these have better spatial resolution. However, colloid scintigraphy is valuable as a second-line technique in patients with equivocal or unexpected results on CT or ultrasound scanning. Liver perfusion scintigraphy, using a single intravenous injection of labelled colloid and a first-pass acquisition technique, has recently been shown to be more sensitive than surgical examination of the liver in the detection of early (occult) liver metastases in patients with colorectal tumours (see Chapter 5). At the time of writing, this technique looks promising but has not yet been widely applied. Liver scintigraphy using gallium citrate and technetium-labelled iminodiacetic acid (IDA) derivatives can have a supplementary role in the differential diagnosis of solitary liver masses (see Chapter 5). Scintigraphy of the spleen is discussed in a later section of this chapter, and scintigraphic localization of functioning tumours of thyroid origin, parathyroids, and adrenal medullary tissue, is discussed in Chapter 7.

The search for tumour-localizing radiopharmaceuticals has been going on for years, but the techniques devised so far have yet to make a major impact on the management of malignant disease. Gallium citrate is taken up by many epithelial tumours and most lymphomas. The method is less accurate than CT in detecting tumours, but may be used as a supplementary procedure or when CT is unavailable. Of the newer methods, the most promising, currently, is the use of labelled antibodies raised against tumour antigens. Most of the antigens found in adult epithelial tumours are fairly non-specific so there is substantial cross-reactivity between the labelled antibody and non-tumorous tissues in the patient. Even when monoclonal antibodies are used, reactivity with tumour antigens is seldom specific and the background reduces the clarity of demonstration of the tumour site. However, a few tumours present much more specific antigens (such as some ovarian cancers in adults and neuroblastoma in children) and in these cases specific uptake of appropriate labelled antibodies leads to a much clearer localization of the lesion. At present these methods are regarded as supplementary to existing anatomical techniques with the exciting potential of being able to offer the delivery of therapeutic doses of radiolabelled antibodies if uptake into the tumour is sufficiently specific. Labelling proteins with technetium presents technical difficulties which have not yet been solved, so studies are presently carried out using iodine-131, iodine-123 or indium-111 as the labelling agent.

The Objectives of Imaging

The Early Detection of Primary Tumours

In patients presenting with symptoms of a tumour in a particular organ, it will be fairly easy to devise an appropriate imaging strategy. Much more difficult is the management of those patients in whom the presenting features suggest malignancy without localizing signs and those presenting with overt metastases but no clue as to the site of the primary. These two categories are discussed in a later section of this chapter.

The value of imaging procedures in oncology depends not only on the precision and reliability of the tests themselves, but also on the natural history of the tumour in question and the effectiveness of available modes of treatment. For example, the value of imaging procedures in the management of testicular teratoma has increased considerably in importance since the introduction of effective chemotherapy for this disease. On the other hand, the introduction of ultrasound, CT and ERCP, although they have enhanced and simplified the diagnosis of pancreatic tumours, have not yet been shown to influence the average survival time from the onset of symptoms to death from this disease. With some tumours there is still not enough time between 'too early to diagnose' and 'too late to cure'.

Logic suggests that the earliest detection of malignant disease should be achieved by screening asymptomatic populations. Imaging procedures have had limited success in this regard, and it is not difficult to see why. When the expected incidence of the tumour in the population being screened is low, very large numbers of patients need to be examined in order to collect an adequate number of positive cases for survival studies. Large groups of control subjects also need to be collected. It may take many years for the effects of medical or surgical intervention in the positive cases to reach convincing levels of statistical significance. A major investment of resources and time is needed to produce what may appear to be a relatively small tangible benefit.

Studies so far suggest little value in the use of routine chest radiographs for the early detection of carcinoma of the lung. The natural history of early gastric cancer, however, has been shown to be amenable to improvement by surgical treatment in patients in whom early lesions were detected by mass population screening in Japan. The incidence of gastric cancer is much smaller in Western populations, and the results of these studies cannot be directly translated to other countries. Long-term studies of the use of mammography in screening for early breast cancer have confirmed its value with a significant improvement in survival in the treated patients at all ages. Recent studies in mass population screening by ultrasound scanning for the detection of early ovarian cancer have also shown promise.

The Diagnosis of Overt Primary Tumours

In the vast majority of cases, the first aim of imaging is to demonstrate that a mass is present. The mass may be visible on plain films of the chest, abdomen or bones, it may be shown on contrast studies, or by ultrasound or CT scanning. Some mass lesions have specific imaging characteristics which can be closely correlated with histological diagnosis. For example, lipomas are unmistakable on CT because of their fat content. Hamartomas in the kidney appear on ultrasound as very sharply defined, rounded lesions within the parenchyma, producing very many echoes but without the acoustic shadowing associated with calcification. The plain film appearance of a soft-tissue mass arising from the pelvis and containing one or two teeth clearly indicates an ovarian dermoid.

The majority of masses, however, are less specific than this, and the next diagnostic step is to identify their site of origin. The range of possibilities is usually narrowed down very considerably once the exact site of the mass is defined. For example, a peripheral opacity on a chest film might represent pathology in the lung, the pleura, or in the chest wall, and the likely diagnosis would clearly be different in each case. If there are clinical pointers to the gastrointestinal or urinary tract, contrast radiology or endoscopy or both will be helpful. In other cases ultrasound, followed where necessary by CT, should provide further information on the site of origin.

For the majority of mass lesions histology will still be required to allow definitive treatment to be planned. Biopsies may be obtained via an endoscope or percutaneously with imaging guidance. Using fluoroscopy, CT, ultrasound or a combination of these techniques, it is possible to obtain carefully directed biopsies of almost any part of the body. If adequate imaging facilities and expertise are available, there is no longer any need to carry out exploratory surgery for diagnostic purposes.

Some tumours, particularly benign lesions of the endocrine glands, present with disturbances of function rather than with mass effects. In these cases the diagnosis is generally made on clinical and biochemical grounds and the role of imaging is to identify the site of the lesion. The identification of functioning tumours of the thyroid, parathyroid and adrenal is discussed in Chapter 7, and the localization of pancreatic islet-cell tumours is described in Chapter 5.

Local Staging

Once a diagnosis has been established, the next step is to assess the local extent of the tumour. This is particularly important as a precursor to surgery as the fixity of the tumour and its relation to vital structures nearby will influence the choice between palliative and curative resection, or may exclude the feasibility of surgical removal. For tumours arising in the bladder or gastrointestinal tract, one of the crucial factors is the depth of penetration through the wall of the organ; this is often difficult to estimate on endoscopic examination. CT shows extramural extension well but may yet be superseded by endoscopic ultrasound. Imaging procedures make a similar contribution to the staging of primary tumours of the cervix, ovary, and renal parenchyma. Imaging the major vascular structures in the upper abdomen is an important prelude to attempts at surgical removal of renal, pancreatic and hepatic tumours. The relation of lung tumours to the chest wall and mediastinum must be clarified before undertaking surgery.

Answering these questions by imaging often involves plain films, contrast radiology, ultrasound and CT, and occasionally, vascular studies. An approach to individual tumours is described on pp. 312–23.

Lymph Nodes

The presence or absence of lymph node metastases has a major influence on the prognosis and management of many types of tumour. The patterns of drainage of tumours in different parts of the body are quite well established, so it is important to direct imaging towards examination of those nodes which are most likely to be involved. With some tumours, metastases may bypass the nearer lymph nodes so the possibility of 'skip' lesions must be borne in mind when searching for nodal disease. Other tumours, e.g. osteosarcoma tend to bypass lymph nodes altogether, metastasizing directly to the lungs; when this occurs lymph node status becomes relatively less important. The anatomy of lymph node metastases varies considerably from one tumour to another. For example, in Hodgkin's disease, affected nodes are almost invariably enlarged, whereas deposits from epithelial tumours may be detected by contrast lymphography in nodes of normal size (Fig. 10.6). The behaviour of diseased nodes also varies with the type of tumour. For example, para-aortic nodal deposits from lymphoma or testicular tumours often reach enormous sizes before causing pressure symptoms of hydronephrosis and vena caval obstruction, whereas much smaller nodal deposits from squamous carcinoma of the cervix may cause ureteric and venous obstruction and painful erosion of the spine.

As mentioned above, contrast lymphography gives the best available demonstration of the inguinal, external iliac, common iliac, and lower para-aortic lymph node groups. Other lymph nodes in the pelvis, abdomen and thorax are not shown by this technique. CT has the great advantage of showing lymph nodes in all areas of the thorax, abdomen and pelvis, and will rarely fail to detect nodes which are significantly enlarged. Moderate degrees of nodal enlargement are less easy to detect by ultrasound, but large lymph node masses should be shown by this method. Radionuclide lymphography using technetium-labelled microcolloid is simpler than conventional lymphography since it does not require lymphatic cannulation, and it also avoids the potential toxicity of oil contrast medium. Although it has advantages as a method for demonstrating lymphatic flow, its reliability in detecting malignant disease in lymph nodes has not yet been established.

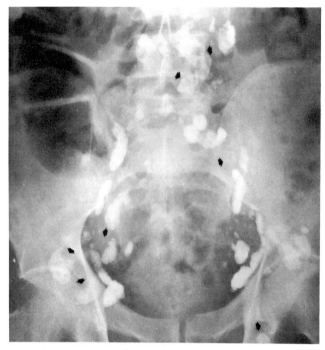

Fig. 10.6 Lymphangiogram of a patient with cervical carcinoma showing numerous small filling defects (arrows) within nodes which are not enlarged. Diagnosis: metastases.

The use of internal mammary lymphoscintigraphy in the staging of breast cancer is discussed below. The plain chest film, supplemented where necessary by CT, is used to demonstrate and follow the progress of nodal disease in the thorax. Axillary lymph nodes are well shown by CT but in assessing cervical lymph node disease, none of the imaging methods has much to offer in advance of careful clinical examination.

Extranodal Metastases

As with nodal involvement, the demonstration of metastatic disease elsewhere is of major importance in assessing the prognosis and deciding the management of a primary tumour. In some cases, patients will present with the local effects of metastatic disease — e.g. fractures from bone secondaries, upper abdominal pain from liver lesions — and appropriate imaging techniques will be directed immediately to the site of the lesion. In the majority of cases, however, imaging for metastases is a screening process carried out in patients who are known to have a primary malignancy but no overt clinical manifestations of distant spread. A plain film of the chest should always be carried out as the first step since, if lung metastases are shown, further staging procedures may be deemed unnecessary. Subsequent procedures should be directed according to the likely route of spread of the particular tumour involved. For example, gastrointestinal tumours arising in the area drained by the portal venous system rarely spread to the lungs without first seeding in the liver; if a careful ultrasonic or CT examination of the liver fails to show metastatic deposits, detailed examination of the chest with whole lung tomography or CT is unlikely to be contributory. On the other hand, renal cell carcinoma is very likely to metastasize to the lungs, so it may be argued that a thorough search for lung metastases should be the first step in the staging of this type of tumour.

The probability of metastases to different organs by particular tumours must also be taken into account when deciding on imaging strategies for staging. For example, treatment of primary tumours which rarely metastasize (such as chordomas and most malignancies of the central nervous system) might reasonably proceed after completing local staging procedures and obtaining a plain chest film. However, there is little justification for carrying out a major dissection or mutilating amputation in patients with melanoma or osteosarcoma unless the absence of lung metastases has been established first by the most sensitive method available. In other cases, the presence or absence of metastases will not influence surgical management of the primary tumour (e.g. testicular tumours) but will have a major impact on prognosis and on postoperative chemotherapy or radiation treatment. Here, staging procedures may be carried out before or after initial surgery.

Lung metastases

In general, CT is the most sensitive method for the detection of lung metastases. It should certainly be used in staging patients with testicular tumours, melanoma, bone and soft-tissue sarcomas, renal cell cancers, hepatocellular carcinoma, and also upper abdominal tumours in children. The case for CT scanning of the lungs in staging most types of head and neck tumours, transitional cell tumours of the urinary tract, gastrointestinal tumours where the liver is normal, breast primaries and carcinomas of the uterine cervix, is less well made.

Liver metastases

Careful ultrasound examination of the liver, supplemented, in cases of difficulty or unexpected results by CT, should be contributory in patients with gastrointestinal primaries. Perfusion scintigraphy provides an interesting new method for detecting occult liver metastases, but the impact of this technique will depend largely on the development of effective forms of treatment of liver lesions. A good case can be made for screening for liver metastases in patients with primary lung tumours before thoracotomy is undertaken. CT is the preferred method since it allows simultaneous examination of the adrenal glands — another common site for secondaries from the lung. Routine liver scanning in patients with early breast cancer is contentious, the incidence of positive results being very low in most centres.

Cerebral metastases

The presence of cerebral metastases carries a very poor prognosis with virtually all types of epithelial tumour and has been used as an argument for routine CT scanning of the brain before undertaking major surgery, e.g. thoracotomy for primary lung cancer. However, in most reported series where this has been done, the incidence of occult cerebral metastases is extremely low. Probably the most productive approach is to obtain CT brain scans only in those patients who have signs or symptoms of intracranial disease.

Scintigraphy

Bone scintigraphy for metastases is accepted as routine in patients with primary tumours which are likely to metastasize to bone, particularly carcinomas of the prostate, lung, thyroid and kidney, skeletal sarcomas, and some types of solid tumour in children. Bone scintigraphy is also indicated in patients with symptoms referable to the skeleton or with suspected local invasion of bone by the primary tumour. However, its value as a routine screening procedure in some other types of tumour is dubious, e.g. in early breast cancer and bladder tumours.

Radiotherapy Planning

The aim of radiation treatment is to deliver an effective dose to the tumour whilst minimizing radiation damage to surrounding normal structures. Radiotherapy planning requires careful correlation between internal anatomy and externally placed skin markers and also adequate imaging of the full extent of the disease. In recent years, CT scanning has come to play a predominant role in this procedure. Not only can precise anatomical localization of tumours be related to skin surface markers but suitable software programmes can allow the direct translation of attenuation measurements in CT slices into radiotherapy simulation computers so that the dose delivered to the tumour can be specified with improved accuracy. Fixed laser positioning devices allow the position of the patient to be reproduced accurately after diagnostic CT scanning. The use of CT guidance for radiotherapy planning has been shown to prolong the survival of patients with bladder cancer when survival time is compared with conventional radiotherapy planning using contrast cystograms. Whether similar improvements will occur in other anatomical areas remains to be shown. Plain films remain invaluable in planning radiotherapy to the chest and also for head and neck tumours.

Monitoring the Response to Treatment

In patients treated by radiotherapy or chemotherapy, imaging of the tumour supplements clinical examination in recording the response to treatment. In general, the preferred imaging technique is the simplest method which shows the tumour adequately. For example, most pulmonary and mediastinal masses can be followed by serial chest films, larger masses in the abdomen and pelvis can be monitored by ultrasound, whilst CT is needed for intracranial tumours and for small masses in the chest, abdomen and extremities.

In the majority of cases, a favourable response to treatment is indicated by a reduction in the size of the tumour. In some cases, however, the size of the tumour may remain constant whilst its internal structure changes. This type of change can often be detected by ultrasound or CT, the commonest transition being from solid tumour to necrosis to liquefaction, and eventually the appearance of a cystic mass. Other tumours appear to melt away completely leaving no visible residue. Alternatively, a period of shrinkage may be followed by persistence of a residual mass, often plaque-like or irregular in shape, which remains unchanged over a long period and may be regarded as inactive scar tissue rather than active disease. Establishing the absence of disease is always more difficult than showing a lesion, and imaging methods cannot be relied upon to exclude the presence of microscopic metastases or tiny tumour residues. With some tumours, the limitations of imaging are more apparent — the most striking example being that of ovarian carcinoma where no imaging technique can yet replace second-look laparotomy as a means of deciding whether residual disease remains after initial treatment.

Detecting Recurrent Tumour

The use of imaging procedures for detecting tumour recurrence broadly follows the lines of primary tumour detection. The effects of chemotherapy, radiation, or surgery usually combine to make the task of diagnosis more difficult than in the initial presentation. Surgical treatment of the primary may result in loss of access for clinical examination (e.g. abdominoperineal excision of the rectum), loss of normal structures and landmarks around the original tumour site (e.g. opaque hemithorax after pneumonectomy), or may introduce artefactual anatomy which makes the interpretation of imaging much more difficult (e.g. the upper abdomen after Whipple's procedure). Radiotherapy to primary tumours may result in extensive fibrosis of surrounding normal structures as well as leaving residual tissue of doubtful significance at the site of the original mass as mentioned above (e.g. pelvic fibrosis following radiotherapy to cervical carcinoma).

Differentiating between inactive residues from a treated primary tumour, surrounding radiation fibrosis, and recurrent or reactivated disease, is impossible by imaging methods in some cases. If there is an urgent requirement to make the diagnosis, needle biopsy may be used; if timing is not crucial then sequential imaging observations will show whether progressive disease is present. In the early detection of recurrent disease, baseline examinations obtained shortly after surgery or a few months after radiotherapy are extremely helpful. The appearance of a mass at the original tumour site or the development of new lesions in other areas may be taken as evidence of active recurrence. Restaging may then be appropriate before deciding how the recurrent lesion should be treated.

Imaging for Specific Tumours

Lymphoma

The staging of lymphoma requires the detection of involved nodes on both sides of the diaphragm together with a survey of the potential extranodal sites of disease. Imaging rarely improves on clinical examination for detecting cervical lymph nodes. Similarly, enlarged axillary and inguinal nodes should be detected clinically, but CT does offer the advantage of accurate measurement of their size. Large mediastinal masses are detected and monitored by consecutive chest films, whereas more subtle degrees of thoracic node involvement require CT which has displaced linear tomography of the mediastinum for this purpose. Contrast lymphography gives an elegant demonstration of pelvic and lower abdominal nodes with the advantage of showing the derangement of internal structure so characteristic of lymphoma. However, CT is preferred since it demonstrates upper abdominal nodes and gives a view of the liver and spleen at the same examination. Although lymphography may show disease in nodes which are not enlarged and thus appear normal on CT, this combination of circumstances is very unusual in lymphoma since most involved nodes are enlarged.

Imaging of the liver is an important, but relatively unproductive, element of staging since lymphoma deposits are often diffuse and microscopic, and large mass lesions are detected only rarely. Imaging the spleen is important but also problematic: large solid tumour masses are relatively uncommon in the spleen where diffuse infiltration is the usual manifestation of involvement. Ultrasound or CT scanning can indicate the size of the spleen and will usually detect mass lesions. Enlargement of the spleen in non-Hodgkin's lymphoma usually indicates diffuse infiltration, whereas in Hodgkin's disease the spleen may be moderately enlarged without being histologically involved. Colloid scintigraphy is sometimes helpful in distinguishing between splenic enlargement due to disease infiltration (where the uptake of colloid is relatively poor for the size of the enlarged spleen) and reactive hypersplenism (where splenic uptake of colloid is increased in proportion to the degree of enlargement). The demonstration of enlarged lymph nodes at the splenic hilum by CT or ultrasound is a strong indicator of splenic infiltration. Extranodal deposits elsewhere will usually be demonstrated by CT (for soft-tissue masses) and bone scintigraphy (for skeletal involvement).

Strategy

A strategy for staging could be summarized as follows: CT scanning of thorax, abdomen and pelvis to examine lymph nodes, to detect extranodal masses in soft tissues and to examine the liver and spleen; bone scintigraphy, colloid scintigraphy for splenic involvement, and possibly also ultrasound examination of the liver and spleen to give the maximum chance of detecting infiltration of these organs; lymphography could be reserved for those patients in whom CT scans show no disease below the diaphragm and the proposed treatment for supradiaphragmatic disease is by local radiotherapy; plain chest and abdominal films (especially after lymphography) will always be useful as a baseline for future comparison.

Checking the response to treatment requires the use of whichever imaging procedures showed the disease best initially. Residual soft-tissue abnormalities often persist after full clinical remission is established and it may be necessary to carry out occasional CT scans to ensure that these areas of disease remain inactive.

Testicular Tumours

These tumours deserve special mention because, typically, they occur in relatively young adults and are eminently curable by surgery, chemotherapy and radiation treatment. Typically both seminomas and teratomas metastasize first to para-aortic lymph nodes and then to adjacent nodes in the abdomen and mediastinum and to the lungs; metastasis to other areas is a late manifestation not commonly seen with current treatment. Detection of lung metastases is crucial, so CT scanning of the chest (or linear tomography of the whole lungs if CT is not available) should be carried out. If the chest film does not show lung nodules then CT examination of the thorax is essential (Fig. 10.7). CT of the abdomen from the diaphragm down to the pelvic brim will encompass the lymph nodes likely to be involved. Pelvic lymph node deposits are very unusual except in patients with overt upper abdominal node disease. If abdominal CT shows no nodal enlargement lymphography is indicated since, in a minority of cases, small intranodal deposits will be demonstrated.

Affected lymph nodes usually shrink back to normal size after treatment.

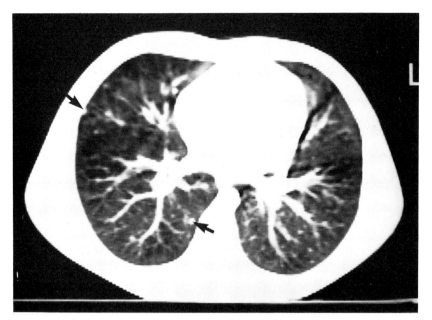

Fig. 10.7 CT through the lungs of a patient with testicular teratoma and a normal chest radiograph. But at least two metastases are shown on this section alone (arrows).

Occasionally, large masses may liquefy leaving cystic residues. In general, these appear inactive on further follow-up, but cases are described in which histology of the resected specimen shows islets of active tumour remaining. Lung metastases occasionally cavitate during treatment and eventually disappear leaving cystic spaces in the lung periphery from which pneumothoraces may develop. However, most lung lesions disappear without trace on treatment.

Whilst most testicular tumours present with an obvious scrotal swelling, a few cases present with large metastatic lymph nodes or lung secondaries at a time when the primary tumour has not yet enlarged the testis. Ultrasound scanning using a high frequency transducer can identify tumours of even a few millimetres in diameter in the testis so this method should be used whenever male patients present with unexplained metastases.

Pelvic Malignancies

Bladder tumours

Bladder tumours are normally treated, at least in the early stages, by endoscopic resection or diathermy, so the contribution of imaging methods to the detection of bladder tumours is limited. Contrast cystography or ultrasound can be used, if desired. Patients with transitional cell tumours in the bladder will, however, require an intravenous urogram to look for tumours elsewhere in the urinary tract since multiple sites are not unusual. Tumours extending deeply into the bladder wall or through it (Stage III and IV) are more likely to be treated by radiotherapy or total cystectomy and in these cases the best demonstration of the extent of tumour penetration is given by CT. At the same time, a search for enlarged pelvic nodes can be made and the CT scan also makes a major contribution to the accuracy of radiotherapy planning. Bladder tumours rarely metastasize outside the pelvis in their early stages, so a plain chest film and (arguably) bone scintigraphy complete the imaging investigations needed for staging and initial treatment.

Tumours of the prostate

The majority of tumours of the prostate are detected clinically and treated endoscopically. Transrectal ultrasound scanning is useful where measurements of the size of the tumour or careful assessment of its extent within the prostate is thought to be desirable. CT scanning through the lower pelvis may be restricted to those patients undergoing radiotherapy in which the major objective is to show whether or not the tumour extends through the prostatic capsule, and also if any pelvic lymph nodes are enlarged. Since prostatic tumours metastasize frequently to bone, and even quite widespread disease can respond to suitable chemotherapy, bone scintigraphy is routine at the time of diagnosis and intermittently during follow-up. Successfully treated bone lesions gradually become quiescent on the scintigram whereas the appearance of new lesions indicates recrudescence of disease. A transient increase in the activity of bone lesions shortly after starting chemotherapy is occasionally seen and this 'flare' phenomenon can be a sign of healing rather than of progressive disease.

Carcinoma of the cervix

Initial staging of the carcinoma of the uterine cervix is carried out by direct examination. Renal ultrasound or scintigraphy, or intravenous urography, can be used to look for evidence of ureteric obstruction. CT scanning of the pelvis is probably of value only in patients with clinical evidence of lateral extension into the parametria or where fixity to the pelvic wall is suspected. Nodal metastases from cervical carcinoma are often small and if nodal spread is suspected, lymphography is necessary in cases where initial CT scanning of pelvis and abdomen shows that the nodes are not enlarged. After local radiotherapy, it can be very difficult to distinguish between radiation fibrosis and recurrent tumour in the pelvis. A new mass shown on CT is a reliable indicator of recurrent disease, but in other cases needle biopsy of the abnormal tissue may be needed.

Ovarian tumours

The ovaries are best demonstrated by ultrasound; the feasibility of mass population screening for the early detection of ovarian tumours has recently been established. In patients presenting with a pelvic mass, plain films of the abdomen may show patterns of calcification which are characteristic for benign fibroid tumours of the uterus, dermoid cysts may be recognized by their fat content and the presence of teeth, whilst extensive peritoneal calcification suggests malignant deposits from pseudomucinous tumour of the ovary or appendix. Ovarian malignancies are usually treated by surgical excision or debulking in cases where the tumour cannot be totally resected, followed by chemotherapy. Ultrasound or CT examination can contribute to the differential diagnosis of clinically overt pelvic masses, in particular distinguishing between ovarian and uterine origin. Distinguishing between benign and malignant ovarian and uterine tumours is by no means easy and the detection of omental, peritoneal and subdiaphragmatic seedlings has not yet been reliably resolved by any imaging technique. The contribution of imaging in these patients is limited largely to postoperative management. Demonstration of residual disease or of new lesions by ultrasound or CT scanning may indicate the need for further chemotherapy or an alteration in the drug regime. When clinical remission is achieved and imaging shows no residual or recurrent disease, second-look laparotomy is still needed to check for small tumour residues at the primary site or in the peritoneum.

Lung Primaries

At the time of presentation, the majority of lung tumours will be visible on a plain chest film, or on bronchoscopy, or by both methods. Population screening for early detection of lung cancer has not yet proved convincing. The use of linear tomography or CT in searching for a potential lung primary when the chest film and bronchoscopy are normal is also often unrewarding, but a solitary lung nodule of dubious nature can frequently be characterized further by tomographic methods. If there is still doubt, needle biopsy, either percutaneous or transbronchial, should allow histological confirmation of the diagnosis.

If thoracotomy is contemplated, it may be helpful to obtain CT scans of the thorax if there is any question from the plain films of invasion of the chest wall or mediastinum. In the latter case, barium swallow can confirm involvement of the oesophagus which would usually be taken as a sign of non-resectability. The role of CT scanning in detecting mediastinal and hilar lymph node metastases preoperatively, and the contribution of mediastino-scopy in this respect, are contentious: current views suggest that nodes which are shown to be enlarged on CT should be biopsied at mediastinoscopy since reactive hyperplasia may coincide with a primary tumour.

Lung tumours metastasize quite frequently to the skeleton so there is a good case for routine bone scintigraphy as part of the preoperative assessment. This technique is also helpful in detecting chest wall invasion, particularly if oblique radiographs of the ribs are equivocal.

Detection of liver metastases by ultrasound or CT scanning is also desirable, the latter technique having the advantage that the same examination will demonstrate the adrenals which are also a common site for metastases from the lung. Patients with cerebral symptoms should also have CT brain scans, since brain metastases confer a very poor prognosis. However, the value of routine head scanning in patients with no clinical evidence of cerebral involvement is dubious.

Radiotherapy planning for lung tumours relies on the plain chest film, supplemented by CT. The detection of recurrent tumour after surgery is often problematic. CT is sometimes helpful: e.g. in patients with a uniformly opaque hemithorax on the plain film, CT will distinguish between fluid and solid components.

Gastrointestinal Tumours

Population screening for early gastric cancer using a limited form of double-contrast barium meal has been used in Japan with success but this approach has not yet been translated to Western countries where the incidence of the disease is much lower. The combination of barium examinations and endoscopy will diagnose the majority of gut tumours at the time of presentation. CT scanning makes a contribution to the assessment of resectability of carcinomas arising in the oesophagus, stomach and rectum. Transrectal ultrasound has recently been shown to be potentially very accurate in determining the submucosal extent of rectal tumours, and endoscopic ultrasound of the oesophagus and the gastric cardia also looks promising. Preoperative detection of liver metastases from gastrointestinal tumours is discussed earlier in this chapter (see p. 309) and also in Chapter 5. Contrast examinations and endoscopy are used as first examinations in the detection of recurrent disease or complications at the site of anastomosis after surgery, with CT as the procedure of choice when surgical alterations in anatomy render the primary techniques impractical. This applies in the pelvis after abdominoperineal resection of the rectum (Fig. 10.8), and in the investigation of upper abdominal symptoms after Polya-type gastrectomy or Whipple's procedure.

Imaging of tumours of the liver and pancreas is discussed in detail in Chapter 5. In general, ultrasound provides a first-line technique, to be followed by CT in cases of technical difficulty or equivocal results. Where there is a high level of suspicion of pancreatic tumour and the non-invasive techniques show normal results, ERCP may be justified. Patients presenting with biliary obstruction will, in general, need direct cholangiography either by percutaneous or endoscopic cannulation. Preoperative assessment of liver tumours requires demonstration not only of the biliary tree but also of the hepatic arterial and venous anatomy, and the inferior vena cava.

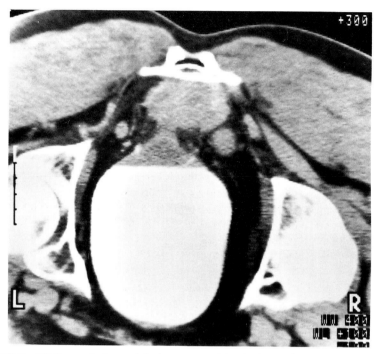

Fig. 10.8 CT through the pelvis (prone position) of a patient with sacral pain nine months after abdominoperineal resection for rectal carcinoma. The bladder is filled with contrast medium. Anterior to the sacrum is an irregular mass of soft-tissue density representing recurrent rectal tumour.

Before undertaking major hepatic resection it may be prudent to exclude pulmonary metastases by CT of the thorax. Similarly, a careful CT examination of the upper abdomen in patients with carcinoma of the pancreas may eliminate the need for further investigation by demonstrating metastases in the liver or lymph nodes, portal vein invasion, or encasement of the coeliac axis, all pointers to non-resectability. Some islet cell tumours will be shown by ultrasound, a few more by CT with contrast enhancement, but others may require arteriography or transhepatic cannulation of the portal system for pancreatic venous sampling.

Renal Tumours

Intravenous urography remains the preferred method for the identification of urothelial tumours within the kidney. Direct opacification of the ureter is also desirable in cases of obstruction by intraluminal tumours, and can be achieved by either retrograde or antegrade injection. The retrograde approach has the advantage of allowing inspection of the bladder at the same time, whereas the antegrade method gives the opportunity to establish percutaneous drainage of the obstructed kidney.

An entirely different approach is needed for renal cell carcinoma. Presentation with haematuria will precipitate an intravenous urogram; if this shows evidence of a renal mass then ultrasound and/or CT should follow automatically. In patients presenting primarily with a mass in the flank, ultrasound will be the first investigation after a plain abdominal film. A plain chest film should be obtained at an early stage since the presence of overt lung metastases may narrow down treatment options to the extent that detailed investigation of the primary becomes irrelevent. Arteriography now plays little part in the diagnosis of hypernephroma, its role being limited to preoperative road mapping and the embolization of large tumours. Biopsy can be carried out conveniently under ultrasound guidance and staging is best achieved by CT which, in addition to showing the local extent of the tumour and its relation to surrounding structures, should allow the recognition of regional lymph node metastases whilst a search for lung metastases may also be made at the same examination. Screening for bone metastases by scintigraphy is of arguable value; it is rare to find bone lesions in the absence of local symptoms since bone deposits from hypernephroma are characteristically aggressive, and sometimes produce scintigraphically 'cold' lesions. Recurrent tumour in the renal bed after nephrectomy can usually be detected by ultrasound, CT being decisive in doubtful cases.

Skin and Connective Tissue Tumours

Imaging procedures need to be tailored to the natural history of the lesion, which varies greatly in this group of tumours — from those which are locally invasive but metastasize only very rarely (e.g. basal cell carcinoma, chordoma) to others which metastasize aggressively and frequently (osteosarcoma, melanoma). Plain films are used in the initial detection and characterization of primary bone tumours (Fig. 10.9) with histology for further characterization. If local resection is planned, CT will often give a helpful indication of the extent of any associated soft-tissue component and also shows the intramedullary extension of endosteal tumours more accurately than plain films. CT is also useful in planning local radiotherapy, measuring the response to treatment and diagnosing recurrent disease. If radical surgery such as block dissection or amputation is proposed, lung metastases should be excluded by CT scanning first. Early detection of lymph node metastases from melanoma and the more aggressive bone tumours remains problematic. Bone scintigraphy adds little to the identification of overt skeletal tumours, but is useful in detecting metastatic disease and also in the early detection of some primary bone tumours which are not visible on plain films at the time of presentation (e.g. some osteoid osteomas). Ultrasound and CT are used as primary techniques in the detection and investigation of retroperitoneal sarcomas where again imaging tests have a role in radiotherapy planning and in the assessment of response to treatment.

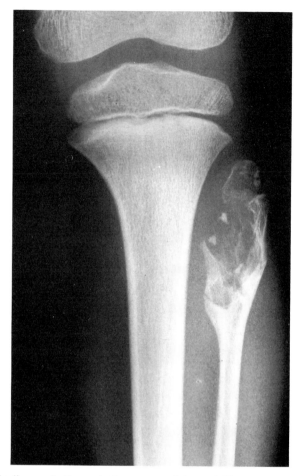

Fig. 10.9 Radiograph through the upper tibia and fibula of a boy with local pain following trauma. Whilst the appearance of the lesion in the fibula strongly suggests benign enchondroma, sarcomatous elements were found at histology.

Paediatric Tumours

Particular efforts should be made to avoid redundancy and duplication of imaging procedures in children. In infants presenting with abdominal masses, plain films of the chest and abdomen should be followed by ultrasound scanning (Fig. 10.10). This will separate solid from cystic masses and, in most cases, distinguish between an intrarenal and extrarenal origin. Intravenous urography has little to add in these cases, since a better estimate of the function of the affected kidney may be obtained preoperatively by one of the scintigraphic techniques. Solid tumours will, in general, require preoperative CT for assessment of local spread and resectability; a search

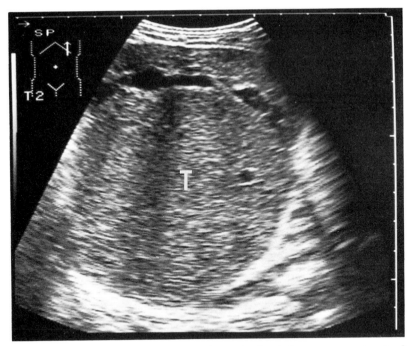

Fig. 10.10 Sagittal ultrasound scan through the left upper quadrant of an infant with a palpable abdominal mass. The tumour (T) is compressing the renal collecting system from behind. Diagnosis: Wilms' tumour.

for metastases in the lungs can be made at the same session. Bone scintigraphy is probably worthwhile, but may require supplementary radiographs of the long bones in children with neuroblastoma since symmetrical metastases can be very difficult to recognize. CT is also needed for planning surgery and/or radiotherapy in children with rhabdomyosarcomas arising in the pelvis or in the head and neck. Investigation of Hodgkin's disease in children follows the same pattern as in adults, but since non-Hodgkin's lymphoma in childhood is almost always treated by chemotherapy, detailed anatomical definition of the extent of disease seems less important. The role of arteriography in childhood tumours is extremely limited, perhaps the only remaining use being in preoperative definition of the blood supply of large upper abdominal tumours whose origin cannot be defined adequately by CT or ultrasound.

Head and Neck Tumours

The role of plain films of the skull has been greatly diminished since the introduction of CT scanning. Only a minority of intracranial tumours produce bone changes, and whilst the progress of bone erosion or destruction can be conveniently followed by repeated plain films, in most circumstances it is preferable to visualize the tumour directly by CT. Intraocular tumours are well-demonstrated by orbital ultrasound, but for orbital mass lesions outside the globe, tumours within the cranial cavity and those arising in the sinuses, nasopharynx and base of skull, CT is necessary. Magnetic resonance imaging (MRI) has a number of advantages over CT in the demonstration of tumours in the posterior fossa, around the base of the brain, and the cervical cord; where available, MRI will be used for tumours in these areas. Arteriography in the head and neck is limited to delineation of vascular tumours in the neck, face and brain, largely, occasionally for diagnosis as a supplement to CT, but more often as a prelude to surgery or radiotherapy planning. The development of techniques for transcatheter embolization of vascular tumours has added a new role for the angiographer.

Thyroid tumours can be identified by scintigraphy and by ultrasound; scintigraphic detection of functioning metastases from the thyroid is discussed in Chapter 7. Tumours of the mouth, pharynx and larynx are generally diagnosed by direct endoscopic visualization. CT is the preferred technique for demonstrating the extent of the tumours, assisting radiotherapy planning and following the response to treatment.

Epithelial tumours of the head and neck generally metastasize late, and do so to the cervical lymph nodes first. Initial assessment for metastatic disease is often limited to careful clinical examination of the neck and a plain chest radiograph. This does not apply to patients with melanoma, lymphoma, thyroid carcinoma or rhabdomyosarcoma arising in the head and neck where the full staging procedures associated with these tumours should be done.

Tumours of the Breast

Since the 1960s when the early population screening studies were carried out using mammography to detect asymptomatic breast cancer, technical improvements have resulted in better images being available at much lower radiation dosage. More recent mammography screening studies have shown a convincing reduction in mortality in younger women as well as in women over 50 years.

As a diagnostic procedure, mammography complements clinical examination of the breast in the detection of dubious mass lesions, demonstrating the extent of the mass, and discriminating benign from malignant lesions. Mammography will detect a proportion of tumours that are impalpable so it is clearly worthwhile investigating patients with vague breast symptoms, lumpy breasts, bleeding or discharge from the nipple. Other applications include screening the opposite breast in patients with established malignancy on one side, looking for recurrences after surgery or radiotherapy, and searching for occult primary tumours in patients with axillary lymph nodes or metastases elsewhere of uncertain origin. Films obtained after opacifying the duct system by contrast injection sometimes reveal intraduct tumours or confirm duct ectasia and are indicated if initial films are inconclusive in the presence of bleeding or discharge from the nipple.

Other imaging techniques have made little impact on the diagnosis of breast masses. Thermography has been unimpressive, ultrasound may be used to distinguish large cystic lesions from solid tumours (but this rarely eliminates the need for biopsy or excision) and specialized CT scanners designed for breast examination have not proven cost-effective. Conventional CT may be useful in radiotherapy planning where measurement of chest wall thickness is important in calculating the absorbed dose.

The choice of staging procedures in breast cancer remains controversial. The incidence of metastases in advanced breast cancer is relatively high so that bone scintigraphy and scintigraphic, ultrasonic or CT scanning of the liver is clearly worthwhile. However, the incidence of positive findings when these tests are carried out in patients with early breast cancer (Stage I or IIa) is very low indeed, so the value of routine screening is in question. However, bone scintigraphy performed near the time of initial surgery is useful as a baseline for comparison in case of future presentation with bone symptoms. Enlargement of the internal mammary lymph nodes is elegantly shown by CT. Lymphatic metastases may also be shown by lymphoscintigraphy carried out after intradermal injection of labelled colloid in the epigastrium, but normal variation in the anatomy of the internal mammary chain makes interpretation difficult.

Related Clinical Topics

Imaging the Spleen

Indications for splenic imaging include clinical evidence of abnormal splenic anatomy (enlarged or displaced spleen, masses in the left upper quadrant, trauma), evidence of disturbed splenic function (hyper- or hyposplenism), and the diagnosis of conditions likely to involve the spleen as well as other organs (suspected lymphoma, portal hypertension). The plain abdominal film should indicate when there is moderate or marked splenic enlargement; calcification in the spleen is seen in hydatid disease, some forms of parasitic infestation, old haematomas or abscesses and some slow-growing tumours. Splenic trauma may result in a number of changes on a plain abdominal radiograph — left-sided rib fractures, loss of the inferior splenic outline, medial displacement of the gastric air bubble, local ileus in the left upper quadrant and elevation of the left hemidiaphragm with or without an associated pleural effusion.

Ultrasound or CT will show the size, shape and position of the spleen as well as detecting mass lesions within it and illuminating the surrounding anatomy. Patients with a mass in the left upper quadrant or a clinically enlarged spleen should have ultrasound in the first instance followed by CT if there is any difficulty in interpreting the result. Tumours in the spleen are often of very low echogenicity on ultrasound scanning and sometimes they may be difficult to differentiate from cysts. The use of contrast enhancement with CT will be decisive in the vast majority of cases. With very large masses, the normal anatomic landmarks are often so distorted that it is difficult to decide from which organ the mass is arising. In these cases, scintigraphy is a useful supplement to the anatomical methods since it will pick out functioning splenic tissue and so indicate whether the spleen is destroyed or simply displaced.

Splenic injuries may be accompanied by a local ileus which can hamper ultrasound examination. CT has proved effective in detecting splenic damage which most often presents as a crescentic subcapsular haematoma. Intrasplenic extensions may occur, usually accompanied by free intraperitoneal haemorrhage. Occasionally, anatomical variants may mimic splenic fracture. CT will usually distinguish between a subphrenic collection and a left basal pleural effusion.

In patients with large spleens, the intensity of uptake of colloid gives a good indication of splenic blood flow and so, indirectly, helps to distinguish between reactive or congestive splenic enlargement (with increased function) and splenomegaly due to infiltration, e.g. by amyloid, lymphoma, Gaucher's disease, where function is not increased in proportion to splenic size. Searching for accessory spleens after splenectomy for trauma, lymphoma or blood dyscrasias is best achieved by scintigraphy using labelled colloid or denatured red cells. Diagnosis of the polysplenia/asplenia syndromes depends on anatomical localization and needs ultrasound or CT. Gallium scintigraphy is occasionally useful as a supplementary technique in sorting out difficult problems in the left upper quadrant, e.g. distinguishing between abscesses and infarcts in the spleen, confirming or refuting dubious subphrenic infection seen on ultrasound or CT.

Pyrexia of Uncertain Origin (PUO)

The investigation of patients with persistent fever presents a challenging diagnostic problem. Where there are localizing features in the history or clinical examination, imaging tests will naturally be directed towards the area under suspicion. Where there are no such features, a systematic strategy of diagnosis by exclusion has to be pursued. The likely causes of a PUO vary according to the age of the patient, recent medical and surgical history, and the possibility of exposure to infections, infestations, toxins and drugs. Imaging procedures supplement clinical, haematological and biochemical tests and may be directed by clues obtained from these other methods. The following check list is offered as a starting point for tailoring an individual approach to each patient.

1.   A chest radiograph should be obtained early on and repeated from time to time.

2. Upper abdominal ultrasound scanning provides the simplest approach to screening for occult tumours in the liver, spleen, kidneys and pancreas, silent infections in the subphrenic and subhepatic spaces, and abnormalities of the biliary tract — particularly cholecystitis which, in elderly patients, may produce surprisingly little in the way of localizing features.

3. Multiple pulmonary emboli may occur without chest symptoms and can be responsible for fevers, particularly in postoperative patients, so V/Q scintigraphy of the lungs is indicated in this group.

4. Echocardiography should detect the rare occurrence of an atrial myxoma and should also be used to look for vegetations on the heart valves which will indicate bacterial or verrucous endocarditis.

5. Bone scintigraphy is suggested to screen the skeleton for sites of infection which, on rare occasions, present without local clinical findings; this test may also reveal asymptomatic bony metastases in patients with occult primary tumours.

6. If good quality ultrasound scans of the upper abdomen are normal, CT is unlikely to reveal a lesion here, but is much more likely to detect occult retroperitoneal infections or tumours. CT of the chest and pelvis will, occasionally, show lymph node enlargement undetected by other means, particularly in patients with lymphoma. Evidence of leaking aortic aneurysms and retroperitoneal fibrosis can also be sought on CT.

7. Whole body scintigraphy with gallium citrate or labelled white cells should be successful in localizing occult inflammatory disease even when the anatomical scanning methods have failed to show an abscess. Because gallium and white cells accumulate in solid inflammatory tissue as well as in the walls of abscesses, areas of phlegmonous inflammation are well shown. Inflammatory bowel disease and infections in bones and joints should also be demonstrated.

8. Polyarteritis nodosa may present as a PUO and, as an alternative to biopsy, selective arteriography of the liver and kidneys should be considered since specific angiographic changes are seen in a substantial proportion of cases.

Metastases from Unknown Primary Tumours

A small proportion of patients with malignant disease present with metastases before the primary tumour has declared itself. In some cases there will be clinical clues to the site of the primary tumour. In others, the primary remains undisclosed after careful clinical history and examination and a good quality chest radiograph. The optimum investigation of this group is contentious. In the majority of such patients, the primary tumour will not be found, whether during life or at autopsy. Consideration must be given to minimizing the discomfort and medical interventions during the short period of life that remains to these patients, whilst ensuring that a treatable tumour is not missed.

Various strategies have been proposed, the more important considerations being as follows:

1. Chemotherapy for testicular tumours is very effective; prostatic and breast tumours may respond to hormonal therapy, and functioning thyroid tumours can often be controlled by radioiodine therapy, so primaries in these sites should be sought. Ultrasound of the testes, mammography, thyroid scintigraphy or ultrasound and perhaps transrectal ultrasound of the prostate should be considered as the first batch of investigations.

2. The site at which metastases present is relevant in deducing their likely primary origin. Lesions presenting with bone destruction should initiate a search for a primary renal cell carcinoma since excision of primary and secondary renal tumours sometimes results in prolonged survival. Similarly, the discovery of lung metastases should precipitate a search for primary tumours in the kidneys, liver or ovaries by ultrasound scanning, once the thyroid and testes have been checked. Silent primaries presenting with liver metastases will usually be found in the pancreas or colon; histology should be obtained because pancreatic islet cell tumours and carcinoids may respond to chemotherapy or transcatheter embolization.

3. Where the presentation is with enlarged lymph nodes, biopsy will be essential. Having excluded patients with lymphoma, the cell type may be helpful in pointing towards the likely primary site. The vast majority of squamous cell metastases in cervical nodes arise from epithelial tumours in the head and neck, for which a careful endoscopic search should be made. Adenocarcinoma deposits, on the other hand, arise from a very wide variety of sources, the most likely primary sites being in the liver, kidneys or pancreas, and the most productive single imaging procedure being CT scanning of the upper abdomen.

4. During the investigation of these patients, the justification for continuing with increasingly unproductive procedures must be constantly reviewed. At some stage humanity must take precedence over technology.

Acquired Immunodeficiency Syndrome (AIDS)

The Acquired Immunodeficiency Syndrome (AIDS) is a new disease of protean manifestations caused by an infective agent, the human immunodeficiency virus (HIV). Whilst serological tests can usually detect the presence of antibodies to HIV in the plasma of infected subjects, the diagnosis of AIDS is based on two major clinical criteria. These are, firstly, the presence of one or more of the infections or malignancies which are commonly associated with deficient cell-mediated immunity, and secondly, the absence of other known causes of impaired resistance such as immunosuppressive therapy or lymphoma. The commonly associated infections are Pneumocystis pneumonia, oesophageal candidiasis, and systemic infections with other fungi, with *Toxoplasma*, or with *Mycobacterium avium intracellulare* (MAI). Other typical manifestations include chronic herpes simplex infections of skin and mucous membranes (occasionally leading to squamous cell carcinomas at these sites), primary lymphoma of the central nervous system, and Kaposi's sarcoma.

The Chest

Pneumocystis pneumonia is the most common pulmonary manifestation of AIDS. Chest radiographs typically show patchy consolidation which, in early stages is perihilar, later spreading towards the periphery of the lungs and also coalescing to form homogenous opacities. Less commonly, the distribution may be unilateral or even lobar; abscess formation and linear atelectasis are other recognized manifestations and, occasionally, the chest X-ray may appear normal in the presence of active infection. Enlargement of hilar or mediastinal nodes is uncommon and should raise the suspicion of tuberculosis or Kaposi's sarcoma. Infection with cytomegalovirus (CMV) may produce an identical appearance on the chest film. Tuberculosis in AIDS usually produces the 'primary complex' type of infection with a patch of consolidation anywhere in the lung periphery together with enlarged hilar nodes.

The presence of enlarged hilar or mediastinal nodes should also raise the suspicion of Kaposi's sarcoma or lymphoma in the chest, although extrathoracic manifestations will usually coexist with these lesions. Kaposi's sarcoma within the lung may produce nodular densities or patchy consolidation similar to Pneumocystis pneumonia and the diagnosis is a difficult one to make, often requiring open biopsy. Lymphocytic interstitial pneumonitis (LIP), a condition which is associated with a wide range of immune deficiencies in adults but is diagnostic of AIDS in children, also produces a range of chest X-ray abnormalities indistinguishable from Pneumocystis pneumonia. The diagnosis usually requires open biopsy for adequate histology.

Gastrointestinal Tract

The gut is a common site for infective manifestations of AIDS. Any part of the GI tract may be affected. Dysphagia is common, usually resulting from oesophageal candidiasis. Double-contrast barium swallow shows mucosal oedema, with diffuse shallow ulceration in later stages. Occasionally, very large areas of ulceration result from CMV infection.

The stomach is less commonly affected, but occasionally shows a diffuse nodular gastritis with CMV infection; this pattern is usually distinguishable from the discrete nodules of Kaposi's sarcoma which are separated by areas of normal mucosa.

Non-specific thickening and ulceration of the duodenal mucosa can result from infection with CMV, MAI or Cryptosporidium, but a very similar pattern may occur with Kaposi's sarcoma or lymphoma of the duodenum. In the small bowel, the same infecting agents produce thickening of the mucosal folds and usually a degree of dilatation and hypersecretion, although mucosal atrophy has also been recognized. CMV infection may present with acute perforation due to local ischaemia of the small bowel.

Infective agents in the colon commonly produce proctitis and periproctitis, but the colitis of CMV infection is unique to AIDS; the pattern on barium examination is that of a granular mucosa with aphthous ulcers and occasionally toxic megacolon.

Lymph Node Disease

Diffuse enlargement of lymph nodes with reactive histology is a major feature of the AIDS precursor which is known under various names including AIDS-related complex (ARC) and persistent generalized lymphadenopathy (PGL). In this condition, retroperitoneal and mesenteric nodes are often larger than 1·5 cm diameter and as full blown AIDS develops, the nodes tend to become smaller. Large nodes with low attenuation centres are more likely to be the result of MAI infection. Needle biopsy, guided by CT, is recommended for distinguishing MAI from *M. tuberculosis* infection, since different treatment regimes are necessary.

Enlarged abdominal nodes in patients with established AIDS are usually caused by lymphoma or MAI infection. Lymphomas in AIDS patients are typically of the non-Hodgkins type and show high grade histology. Extra-nodal lesions occur in the majority of cases, the usual sites being liver, spleen, kidneys, gastrointestinal tract and brain; thoracic lymphoma is much less common. Most patients with Kaposi's sarcoma have lesions in the skin and mucous membranes; lung and mediastinal node deposits are common, as are gastrointestinal lesions which take the form of submucosal nodules, thickened folds or polyps. Kaposi lesions are less common in the liver and spleen.

As with the infective manifestations of AIDS, the radiographic features of AIDS-related lymphomas and Kaposi sarcomas are often non-specific and cytological or histological evidence is necessary to confirm the diagnosis.

Central Nervous System

Abnormalities in the CNS which are commonly seen in AIDS patients may be the result of infection or of neoplasia or of both. Granulomatous meningitis is the usual manifestation of *Cryptococcus* infection and meningeal infiltration is the most common intracranial result of systemic AIDS-related lymphomas. CT and magnetic resonance (MR) scans are often normal when disease is limited to the meninges in these conditions. Mass lesions within the brain, usually of low density with patchy contrast enhancement, result from infection with *Toxoplasma* or *M. tuberculosis* and similar appearances are seen with primary lymphoma of the brain, a condition almost peculiar to AIDS patients. Focal inflammatory lesions are also shown with CMV or herpes simplex infection, whereas more chronic encephalitides are probably the result of infection with the HIV agent itself. The demyelinating condition, progressive multifocal leukoencephalopathy (PML), also produces recognizable changes on CT and MR scans related to patchy local increases in the water content of white matter. MR is said to be more sensitive than CT in the early detection of both infective and neoplastic brain lesions.

An Approach to Imaging in AIDS

The manifestations of AIDS are protean and investigation needs to be tailored to the clinical presentation in each individual patient.

In the chest there are no specific radiographic features of AIDS. Plain chest films will always be useful, although transbronchial biopsy and bronchial lavage will usually be necessary to achieve a specific diagnosis of Pneumocystis pneumonia, tuberculosis and most of the other types of infection. Open biopsy may be needed for confirmation of LIP and Kaposi's sarcoma. CT of the thorax will be useful in revealing mediastinal disease and distinguishing between pleural and peripheral lung lesions. Where pulmonary involvement is suspected in the presence of a normal chest radiograph, gallium scintigraphy has been advocated as a more sensitive means for detecting early infections, but the results have a low specificity. The recognition of hilar or mediastinal adenopathy should prompt a search for tuberculosis or malignancy. Where lung consolidation is segmental or lobar the possibility of superimposed bacterial infection should be pursued since such a complication may be readily treatable.

Disease involving the central nervous system will usually be detected by CT scanning, but for mass lesions in the posterior fossa, for demyelinating lesions, and probably for cerebral inflammatory disorders, MR imaging is likely to be more sensitive.

Double-contrast barium examinations, together with endoscopy, provide an approach to the diagnosis of infections and neoplasia in the gastrointestinal tract. Lesions in the liver, spleen and kidneys will usually be detected by ultrasound, but probably the most useful single investigation for abdominal disease will be CT since the whole of the abdominal contents can be assessed at a single examination. As in the thorax, biopsy of abdominal lesions will often be required for specific diagnosis and this may be achieved either endoscopically for gut lesions, or with imaging guidance using ultrasound or CT for lesions in the lymph nodes and solid organs.

Further Reading

Ackery, D. and Batty, V. (Eds.). (1986). Nuclear medicine in oncology. *Clinics in Oncology* 5, **1**. W.B. Saunders & Co., Eastbourne & Philadelphia.

Glazer, G.M. (Ed.). (1986). *Contemporary Issues in CT — Staging of Neoplasms*. Churchill Livingstone, Edinburgh.

Goldberg, B. (Ed.). (1981). Ultrasound in cancer. *Clinics in Diagnostic Ultrasound* Vol. 6. Churchill Livingstone, Edinburgh.

INDEX